WA 1165521 6

D0302901

The CM Contracting System: Fundamentals and Practices

C. Edwin Haltenhoff

Consulting Engineer

Distinguished Lecturer Emeritus
Michigan Technological University

PRENTICE HALL
Upper Saddle River, New Jersey 07458

Library of Congress Cataloging-in-Publication Data
Haltenhoff, C. Edwin.
The CM contracting system : fundamentals and practices / C. Edwin Haltenhoff. — 1st ed.
p. cm.
Includes index.
ISBN 0-13-744400-1
1. Construction industry—Management. I. Title.
HD9715.A2H32 1998
624'.068—dc21 98-4729
CIP

Acquisition Editor: Bill Stenquist
Editorial/Production/Composition: Interactive Composition Corporation
Editor-in-Chief: Marcia Horton
Assistant Vice President of Production and Manufacturing: David W. Riccardi
Managing Editor: Bayani Mendoza de Leon
Manufacturing Manager: Trudy Pisciotti
Full Service/Manufacturing Coordinator: Donna Sullivan
Creative Director: Jayne Conte
Cover Designer: Bruce Kenselaar
Editorial Assistant: Meg Weist
Copy Editor: Lori Stephens

Printed in the United States of America

10 9 8 7 6 5 4 3 2 1

ISBN 0-13-744400-1

Prentice-Hall International (UK) Limited, *London*
Prentice-Hall of Australia Pty. Limited, *Sydney*
Prentice-Hall Canada Inc., *Toronto*
Prentice-Hall Hispanoamericana, S.A., *Mexico*
Prentice-Hall of India Private Limited, *New Delhi*
Prentice-Hall of Japan, Inc., *Tokyo*
Simon & Schuster Asia Pte. Ltd., *Singapore*
Editora Prentice-Hall do Brasil, Ltda., *Rio de Janeiro*

To C.A. Richardson, the consummate Mentor
and
Harriet, the consummate friend, spouse and Mother

Contents

Figures, Examples, and Lists

Introduction

The CM contracting system was developed in the 1970s in response to owner requests for improved construction industry performance. Projects commonly exceeded owner budgets and fell behind schedule; construction quality and productivity declined; design services fell short of expectations; contract disputes were common; and owners generally perceived that they were not receiving fair value for the money they spend on new facilities.

Several factors precipitated negative conclusions by construction users. There was a rapid increase in the number of both public and private projects, new and innovative construction products innundated the marketplace, larger and more complex facilites were demanded, trade contracting became more specialized, merit shop contractors increased in numbers, and many new government regulations were promulgated. These events further divided an already fragmented construction industry and frustrated its performance.

The contracting system predominantly used to satisfy the needs of owners was general contracting, a system that evolved during the previous century. The design–build system, the heir apparent to the master builder concept, was available at the time but not widely used.

The general contracting system, with its design–bid–build project-delivery sequence, conformed to the legal bidding and contracting requirements of public owners and was their political choice.

The design–build contracting system, with its bid–design–build sequence, was contrary to public contracting requirements but successfully competed with general contracting in the private sector.

In the late 1960s, many owners voiced displeasure with the general contracting system and strongly suggested change. Unfortunately, the structure of general contracting and the attitude of most general contractors was too rigid to satisfactorily accommodate the needs of the owners.

In order to extract more from the new construction enironment and dispel owner concerns, new procedures would have to be added and older ones modified. An alternative contracting system—one that retained the positive attributes of general contracting and included the positive features of design–build contracting but provided more contracting flexibility—proved to be the answer.

This alternative, developed by coincidental consenus of many construction industry participants, became known as Construction Management or CM. The CM system retained the design–bid–build sequence and competitive bidding attributes of general contracting and included several of the virtues of design–build contracting.

CM enhanced the primary design and contracting components of the project-delivery process and facilitated increased owner involvement in both design and construction. This was accomplished by adding an advisor/manager (a construction manager, or CM) into the contracting structure and eliminating the traditional management role of the general contractor.

The basic form of CM contracting became commonly known as Agency CM or ACM, a descriptor that specifically identifies the role of the CM as a fiduciary agent of the owner. In the ACM system, the CM, A/E, and owner operate as a team from the start of design until the end of construction and share a common goal: the satisfaction of the owner.

Other forms and variations of CM developed over the years, but ACM stands as the root system, the one which must be understood by providers and users of all CM forms and variations if they are to understand CM at all. ACM unequivocally meets the overt, fair, competitive, and legal requirements on public projects, and when emulated in the private sector, provides the same positive results.

While the GC contracting system evolved over a period of a century or more, CM contracting was basically formulated and instituted within approximately fifteen years. The CM system developed independently but simultaneously in many parts of the country without guidance from a coordinating group and with little communication between pioneer CM practitioners.

Remarkably, many of the CM procedures developed in the early 1970s are in use today without even minor modifications, confirming the fact that CM development was ideologically consistent wherever it occured.

Perhaps just as remarkable is that many collateral benefits which emerged during CM's early practice were quickly incorporated into the CM menu of services. They have significantly increased the efficiency of the CM system and added substantial credibility to CM services. Currently it is difficult to determine which benefits provide the most value—those that were originally built into the CM contracting system or those that surfaced during CM's early use and became part of CM Services over the years—but all are important to owners.

The CM system did not develop without controversy. It must be remembered that the construction industry, prior to the advent of CM, was primarily the domain of general contractors. They formulated the GC contracting structure, instituted its operating procedures, and established the image of the construction industry.

At first, general contractors considered CM an intruder, a miscreant system, a fad, a buzzword that would soon fade away. CM proved otherwise and, although general contractors and their advocates continue to search for ways to compete with the positive attribues of CM, many general contractors have added CM to their contracting services.

Several construction industry associations or societies have made moves to establish themselves as the consensus CM authority. Unfortunately, provincialism and market share have been the motivation, not concern for the problems and well-being of the construction industry or the needs of owners. None have unequivocally earned the respect of the construction industry and its users or moved

CM practice significantly closer to a professional practice on par with architecture and engineering.

One group, the Construction Management Association of America (CMAA), was formed in 1981 to provide a home exclusively for CM practitioners. Other groups recognized CM as an adjunct to the primary purpose of their group. The CMAA postured itself as an industry association, with a structure similar to the Associated General Contractors of America, Inc. (AGC). In recent years, however, the CMAA has altered its direction and is leaning toward improving CM technical skills and attaining professional status for CM practitioners. If sustained, this shift in direction should eventually position the CMAA as the recognized CM authority.

CM's tenure, popularity, and quiet divisiveness has created a need for documentation and discussion. CM is no longer in search of an identity or definition; it has clearly defined itself through use over the years. However, to the author's knowledge, this is the first book-length publication devoted to CM fundamentals and practices to date.

If the system is to retain its position as a practical alternative to general contracting and design–build, its practice and level of practice will have to improve, yet there are signs that CM practices are actually waning. Services are being diluted and sundered by the willingness of CM practitioners to engage in price-based selection competition rather than fostering qualification-based selection. Through a lack of any influences to the contrary, owners have formed the opinion that CM services are more akin to independent contractor services than to professional services.

This book is written within the context of CM's self-definition from its beginning to current practices. It does not favor the provincial opinions of any one organization, association, or society. It was written on the basis of what the author has learned as a CM pioneer and thirty years of experiences, learning what does and does not work in actual CM practice.

The purpose of the following chapters is to explain the CM system as completely as possible; to provide insight to its philosophy; to explain its fundamentals, practices and procedures; and to provide a benchmark for understanding CM as it is, has been, and can be practiced.

Small- to medium-large-size projects for public sector owners below the federal level provide the prime context for this book. Small- to medium-large projects in the private sector that emulate public sector contracting requirements qualify as well. (This is not to say that large projects, federal level owners, and private sector projects that do not emulate public sector contracting requirements will not benefit from the information provided.) However, the heart of our nation's construction industry should be located where most of the projects occur. Consequently, this book relates to the large number of construction industry participants who are engaged in the small- to medium-large-project market in both the public and private sectors.

Early chapters concentrate on Agency CM, the accepted root form of the CM system, and describe other CM forms and variations. Later chapters cover CM knowlege requirements and CM fundamentals and practices.

BEFORE STARTING CHAPTER 1

To gain the most from this book, it is suggested that the reader:

1. suppress all preconceived ideas about the construction industry and its current practices
2. agree that "construction managment" and "the management of construction" are not synonymous
3. understand that CM's development was fueled by owners who wanted a more equitable return on construction investments
4. realize that CM has become competitive to a point where not all CMs offer the same level, range, or quality of services, and
5. understand that *macro*-planning and *macro*-managing plays only a minor role in a CM's performance; the success of CM is vested in the CM firm's ability to *micro*-plan and *micro*-manage every activity.

After digesting the content of the book, an owner, practitioner, or student may conclude that it is impossible to provide CM services to the extent and degree described. Others will feel compelled to add services and procedures to make them more effective and complete. Whatever the reaction, this book will have achieved its purpose; readers will know more about CM than they knew before.

Finally, the reader can be assured that all practices and procedures in this book have been used by construction managers on actual projects. None are fictitious or purely academic.

CHAPTER 1

The Fundamentals of the Root Form of CM (ACM)

The CM system of contracting (CM) uses the same construction industry resources as the other contracting systems, and requires the same services to complete a project. The differences between the systems are the contractual ties and responsibilities of the parties involved. The essential differences are determined by the responsibilities of each party, the contracts within the system, and the legal performance requirements of each party.

The fundamental form of CM is Agency CM or ACM; it is the root form of all other forms and variations of the CM system. Understanding ACM is essential for understanding all forms of CM consequently it is presented first. The other forms and their variations of CM are covered in Chapter 5, CM System Forms and Variations.

1.1 THE SIX ESSENTIAL SERVICE ELEMENTS

Every project, without exception, requires the same six service elements for its complete execution:

1. Project Management
2. Design
3. Contracting
4. Construction
5. Contract Administration
6. Construction Coordination.

The identification of the six essential elements of service and the contracting structure that allocates the responsibility for providing the elements, clearly defines each system and establishes the differences between CM contracting, general contracting, and design–build contracting.

The six essential elements are defined as follows:

Project Management: Guidance of project-related activities from design and construction to occupancy.

Design: The solution to the owner's project needs in the form of contract documents from which cost estimates from contractors can be obtained and the project constructed.

Contracting: Arranging for or the holding of contracts for the services required to produce the project.

Construction: The hands-on work of constructors; the performing contractors who build the project with workers on their own payroll.

Contract Administration: The servicing of contracts for construction between the owner and constructors.

Construction Coordination: The orchestrating of constructor activities during the building of the project.

1.2 ASSIGNMENT OF RESPONSIBILITIES

Responsibility for the performance of the six elements must be contractually assigned to one or more of the parties included in the contracting structure.

For example, when using the **General Contracting** (GC) system,

- the Architect/Engineer (A/E) is assigned responsibility for Project Management, Contract Administration, and Design;
- the General Contractor (GC) is assigned responsibility for Contracting (with trade contractors), Construction, and Construction Coordination;
- the Owner retains responsibility for Contracting (with the GC and the assistance of A/E).

When using the **Design–Build contracting** (D–B) system,

- the D–B contractor is assigned responsibility for Project Management (in part), Design, Contracting (with trade contractors), Construction, Contract Aministration (in part), and Construction Coordination;
- the Owner retains responsibility for Project Management (in part), Contract Administration (in part), and Contracting (with the D–B contractor).

We will see that when using the **Agency Construction Management** (ACM) system:

- the A/E is assigned responsibility for Design, Project Management (in part), and Contract Administration (in part);
- the CM is assigned responsibility for Project Management (in part), Contract Administration (in part), and Construction Coordination;
- the Contractors are assigned responsibility for Construction; and
- the Owner retains responsibility for Contracting (with the help of the CM and A/E).

1.2.1 Shared Responsibilities

Responsibilities for providing some of the elements are shared between two parties. When this is the case, each contract must clearly define the shared responsibilities. This could lead to more wordy contracts, something the construction industry can well do without. Chapter 16, Project Management, offers a contract management device called a *responsibility chart* which provides solutions to the problem of contractually assigning co-responsibilities.

Figure 1.1 shows the distribution of responsibilities in matrix form.

	Project Delivery Element					
Contracting System	Design	Project Management	Contracting	Construction	Construction Coordination	Construction Administration
General Contracting	AE	AE	GC	GC	GC	AE
D–B Contracting	DB	DB/O	DB	DB	DB	DB/O
Agency CM (ACM)	AE	AE/CM	O	C	CM	AE/CM

CM: Construction Manager O: Owner GC: General Contractor
AE: Architect/Engineer C: Contractors DB: Design–Build Contractor

FIGURE 1.1 Responsibility Distribution; ACM, GC and D–B systems.

1.3 CONTRACTING STRUCTURES

The contracting structure of the three systems is the second distinguishing characteristic, and Figure 1.2 depicts the contract structure for the traditional GC system of contracting. It shows the contractual relationships of the owner, A/E and GC (note that the A/E and GC have no direct contractual ties) and the several contracting tiers or levels. It also differentiates between contractors and constructors or performing contractors and the type of contract that exists between each party.

1.3.1 Agents and Independent Contractors

Figure 1.2 shows how the A/E has an agency agreement Ⓐ with the owner and the GC has an independent contract Ⓢ with the owner.

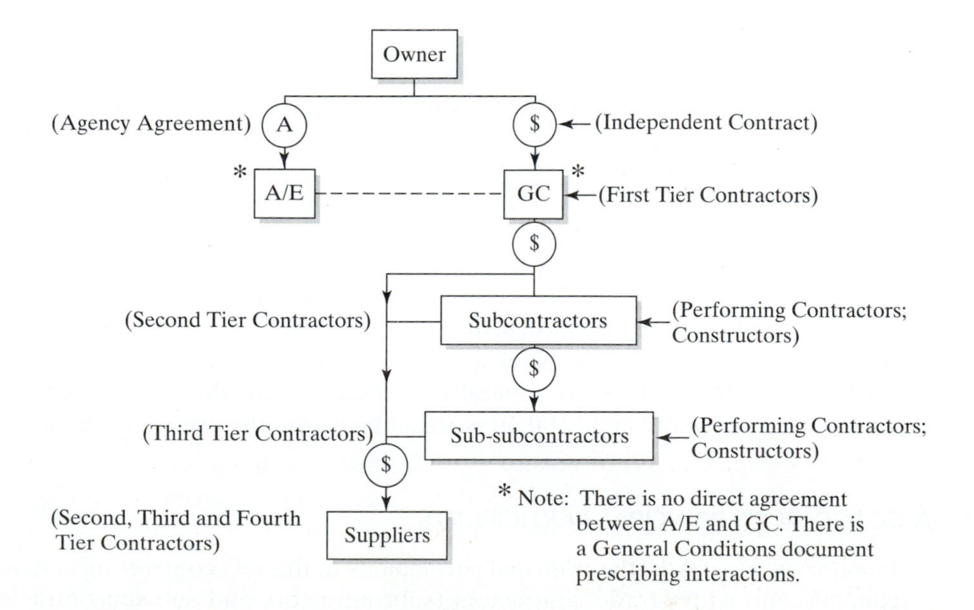

FIGURE 1.2 The GC contracting structure.

It is important to realize the difference in performance responsibilities between an independent contractor and an agent. Our legal system sets different performance standards for each and their unique status significantly affects the way contracting systems work. When providing services to an owner, an agent is bound by law to make all decisions affecting the owner to the benefit of the owner. An agent is essentially an extension of the owner—a representative who can act in behalf of the owner to the extent stated in their contract.

If the owner perceives an error in judgement on the agent's part, the degree of the agent's negligence, if any, would be measured against the standard of care that agents are expected to provide. If it can be shown that other agents, under the same conditions and circumstances, would have acted differently, a charge of negligence against the agent by the owner might be appropriate.

The opposite is true of independent contractors. Independent contractors are not only permitted but expected to make decisions to their own benefit, so long as their decisions are consistent with the terms of their contract with the owner. If a dispute is taken to court, a judge will turn to the wording of the contract to determine if a breach occurred.

This single difference between an independent contractor and an agent is significant when comparing, from the perspective of an owner, the three basic contracting systems.

1.4 CONTRACTORS AND CONSTRUCTORS

To understand the workings of a contracting system through its structure and to accurately compare contracting systems, it is necessary to use precise descriptive terminology. To this end, the following definitions are used throughout this book.

The term **contractor** is reserved for a business entity that, typically, subcontracts all of the work for which it is responsible to other contractors and with few exceptions, does not perform work with employees. A general contractor who subcontracts 100% of the work for which he is responsible fits this definition and is often referred to as a "broker general contractor."

The term **constructor** is reserved for a business entity/individual that, with very few exceptions, does all the work for which it is responsible with employees, and with few exceptions, does not subcontract work to others. A constructor is also referred to as a performing contractor. Trade contractors, subcontractors and sub-subcontractors usually fit the definition of constructors or performing contractors.

The term **contractor/constructor** is reserved for a business entity that subcontracts work to others but also does work with its own employees. A GC who does not subcontract 100% of the work generally fits this definition. Most GCs who build buildings subcontract about 80% of their work and perform the other 20% with employees.

1.5 A GC PROJECT'S PRINCIPAL PARTICIPANTS

In addition to an A/E, the principal participants in the GC contracting system include general contractors, trade contractors (subcontractors and sub-subcontractors), and suppliers.

A general contractor is the **prime contractor**, the business entity that holds a contract directly with the owner, and is referred to as a **First Tier Contractor** in the GC contract structure. **Trade contractors** are business entities that specialize in performing specific portions of the work on a project. Trade contractors are constructors hired as subcontractors or sub-subcontractors.

Subcontractors are business entities that have a contract with a prime contractor or First Tier contractor, and are considered **Second Tier Contractors** in the GC contract structure. Trade contractors hired by GCs are subcontractors.

Sub-subcontractors are business entities that have contracts with subcontractors or Second Tier contractors, and are considered **Third Tier Contractors** in the GC contract structure. Trade contractors become sub-subcontractors when they are hired by a subcontractor.

Suppliers are manufacturers, wholesalers and retailers who sell raw, processed, fabricated and finished material and equipment to contractors at all levels. Suppliers are contractors at any tier in the GC contract structure.

1.6 GC PARTICIPATING BUSINESS ENTITIES

In addition to the principal participants, all contracting systems are supported by business entities that make the systems work. In fact, it would be impossible to have a functioning construction industry without these supporting participants.

Figure 1.3 shows how the principal and the supporting business entities fit into the GC system. Note that the diagram includes an associate A/E firm and a GC joint venture partner. Although the involvement of these entities are not always necessary, they provide an important service on projects that can use their involvement.

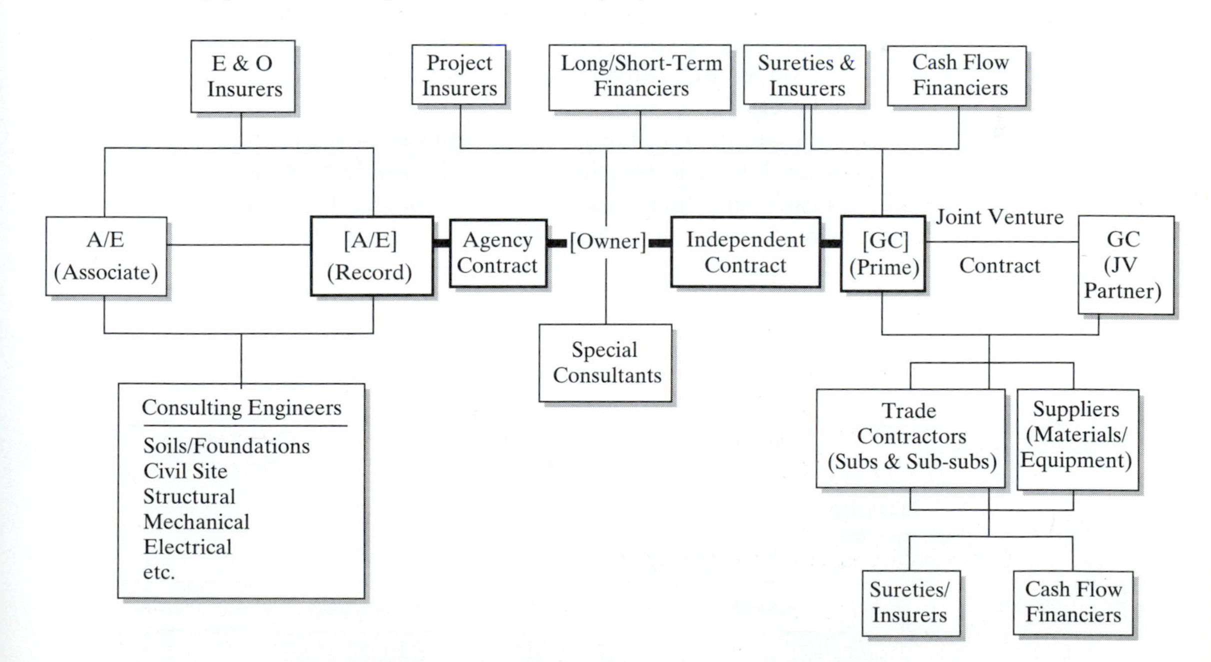

FIGURE 1.3 Participants in the GC contracting structure.

Architect/Engineer: A design professional who is qualified by license to design facilities. Architects design facilities that people occupy (architectural projects) and engineers design facilities that serve people (engineering projects). The A/E of record is in responsible charge of the project and holds the contract with the owner. An associate architect could be a limited partner or joint venturer of the A/E of record for a specific project.

Consulting Engineer: A design professional who is qualified by license to design specific elements or portions of a facility. They are either a part of the A/E organization or have contracted their services to the A/E on a project-by-project basis.

Financiers: Lending institutions that provide capital in the form of short-term (construction) loans, long-term (mortgage) loans to owners, and long- and short-term loans to contractors.

Insurers: Providers of a variety of insurance coverage such as errors and omissions, builder's risk, worker's compensation, public liability and property damage, and other special coverage.

Special Consultants: Business entities or individuals with unique expertise who can assist owners in making use-oriented decisions for the project.

Sureties: Financial backers who provide bid bonds, labor and material bonds, performance bonds, and other special bonds to contractors. The owner or another contractor is the obligee in the event of default.

1.7 A GC BUILDING PROJECT'S NINE PHASES

A construction project moves to occupancy by undertaking and completing a sequence of activities or phases that are unique to the contracting system being used. General contracting, when used on an architectural or building project uses a design-bid-build sequence as shown in Figure 1.4.

For an engineering project (such as a bridge, highway, or waste treatment plant) rather than a building as shown, design phases 2, 3 and 4 would be combined into two phases; preliminary design and final design. The activities of the phases would not change, the number would simply be reduced by one. For an overview of the activities during each of the phases, see Appendix E.

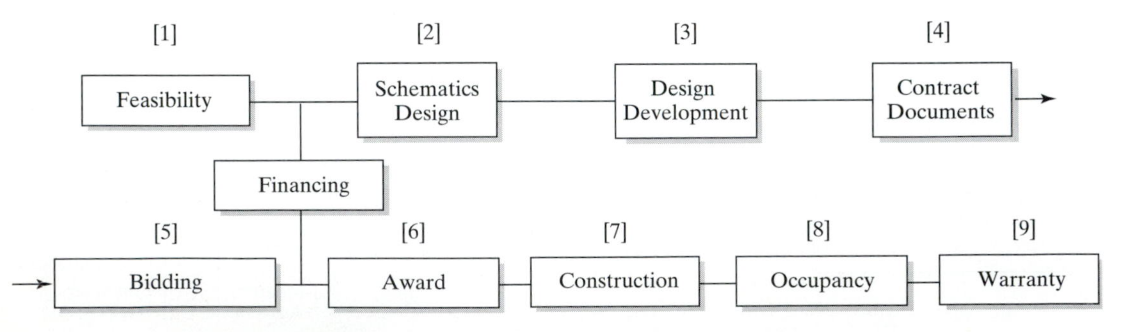

FIGURE 1.4 The nine phases of a GC building project.

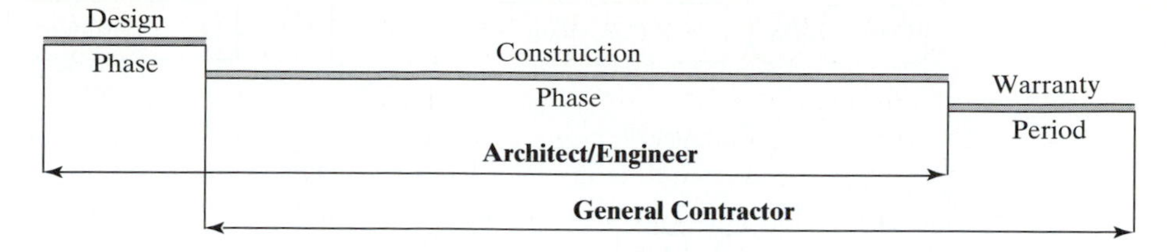

FIGURE 1.5 The tenure of the GC and A/E on a GC project.

1.8 GC TENURE

Another important aspect of contracting systems is the length of time for which each contracting party is involved in the project. The tenure of the A/E in all three systems extends from the start of design to the end of construction. The tenure of the GC is from the signing of the contract for construction to the end of the stipulated warranty period. The tenure of the A/E and GC is shown in Figure 1.5.

1.9 DESIGN–BUILD CONTRACTING

The simplest contracting structure to visualize is the D–B system. There is one prime contract that between the owner and the D–B contractor. It requires the D–B contractor to provide both design and construction services. Other parties who provide any of the six essential elements, in whole or in part, are contractual to the D–B contractor.

The contracting structure is depicted in Figure 1.6. Refer to Figure 1.1 to correlate responsibility assignments with the contracting structure.

1.9.1 Alternative D–B Contractor

The box in the upper left corner of Figure 1.6 is an alternative D–B contractor arrangement. Some owners prefer this arrangement because it provides a consequential separation of the design and contracting entities of the D–B contractor.

This separation may seem counter to the fundamental justification for using the D–B system in the first place, but from the perspective of design-construction checks and balances, this alternative provides the owner a margin of protection against design-contracting overcomplicity.

Although the A/E is the GC's agent and consequently is required to act in the GC's best interests in design decisions, the A/E's professional ethics provides a solid platform from which to make decisions that consistently meet the standard of care for a design professional.

1.10 GC VS. D–B DIFFERENCES

One major difference between the GC and D–B systems is that in the D–B system the owner has no direct contract with the A/E, who designs the project and produces the contract documents. The purported strength of the D–B contracting structure is based on the owner having a single point of responsibility; one contract instead of two. The

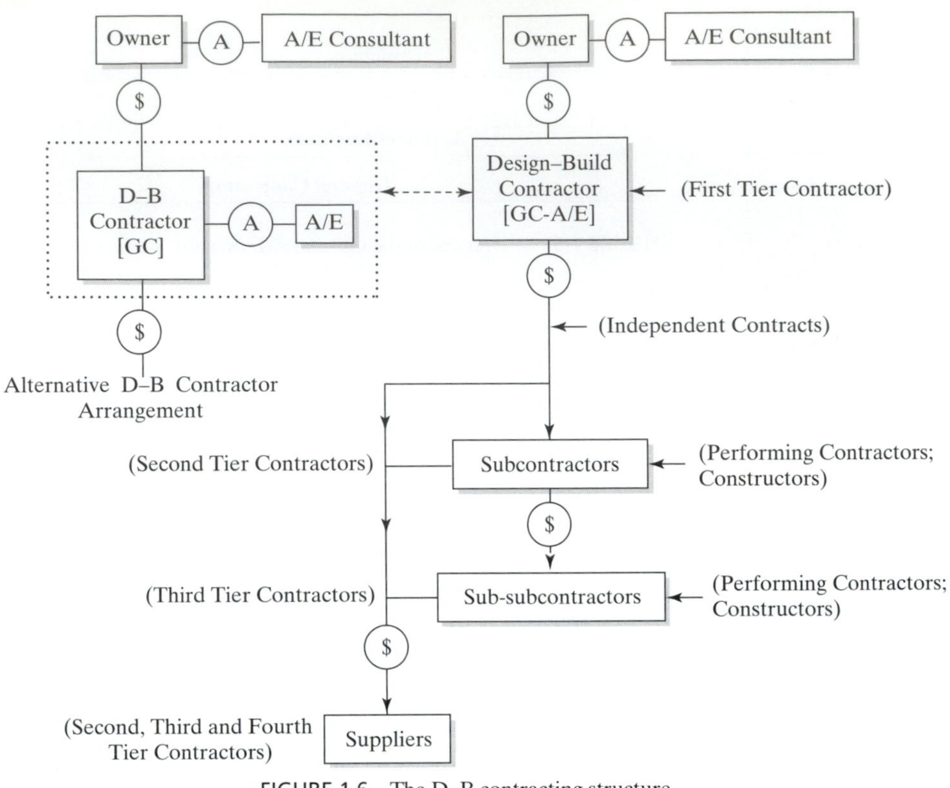

FIGURE 1.6 The D–B contracting structure.

D–B contractor is entrusted to look after the owner's interests in the design and construction of the project and is accountable to the owner for both.

The pros and cons of the D–B system are covered in Chapter 3, The Development of the CM System. The option of fast-tracking is mentioned in Chapter 15, Value Management.

1.11 D–B PARTICIPATING BUSINESS ENTITIES

Figure 1.7 shows how the various construction industry business entities fit into the D–B system of contracting.

Aside from the fact that design and construction are under one contract to the owner, the function of a GC and D–B contractor during construction is practically identical, and the participants the same.

1.11.1 A D–B Building Project's Eleven Phases

Another major difference between the GC system and D–B system is the sequence and number of phases required to execute the project. D–B contracting has a bid–design–build sequence as compared to the GC sequence of design–bid–build. The progression of activities required to deliver a building project using the D–B contracting system is shown in Figure 1.8. An overview of the phase activities is shown in Appendix E.

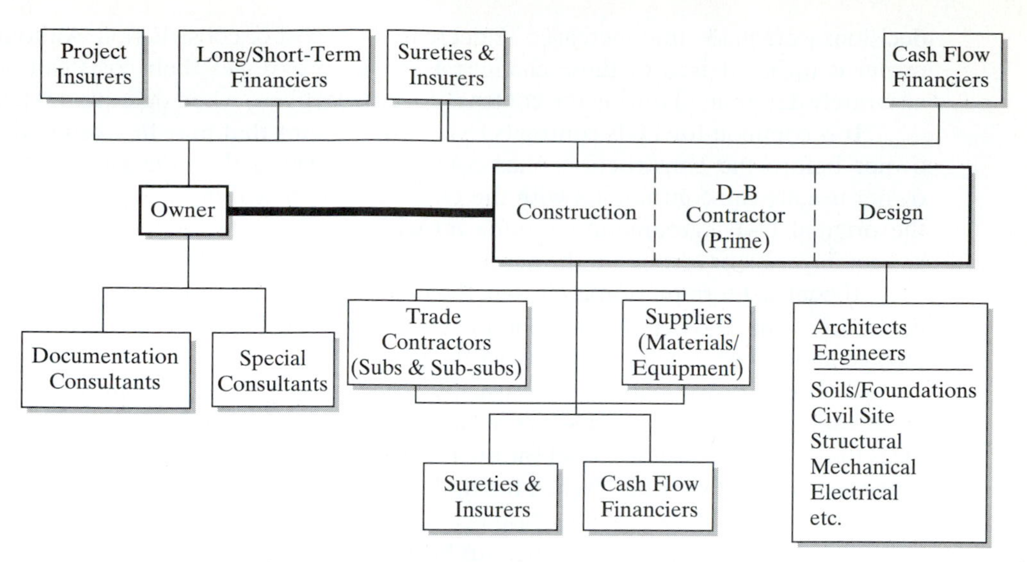

FIGURE 1.7 Participants in the D–B contracting structure.

The two additional phases in the D–B system, the documentation phase and the cost confirmation phase, are essential to the system. The documentation phase is required to develop adequate parametric design definition (schematic type drawings and outline specifications) to provide project scope and quality parity in the proposals of competing D–B contractors. The more complete and detailed the documentation, the better the chances that the owner will get the desired end result.

The cost conformation phase is required to develop an estimate of construction cost, based on completed or almost completed contract documents, before proceeding with the construction phase. It is probable that during the design phase, many mutual

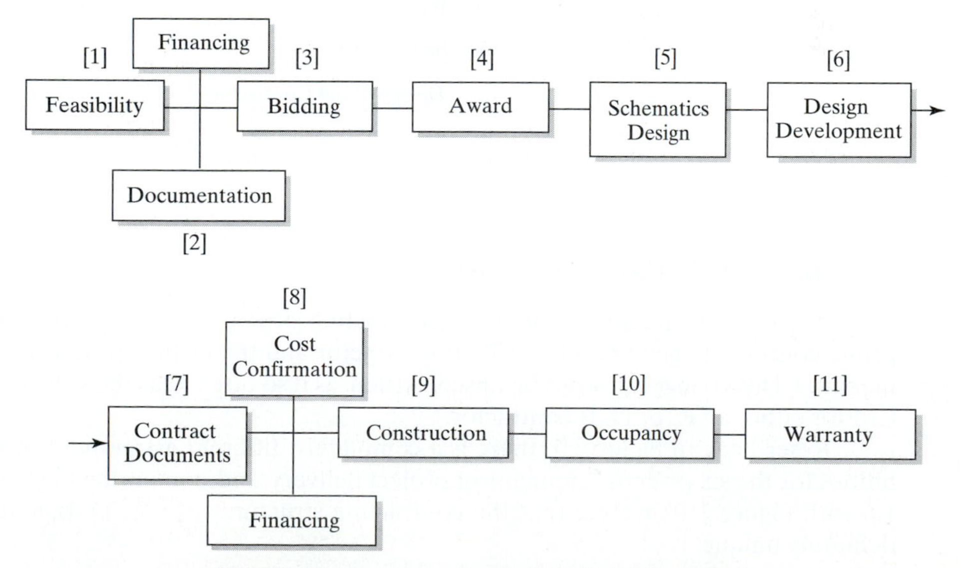

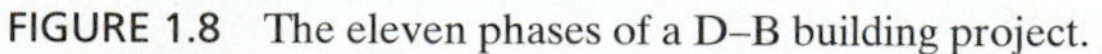
FIGURE 1.8 The eleven phases of a D–B building project.

decisions were made that increased or decreased the cost of construction. Although the owner is to be advised of these changes as design progresses their costs can only be accurately determined during the contract documents phase when definition is peaking.

It is common for D–B contracts to contain options that may be exercised by the owner before the construction phase. After establishing the construction cost, the owner usually can continue on with the D–B contractor within the economic terms of the original D–B agreement, continue on with the D–B contractor under amended economic terms, or terminate the agreement with the D–B contractor.

If contract termination is chosen, the owner has the option to not build the project or have the D–B contractor complete the contract documents so the owner can engage another contractor to construct the project.

D–B contract documents are usually structured to accommodate these options by dividing the services of a D–B contractor into design and construction. Design services under the contract are paid for on a cost-plus-fee arrangement, with or without a maximum amount specified. When the updated cost of the project is determined in the cost confirmation phase, construction services can be provided as a lump sum, a guaranteed maximum price, or on a cost-plus basis.

1.11.2 D–B Tenure

Figure 1.9 compares the tenure of a D–B contractor with that of a GC contractor. The significance of each party's tenure in a contracting system is covered in later chapters. Tenure is very important when evaluating the potential effectiveness of different project delivery systems.

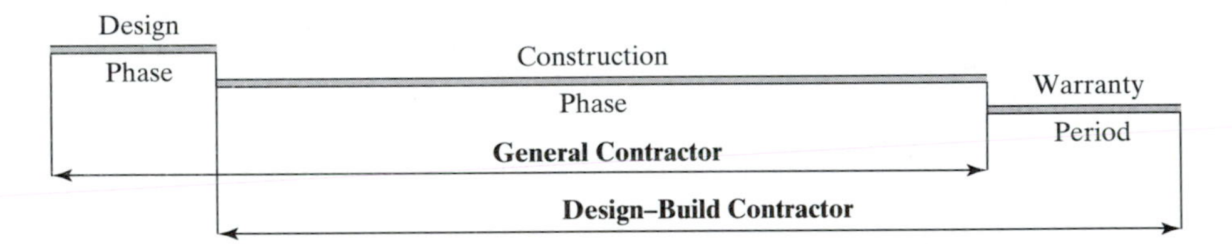

FIGURE 1.9 A comparison of the tenure of a GC and D–B contractor.

1.12 THE ACM CONTRACTING STRUCTURE

The obvious characteristic of the ACM contracting structure is the absence of a single prime contractor such as a GC or D–B contractor and the inclusion of a construction manager. This change must not be misconstrued, as it so often is, as the substitution of a CM for either a GC or D–B contractor.

Refer back to Figure 1.1; there is a completely different assignment of responsibilities for the six essential elements of project delivery, and, comparing Figures 1.2 and 1.6 with Figure 1.10, we see that the contracting structures of GC, D–B, and CM are definitely unique.

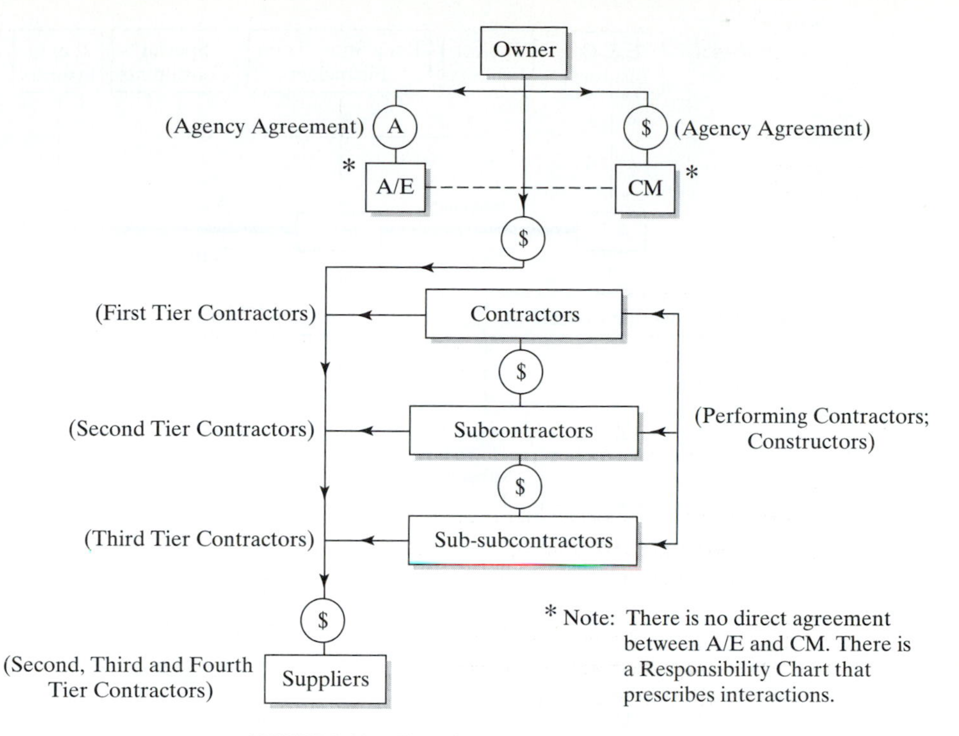

FIGURE 1.10 The ACM contracting structure.

1.13 ACM PARTICIPATING BUSINESS ENTITIES

Figure 1.11 shows how construction industry business entities fit into the CM contracting structure. The GC is conspicuously absent. However, GCs (contractors/constructors) that normally construct parts of their projects, doing work such as concrete, carpentry, excavation and masonry, can be involved as trade contractors on CM projects in those categories. The GC is excluded in the subcontracting and managing of the project. Additionally, many GCs have responded by providing CM services in addition to general contracting services. D–B contractors have unlimited opportunities in the ACM contracting structure as A/Es, CMs and Constructors.

An overview of the activities included in each phase of the ACM contracting structure can be found in Appendix E.

1.14 CM TENURE

The tenure of a CM's services on a project exceeds that of a GC's services and is no less than the tenure of a D–B contractor's services. This important feature of the CM contracting system is evident in Figure 1.12. A major portion of the CM's efforts in behalf of the owner occur during the design phase in respect to designability, constructability, and contractability services.

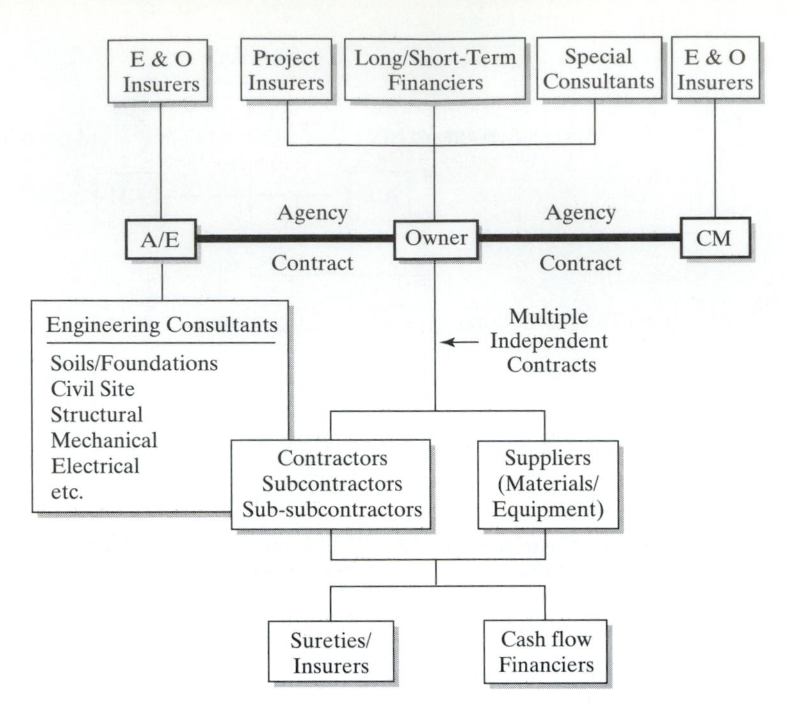

FIGURE 1.11 Participants in the ACM contracting structure.

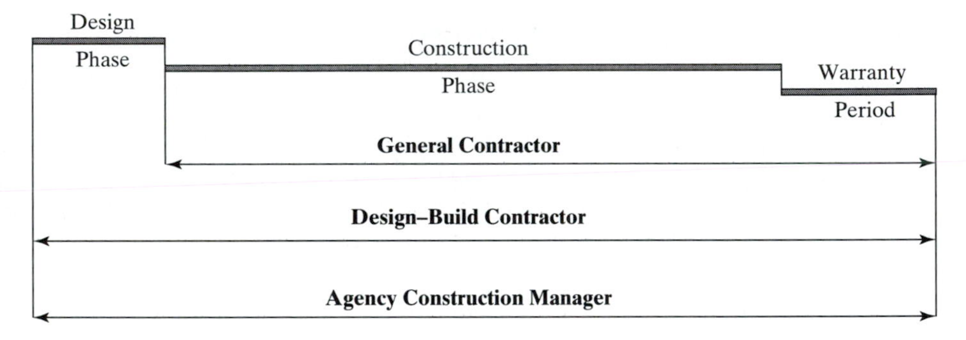

FIGURE 1.12 The tenure of GC and D–B contractors and the CM.

1.15 AN ACM PROJECT'S NINE PHASES

The phases of an ACM project, as shown in Figure 1.13, are identical to those of the GC system unless the ACM option of fast-tracking is used. On a fast-tracked ACM project, the design–bid–build sequence is repeated for each phase that is bid. The design–bid–build fast-track procedure, unique to ACM projects, is discussed in Chapter 15, Value Management.

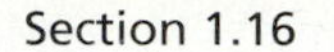

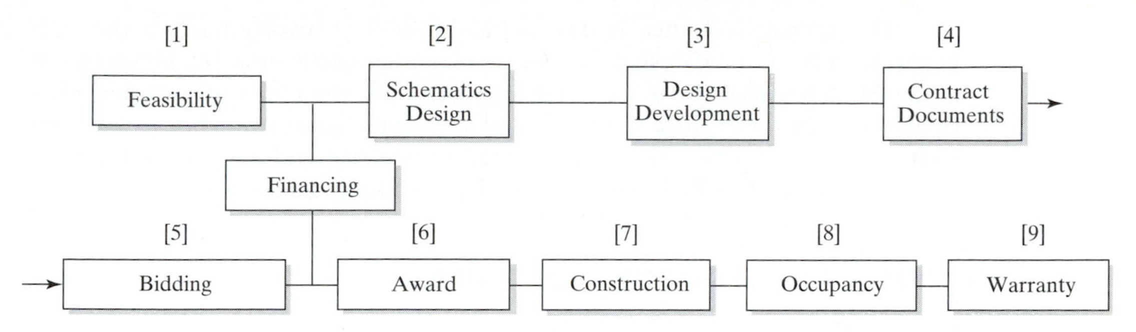

FIGURE 1.13 The nine phases of an ACM building project (identical to Figure 1.4 for a GC building project).

1.16 PRELIMINARY DEFINITIONS FOR THE THREE SYSTEMS

Based on the information provided so far, the three systems can be defined on the basis of their most obvious characteristics. More comprehensive definitions will emerge as additional information is provided in later chapters.

General Contracting: A contracting structure that consists of two prime contracts: (1) an agency contract between an owner and an A/E, who has responsibility for design, project management and contract administration services; and (2) an independent contract between an owner and a GC, who has responsibility for construction, contracting, and construction coordination services. The selection, administration, coordination and performance of trade contractors is the responsibility of the GC contractor.

The project sequence is design–bid–build. The A/E's involvement begins with design and ends with owner occupancy. The GC's involvement begins with the start of construction and ends when warranty periods expire. Fast-tracking is *not* an option.

Design–Build Contracting: A contracting structure that consists of one prime contract; an independent contractor contract between an owner and a D–B contractor, who has responsibility for design, project management, contract administration, construction, contracting, and construction coordination services. The selection, administration, coordination, and performance of trade contractors is the responsibility of the D–B contractor.

The project sequence is bid–design–build. The contractor's involvement begins with the start of design and ends when the warranty period expires. Fast-tracking *is* an option.

Agency Construction Management Contracting: A contracting structure that consists of several prime contracts: (1) an agency contract between an owner and an A/E, who has responsibility for design and partial responsibility for project management and contract administration services; (2) an agency contract between an owner and ACM, who has sole responsibility for construction coordination, and partial responsibility for project management, contract administration and contracting services; and (3) several independent contractor contracts between an owner and trade contractors, who have individual responsibility for the construction services they provide.

The project sequence is design–bid–build. The involvement of the A/E begins with the start of design and ends with owner occupancy. The involvement of the ACM begins with the start of design and ends when the warranty period of the last contractor expires. The involvement of trade contractors begins with the start of the construction they are engaged to perform and ends when the warranty period covering their work expires. Fast-tracking *is* an option.

1.17 DEFINITIONS OF A GC, D–B CONTRACTOR, AND A CM

General Contractor: A general contractor is a business organization that contracts with owners as an independent contractor to construct facilities that are designed and documented by qualified design professionals.

Design–Build Contractor: A design–build contractor is a business organization that contracts with owners as an independent contractor to design, document, and construct facilities for owners.

Construction Manager: A construction manager is a business organization that contracts with owners as an agent to manage the process that produces facilities for owners.

The definition of a design–build contractor can be expanded to include additional services such as furnishing a site, financing the project, commissioning the project, training owner personnel, and operating the facility after it has been completed.

The definition of a construction manager can be expanded to include a number of optional forms and variations, all of which are covered in Chapter 5, CM System Forms and Variations. The definition provided here is for an CM firm providing ACM services. Regardless of the CM form and variation used on the project, ACM services are always part of a CM project.

1.18 ACQUISITION OF SERVICE CONTRACTS

How owners acquire construction industry services depends on the legal status of the contract being entered into and whether the owner is operating in the public or private sector.

Public sector owners commonly use a qualification based selection (QBS) process to acquire the services of agents such as A/Es and CMs and do so with minimum interference from laws and regulations. A QBS process stresses consideration of a competitor's qualifications as well as the cost of services.

A price-based selection (PBS) process, where selection is based on the cost of services, can be used as an option but is seldom mandated. The acquisition of independent contractors providing contracting and construction services generally stipulates that a PBS procedure be used.

1.18.1 Public and Private Sector

In the private sector, the laws and regulations that control the public acquisition of service contracts do not apply. The services of contractors and agents can be acquired according to individual owner preferences.

Public sector criteria and procedures were established for the GC system to be as equitable as possible to owners and to those competing for services. Unfortunately, the criteria and procedures used by some private sector owners are less than equitable to those competing for project services. The unbalance stems from the leverage that private owners can exert by "dangling" a contract in front of competing service providers. Leverage is difficult to apply in the public sector by virtue of public laws and regulations.

Chapter 2, The Reasons for a Third System of Contracting, points out that public sector acquisition procedures are not without problems. However, private sector owners would have better experiences with the construction industry if they emulated the public sector by at least eliminating leverage regardless of the contracting system they use.

The contracting structure of the CM system eliminates the application of leverage that is detrimental to the owner in the private sector in favor of equitable competition. The GC and D–B contracting structures unwittingly invite this owner practice.

1.18.2 Acquiring Design–Build Services

PBS procedures, applicable to GC selection in the public sector, cannot be used for D–B contractor acquisition without significant modification. While this is readily accomplished at the impulsive federal level, it is not the case at the staid state, county, city, township and school district levels.

A combined QBS/PBS procedure for the D–B system has been made viable at the federal level, but combined procedures are burdensome, complicated, and slow in the lower public echelons, where acquisition procedures must legally bend existing regulations.

QBS is based on the past performance and current abilities of the D–B contractor, particularly in the area of design. PBS is based on an acceptable cost competition. An unbiased formula combines QBS and PBS results for final selection.

Trade contractors hired by the D–B contractor as subcontractors are selected by the D–B contractor on an subjective basis. Laws and regulations that govern D–B selection on public projects do not apply to the hiring of subcontractors.

The use of the D–B system is more common in the private sector where public acquisition restrictions do not apply and where industrial owners can use the system to advantage on a performance, rather than a prescription, specification basis.

1.18.3 Acquiring General Contracting Services

Current PBS procedures were established to accommodate the use of general contractors on public projects. Unlike the D–B system where design is conceptual at the time of bidding, GCs can accurately quantify the work and estimate its cost from drawings and prescription specifications before submitting proposals to owners. It is legally mandatory that all bidders have identical bidding documents to work from and all proposals be submitted at a single location prior to a specified date and time.

The separation of the design contract from the construction contract facilitates public bidding laws and allows QBS procedures to be predominantly used for the selection of an A/E. A/E QBS is based on past performance and current potential abilities. A/E selection is ultimately determined from a face-to-face interview process.

Trade contractors hired by the GC as subcontractors are selected by the GC on a subjective basis. Laws and regulations that govern GC selection on public projects do not apply to the hiring of subcontractors.

The use of GC contracting is popular in both the public and private sectors. However, its compliance to laws and regulations influences its use in the public sector.

1.18.4 Construction Management Services

CM firms hired by owners to perform ACM services are agents and typically compete for projects through QBS procedures similar to those used by owners to select A/Es. The use of QBS for ACM services complies with most service acquisition laws and regulations in all jurisdictions.

The separation of the design contract from construction contracts facilitates public bidding laws and allows QBS procedures to be predominantly used for the selection of an A/E. The A/E QBS selection procedure is similar to that used in the GC system.

Contractors needed to construct an ACM project are obtained through a competitive PBS process similar to that used by owners to acquire the services of general contractors. The contractors are trade contractors and are prime contractors in the ACM system rather than subcontractors as they are in the GC and D–B systems.

ACM contracting is an alternative to GC contracting in either the public or private sector. The flexibility of other forms and variations of CM position it as an alternative to D–B contracting in both public and private sectors.

CHAPTER 2

The Reasons for a Third System of Contracting

Chapter 1 explained the ACM contracting system in terms of contracting structure and assignment of responsibilities and compared some of its attributes with those of the general contracting and design–build systems.

Comparisons show that, with the exception of the addition of a construction manager and the elimination of the general contractor's traditional role, project participants are the same in all three systems. Only contractual relationships, assigned responsibilities, and time of involvement in a project differ.

In this chapter, the reasons for developing a third system of contracting after years of using the GC system (and to a lesser degree, the D–B system) are explained. In essence, the ideological development of the CM system and its processes is based on a combination of owner concerns for certain restrictive GC procedures and several questionable GC practices.

2.1 OWNER CONCERNS

Several areas of concern were informally but repeatedly expressed by owners in the late 1960s, most of which could not be corrected without a synthesized change in the way construction projects were being delivered.

These concerns included:

1. Projects commonly exceeded budgets
2. Projects consistently fell behind schedule
3. Construction quality was not as expected
4. Design services fell short of expectations
5. Owners had limited participation
6. Contracting flexibility was minimal
7. Contract disputes were common
8. Fair value was not being produced.

Until the late 1970s, owners had no organized access to the construction industry. This state of affairs slowly changed in the next decade with the proliferation of local and national owner organizations and construction-user councils. These groups provided access to construction industry participants and facilitated an ongoing dialogue between them and the contractors.

Owners had been vocal with their concerns prior to the formation of these groups, but without association or a forum it was difficult to impress the members of what had become a very autonomous and self-centered construction industry, to change in their behalf. Their option was to vent individual problems and concerns to the A/Es and contractors who designed and constructed their facilities. With few exceptions, owner complaints were considered part of doing business as an A/E or contractor.

Over a period of years, continuing owner grievances became a more frequent topic of conversation among construction users. This eventually prompted media attention, the initiation of owner collectives, and eventual recognition of the problems by some of the more concerned construction industry practitioners. Although most simply countered the complaints with defensive comments and denials, others interpreted owner persistence as credible and responded with action.

Using present-day terms, the major concerns of construction users fell under five specific headings: Designability, Constructability, Management, Contractability, and Questionable Industry Practices; and addressed four broad problems: Cost, Time, Quality, and Business Interruption.

2.1.1 Designability

A major owner concern was the A/E's interpretation of the owner's needs in the drawings and specifications. More owners are first-time builders than repeat-builders. They are capable of explaining their project requirements when properly prompted to do so, but cannot readily determine if their needs have been included in the A/E's design by looking at drawings and reading technical specifications.

Many cases were cited where owners were upset by the inadequacy or overprovision of the sizes and relationships of spaces. The two-dimensional drawings and the specifications that the owner approved during the design phase were acceptable to untrained eyes, but after construction and three-dimensional inspection were often questioned and sometimes unacceptable.

Owner comments to the effect that A/E firms often designed facilities for their own edification rather than for the owner's specific use and purpose were fairly common. In essence, owners felt that design professionals had distanced themselves from the realities of owner needs and were treating interior and exterior architecture as more of an artform than a pragmatic solution to the owner's needs.

The term *designability* was introduced to remind A/Es that satisfying the owner's pragmatic needs, and doing so economically, has a much higher priority than including architecture for the sake of architecture. Although both are a necessary part of good design, the balance between the two must favor the owner's needs.

2.1.2 Constructability

A serious deficiency of the GC system, as perceived by construction users, was inadequate pragmatic construction input during design. Few A/E firms require construction site experience as a prerequisite for their design personnel, and those that do have their sights set too low. It was common for awkward construction details and inappropriate materials and equipment to be incorporated into design.

The term *constructability* was introduced to describe the level of cost-related features designed into projects. Good constructability referred to a good economic match between what had to be constructed and what needed to be done to construct it.

Sound knowledge of hands-on construction is a prerequisite for assessing and improving constructability. The GC system inherently excludes contractors, the unchallenged constructability experts, from participation in a project until design is complete. Constructability responsibility is vested solely in the A/E.

A solution that surfaced was to hire a third party to perform a constructability review of the drawings and specifications prior to bidding. This solution not only proved costly because of the expense of extensive redesign but also raised questions regarding the A/E's competency.

Another proposed solution was to keep constructability expertise available during the design process so that changes could be incorporated as the design proceeds. This proved less costly and more acceptable. Design, review, and redesign became an ongoing less disruptive process, and also eliminated the trauma usually connected with a post-design review. This solution allowed time for the A/E and constructability reviewer to gain each other's respect during the process.

2.1.3 Management

Many owners' concerns were management related. They were aware that subcontracting was increasing—that the role of the general contractor had switched from constructor to contractor. This was especially true in the construction of buildings where it could be fairly stated that trade contractors, not general contractors, build buildings. A general contractor's main function was that of a manager; his main responsibility was the supervision of subcontractors.

The traditional GC bidding process provided absolutely no assurance of a general contractor's management ability. Owners simply assumed that all general contractors were capable managers, when in fact their ability and dedication as managers varied greatly. The awarding of contracts in the public sector was on a low-bid basis. The conventional GC bidding process only attested to a contractor's ability to be the low bidder and did not attest in any way to a contractor's capability to manage.

Owners realized that the keen price competition created by the GC system placed general contractors in a compromising position when rationalizing the amount of management costs they could afford to include in a competitive proposal. They couldn't lower their bids by cutting out a yard of concrete or a ton of steel, but they could pare monies allocated to management to a bare minimum. The amount and quality of concrete and steel are specified; the amount of management applied to the project is not.

Consequently, the proper management of the project during the construction process emerged as a very serious owner concern when it was realized that even a minimum level of management might be further pruned from GC dollar proposals as a result of the intense competition in the bidding process.

The fact that GCs contractually agreed to complete a project on time and at specified quality no longer provided sufficient assurances to owners that those agreements would be kept. While a contract could mitigate financial loss to the owner if these promises were not kept, a contract could not compensate the owner for the business interruption contingent to the resolution of claims or a breach of contract.

Owners realized that if they were to get fair value for dollars spent on construction, a way had to be found to guarantee high-level management without sacrificing earnest competitive bidding. Two alternatives emerged. They could explicitly specify management requirements as they did steel and concrete, or they could isolate management from the competition for contracting services. Either approach would likely produce the desired results.

2.1.4 Contractability

Contractability was originally part of constructability but experience in both areas revealed that the expertise needed for each was entirely different.

Constructability is the economic optimization of the construction aspects of a project. *Contractability* is the economic optimization of the contracting aspects of a project. Constructability references contract drawings and technical specifications. Contractability references, contract structures, contract documents and general conditions.

As owners became familiar with the inner workings of the construction industry, they showed increasing interest in the contractual arrangements of the parties involved. It was apparent that traditional contracting structures had to be modified in order to accommodate the emerging needs of owners.

The GC contract structure was straightforward but inflexible; there was little room for contracting innovation. The D–B contract structure was even more straightforward but even less flexible. Both had been used for many years and were codified by strict procedures.

Owners were ready to depart from traditional contracting and try something other than what the construction industry then had to offer. This bold attitude provided the license for responsive construction industry participants to develop the CM contracting structure.

2.2 QUESTIONABLE INDUSTRY PRACTICES

Owner concern in the areas of designability, constructability, management, and contractability spearheaded the changes. However, two questionable GC system practices influenced the direction that the changes should take (both involving the bidding process): the manner in which trade contractors were selected to be subcontractors by GCs, and the manner in which GC proposals were assembled.

2.2.1 Trade Contractor Selection

As owners realized that subcontractors, not general contractors, actually built buildings and that general contractors essentially provided supervision of the construction phase, more importance was focused on trade contractors, and increased their value to owners in the contracting process. These new perspectives influenced the direction of any new contracting system.

In the GC system, trade contractors became subcontractors only at the discretion of general contractors. As the prime contractor, the general contractor autonomously selected the trade contractors to whom work would be subcontracted. Trade contractors were dependent on general contractors for their employment and, in many cases, their success in the construction industry.

This dependency gave general contractors the license to individualize their trade contractor selection process and tailor subcontract provisions to their own advantage. Consequently, a broad range of subcontracting policies and practices coexisted. From a trade contractor's perspective, some of these practices were fair and equitable and others were partial and demanding. (See Appendix F.)

Subcontractors persistently petitioned general contractors for relief in three areas:

1. Payment for work performed paid directly by the owner rather than a pass-through payment from the general contractor.
2. Relief from the general contractor's practices of pre-bid and post-bid shopping.
3. Elimination of "pay if paid" clauses in general contractor's contracts with subcontractors.

These three requests remained on subcontractor association legislation agendas year after year. Subcontractors could not get relief from general contractor associations, so they petitioned the federal and state governments to enact legislation that would ease their burden on public sector projects.

Not all general contractors condoned these practices, and many who did were not equally overbearing. However, sufficient complaints existed to motivate subcontractors to actively seek relief wherever they could find it.

2.2.2 Direct Payment to Subcontractors

Subcontractors claim that general contractors do not make prompt progress payments to them after the general contractor has been paid by the owner for the work the subcontractors performed in the general contractor's behalf. This payment lag adversely affects the subcontractors cash flow and forces additional financing until payment is received. Conversely, it enhances the cash flow of the general contractor.

Subcontractors compensate for slow payments by inflating their proposals when dealing with general contractors known to use this practice. The inflation effect of this counter-practice by subcontractors eventually reaches the owner. Subcontractors claim that owners would save money if their progress payments were made on time. They suggested direct payment by owners to subcontractors as the ultimate solution.

2.2.3 Pre-Bid Shopping

During the GC system's bidding phase, general contractors solicit proposals from trade contractors for the work they have decided to subcontract. Proposals received from trade contractors are in fact offers to do portions of the work for a stipulated amount, and they are binding on the part of the trade contractor but not binding on the part of the general contractor.

Cost proposals are assumed to be confidential between the subcontractor and general contractor. However, to intensify the competition in the trade contractor categories, general contractors sometimes reveal actual or false price information to competing trade contractors. The intent of this action is to coerce trade contractors into reducing their dollar proposals or to increase the scope of work included in their proposals without increasing their price. This practice is called *pre-bid shopping*, and in the latter case, *scope enhancement*, which is considered by general contractors as a less intimidating descriptive than bid-shopping.

Trade contractors are well aware of pre-bid shopping and do what they can to protect their proposals while attempting to remain competitive based on the cost information being bandied around by general contractors. One protective action is to inflate their initial dollar proposal so there is room to absorb price-cutting or scope enhancement when pre-bid shopping begins.

Pre-bid shopping benefits the owner; it tends to lower general contractors' proposals prior to their submittal to the owner. However, its deviousness detracts from the intended credibility of the bidding process established for public projects. Subcontractor prices reflect defensive strategies rather than the true market value of their work.

2.2.4 Stress on GC Estimators

The sizable number of trade contractors required on building projects has turned the bidding phase in the GC system into a period of high stress for GC estimators. Literally hundreds of pieces of new and modified information from trade contractors and suppliers must be received, sorted, and analyzed before a GC's proposal is finalized.

Although bidding periods are several weeks long, the best trade contractor proposals are not received by GCs until the last day of the bidding period. The reason for this is that trade contractors do not want to reveal their best prices until it is too late for GCs to shop them. As a result, a good deal of trade contractor information is received by the GC in a brief period of time, and in some cases not in sufficient time to be included in the proposal. It can be accepted as a certainty that on most projects the low bid GC does not have all of the best trade contract prices that were available prior to bidding.

The chance of an error in the GC's dollar proposal becomes significantly higher as the owner's bid deadline approaches. Whether large, medium or small, errors do occur on practically every proposal submitted to the owner.

Low bidders often stumble into a contract award because the error they made was a "good" error, one so-called because the second bidder's proposal was closely higher (a "bad" error is when the second bidder is significantly higher). It is not unusual for the low bid contractor to recheck the proposal's content to locate where errors might have occurred.

2.2.5 Post-Bid Shopping

Post-bid shopping occurs after a general contractor receives a contract award from the owner. As previously stated, proposals received from trade contractors are offers that can be accepted or rejected by the general contractor for any reason at any time. After an award is made, the general contractor is free to search for better proposals from trade contractors, whether they submitted a proposal to the GC during the bidding period or not.

Post-bid shopping provides no benefit to the owner; only the general contractor, who pockets the difference when trade contractors lower their pre-bid proposals, benefits from post-bid shopping.

During periods of heavy competition for GC work, it is common for general contractors to bid projects at their bare cost with the intention of extracting profit through post-bid shopping. To accomplish this, all pre-bid offers from trade contractors are shopped. When the GC resorts to this practice, it is logical to assume that the GC will probably select subcontractors on the basis of their price rather than on the basis of their potential performance.

Trade contractors protect themselves from post-bid shopping in much the same way that they protect themselves from pre-bid shopping: by increasing their proposals to general contractors who they have identified as post-bid shoppers from experience. In many instances, final trade contractor offers are more dependent on their counter bid-shopping strategies than on the actual cost of the work.

Experienced trade contractors submit their offers to GCs very prudently. They do not submit proposals to all GCs nor do they give the same dollar figure to every general contractor they make an offer to. Based on this selectivity, it can be safely assumed that low-bid general contractors rarely have all of the lowest trade contractor proposals that are available when their bids are submitted to the owners.

2.2.6 Pay-If-Paid Clauses

The commanding role of general contractors in the GC contracting structure gives them license to use subcontract documents that lean heavily in their favor. One such contract provision is referred to as the *pay-if-paid* clause. This provision states that subcontractors will only be paid by the general contractor if the general contractor is paid by the owner, regardless of the amount of work done or expense incurred by the subcontractor.

When GC projects are few and far between, general contractors sometimes overextend their normally aggressive pursuit of new work and (wittingly or unwittingly) get involved with owners who are not financially sound. Anxious to get the work themselves, trade contractors sign subcontractor agreements with the GC and proceed as usual to perform their portion of the work. If, during the course of construction, the owner defaults and cannot pay a general contractor's progress payment, the general contractor has no obligation to pay the subcontractor if a pay-if-paid clause is included in the subcontract, even if the subcontractor has done the work included in the progress payment request.

To a lesser degree, on projects that are properly funded, subcontractors often perform work on contract changes and extra work requested by the general contractor or the A/E. If the change or extra work is not approved as such by the owner, and payment is not made to the GC for the work done, and if a pay-if-paid clause is in effect the subcontractor who performed the work is out of pocket without contractual recourse.

2.3 GC SYSTEM BIDDING PROCEDURES

The bidding of public projects under the GC system is a very organized process—at least on the surface. Bid forms are supplied to each bidder with instructions as to how to fill them out and exactly when, where, and how to submit them. Bids not received on time, regardless of cause, and those improperly filed are deemed unresponsive and not accepted. Bids properly filed are open and read publicly. The contractor with the lowest bid is the apparent low bidder and will be awarded a contract unless the owner rejects all bids for cause, and if no latent irregularities surface during subsequent reviews by the owner and A/E.

Behind the scenes, the GC bidding procedure, regardless of its longevity, has shortcomings that adversely affect all participants—owner, general contractors, subcontractors, and suppliers. This is especially true on architectural projects (where a greater number of subcontractors are involved) as opposed to engineering projects where the general contractor is essentially a constructor and performs the major portion of the work with on-payroll personnel.

In addition to planning an approach to constructing the project, bidding consists of gathering strategic data and pricing information from which a firm cost and time commitment can be made by the general contractor to the owner. Pricing is the main component of a general contractor's bid.

2.3.1 Sorting Bidding Information

A large amount of pricing information is gathered and processed by a general contractor while preparing a bid. The number of information pieces probably exceeds 500 on an uncomplicated project such as a school and reaches 1000 on a hospital. More important than the number of pieces of price information to be processed is the time in which processing must be accomplished.

The length of bidding periods depends upon the size and complexity of the project. However, regardless of the time allocated by the owner for bid compilation, the last day (the last hour) on every project is a hectic period for a general contractor's estimator.

As noted above, trade contractors and suppliers withhold their best price from general contractors until late in the bidding period. Consequently, general contractors must do everything possible to take advantage of last-minute prices. To this end, the final bid amount, and other required information relating to the final bid amount such as the names of major subcontractors and suppliers, is not entered on the bid form until the last possible moment.

2.3.2 Submitting GC Proposals

Local bidders and bidders from out of town have their own timing strategies. Local bidders gauge the time it will take to enter the final number and deliver the proposal to the location where they are to be received. To "level the playing field," bidders from out of town send a person to the location where proposals will be received with a bid form that is complete except for the bid amount. A phone call from the out of town bidder's office to his representative, stationed at a telephone close to where bids will be received, provides the final price information at the last possible moment. The representative enters the final bid number on the proposal form and delivers it just in time.

This scenario probably sounds ludicrous to those unfamiliar with the inner workings of the construction industry, but be assured that it is a common occurrence. FAX machines eliminate the manual entry of the final amount and the personal delivery (providing that a FAX delivery is listed as an acceptable method of submitting bids in the instructions to bidders). The use of a FAX may require that the bid amount is not revealed. In that case, a stand-in bid amount is entered on the proposal form before it is submitted, and the last-minute FAX only contains a plus or minus adjustment figure.

2.4 OWNER ASSESSMENTS OF THE GC SYSTEM

As owners became familiar with the intrinsic workings of the construction industry they realized that its business style was archaic in many ways when compared to their own. While blanket assumptions were not fair because each industry is unique, there were areas where counterpart comparisons were appropriate.

For example, owners in the manufacturing industry had difficulty understanding why the level of quality on a construction project always seemed to be controversial. Why was the quality goal in construction "minimum defects" when the quality goal in manufacturing was "zero defects"? However, once the construction process was thoroughly understood, they accepted the fact that construction quality could never simulate manufacturing quality. Construction's uncontrollable production environment, constantly changing work force, rapid mobilization, and one-of-a-kind product lines were overwhelming deterrents. Owners came to understand that quality in a construction project is the result of the overall project management rather than specific quality-control measures.

2.4.1 The GC Bidding Procedure

Private owners were especially critical of GC bidding procedures. Many of them represented industries where preparing and submitting competitive bids was routine. They quickly pointed out that neither the philosophy nor the procedures used in the GC system would be acceptable in their own industry.

Their areas of concern centered on the GC's cost-gathering process and the confusion created by sorting out volumes of information in a brief period of time. Owners agreed that intense competition was essential to favorable construction economics but felt that competition for contracts should be direct, fair, and as free from time pressures as possible.

They observed that immediately after GC bids have been received, the owner: (1) knows the name of the low bid contractor, (2) has a price on which to base a contract award, (3) has a promised date of completion, (4) doesn't know who the trade contractors are that will build the project, and (5) is aware that the low bid did not include all of the lowest trade contractor and supplier prices available. Owners felt that more and better information should be drawn from a bidding process.

2.4.2 Direct Payments to Subcontractors

Another area of concern for private sector owners was the pass-through payment arrangement for subcontractors. One of the anxieties of private sector owners is mechanic's liens filed against their property during construction. (Public sector owners are spared this anxiety because most public properties are exempt from liens.)

Owners saw direct payment to subcontractors as a way to automatically reduce the occasions to have liens filed on their property. The pass-through arrangement can and does delay payment by GCs to Second Tier parties who, along with Third and Fourth Tier parties are usually the ones who have reason to file liens.

The owner had no way of knowing if and to whom a GC was dispersing progress payments by the owner. Direct payments to Second Tier parties would eliminate owner concern with an entire tier of contractors and suppliers.

2.4.3 Bid Shopping

In the GC system, pre-bid shopping ambiguously benefits the owner and post-bid shopping indisputably does not. Unaware of the extent and severity of bid shopping, owners quickly realized through conversations with trade contractors and from their own experience in purchasing that its practice (especially post-bid shopping) was deleterious and should be replaced in the contracting process.

Some owners initially thought bid shopping was a practice equally accepted by both GCs and trade contractors—a business procedure used by the construction industry to provide fair competition for the work to be done. They were not aware that trade contractors vehemently opposed bid shopping and considered it a means of unfair competition.

Private sector and public sector owners indicated that they preferred to pay a fair, competitively-derived price rather than to coerce contractors and suppliers into a blind auction that could adversely affect the quality and caliber of their performance under lump-sum contracts. This coincided with trade contractor preferences to provide their best price, once, in a fair and open competition. (See Appendix F.)

CHAPTER 3

The Development of the CM System

Chapter 2 covered the reasons for developing a third system of contracting, pointing out owner concerns of the day and the frailties of the GC system as it developed and was practiced. This chapter provides insight into the development of the CM system as it occurred in the early 1970s, in separate parts of the United States.

CM's development was not a coordinated effort under the guidance of a group or association. Rather, it was the product of independent efforts by individual firms who recognized that inherent problems existed with the GC and D–B systems, and coincidentally found surprisingly similar ways to solve them.

The point that CM's ideological development and its practices were not a coordinated effort is an exceptionally important one. The fact that similar problems were recognized without collaboration, and that similar solutions were installed without association, helped to legitimatize CM and validate its solutions.

There is a major difference between the development of the GC and D–B systems and the CM system. The GC system, as we know it today, evolved over a long period. Throughout the last century, it was adjusted to accommodate the ongoing changes in the construction industry and the developing needs of private and public owners.

The D–B system, an adaptation of the GC system, developed in the private sector to serve manufacturing industries who had competitive production schedules to meet. Combining design and construction into one contract permitted construction to begin before design was completed, saving considerable time in the total length of the project delivery process.

In the public sector, the D–B system was used on a large scale during World War II to accelerate the construction of munitions manufacturing facilities. The strategy was to quickly catch up to and then outproduce the axis nations, who were far ahead of the allies in weapons and ammunition inventories. There was no faster way to create a producing facility than by using the D–B process. The federal government's use of the system under those threatening circumstances was tremendously effective and served its purpose.

If the development of the GC and D–B systems was an evolution, the development of the CM system was by time comparison a revolution. CM services were being provided by fledgling CM firms in various parts of the country in the late 1960s and early 1970s, but by 1975, the first standard contract documents covering the CM system were in print.

A number of proprietary CM documents, fashioned after the existing standard documents used in the GC system, were forerunners of the 1975 documents. Prior to

1975, pioneer CM firms chose to modify standard GC documents to fit the CM system, because owners and their attorneys felt comfortable with the wording of standard documents and had less difficulty understanding the CM amendments than digesting a complete new set of documents.

CM's rapid maturation was influenced by the owners' enthusiastic acceptance of the new contracting concept. CM's potential as the system they were looking for, and the possibilities that surfaced as the system was used, generated a synergistic effort by owners and CM to develop the system as quickly as possible.

Once CM's basic concepts and initial procedures were installed and in working order, attention shifted to the numerous collateral benefits of the system that emerged; these too were quickly refined and incorporated as procedures. Today it is difficult to determine which are more beneficial to the owner, those that formed the basis for CM, or those that are the product of CM.

A point worthy of note is that the ACM form as practiced by many firms prior to 1975 is still practiced with little change today. ACM practices and procedures were so obvious and straightforward when first developed that few modifications were required as experience with ACM was gained. The only real changes were in the tools used to get the job done (i.e., personal computers, FAX communication, and teleconferencing). This stability speaks well for the system's early credibility and its present integrity.

Problems that persist today are the lack of uniformity in the extent and quality of services provided by CM firms and the availability of expertise to properly provide CM services that serve the owner's best interests. The expected level and range of CM services will be clearly explained in following chapters.

3.1 ACM SYSTEM DEVELOPMENT

The process most effectively used by firms when developing CM services was to consider how to retain the positive and eliminate the negative attributes of the GC and D–B systems. The governing criteria for change were: (1) the result should use existing construction industry resources and be consistent with its basic practices, and (2) it should improve the cost, time, quality, and business interruption considerations from the perspective of the owner.

The following is a list of the positive and negative attributes of the GC and D–B systems. The GC list assumes building projects competitively bid in the public sector. The D–B list assumes building projects not competitively bid in the private sector. Lists are not prioritized.

3.1.1 Positive Attributes of the GC System as Used in the Public Sector

P01) Firm commitment to cost, time, and quality of construction.
P02) Competitive bidding of a prescribed work-scope.
P03) Single construction contract to administer.
P04) Single contract responsibility for construction performance.
P05) Minimum owner involvement required during construction.

P06) Established contracting process, familiar to participants.
P07) Contract documents which have stood the test of time.

3.1.2 Positive Attributes of the D–B System as Used in the Private Sector

P08) Firm commitment on cost and time of construction.
P09) Single design/construct contract to administer.
P10) Single contract responsibility for design and construction.
P11) Accommodates fast-track construction.
P12) Construction expertise availability during design.
P13) Continuity between design and construction.

3.1.3 Positive Attributes Common to Both the GC and D–B Systems

P14) Sheds the majority of owner risks connected with construction.
P15) Contractor's performance secured by surety bonds.
P16) The contractor is an at-risk stakeholder in the project.
P17) Uses Existing Industry Participants.

3.1.4 Negative Attributes of the GC System as Used in the Public Sector

N01) Insufficient construction expertise available during design.
N02) Insufficient contracting expertise available to the owner.
N03) Lack of owner decision-making involvement during construction.
N04) No measure of a bidder's management ability prior to award.
N05) Limits owner involvement in the selection of trade contractors.
N06) Insufficient continuity between design and construction.
N07) The stressful nature of the proposal assembly period.

3.1.5 Negative Attributes of the D–B System as Used in the Private Sector

N08) Competition is provisional without complete contract documents.
N09) Designed quality directly affects the contractor's profit.
N10) Owner must have high-level construction industry expertise.
N11) Potential conflict of interest with combined services contract.
N12) No decision-making checks and balances.
N13) Cannot be used in the public sector without modification.
N14) Non-synergistic approach to problem solving.

3.1.6 Negative Attributes Common to Both the GC and D–B Systems

N15) Long retention time on early subcontractor completions.
N16) Covert assembly of general contractor's cost proposal.
N17) Adversarial relationship exists between contractor and owner.
N18) Pass-through progress payments to performing contractors.
N19) Questionable competition between trade contractors.
N20) Prepackaged risk identification and management.
N21) Questionable supervision of trade contractors.
N22) Does not permit direct contact with trade contractors.
N23) Delegates authority for determining contract change costs.

3.2 KEEPING THE POSITIVE ATTRIBUTES

3.2.1 Firm Commitment on Cost, Time and Quality (P01)

The GC contract for construction provided this three-pronged attribute because of the definitive work-scope provided by the complete drawings and specifications provided by the A/E prior to the award. The D–B system could not equal the requirement because of the bid–design–build sequence it followed.

However, having a commitment on cost, time, and quality in the form of a contract does not guarantee a trouble-free end result. In the event a general contractor did not meet the terms of the contract, the owner had legal recourse to recover damages for shortcomings but had no way to avoid the expense and inconvenience of recovery.

A contractual commitment from a contractor on cost, time, and quality was not providing the level of assurance that owners wanted in these areas. They wanted minimum interruption of their business, both during the project and after its completion.

3.2.2 Competitive Bidding of Prescribed Work-Scope (P02)

From the perspective of the public sector, open bidding on a competitive price basis is a legal prerequisite for any contracting system. Equally fair but more complex bidding formats have been used with some measure of success in the public sector. Bidders are asked to include other value criteria, which converts to cost and modifies the dollar proposal. However, most CM system developers opted for the simplicity and decisiveness of comparing dollar proposals only. It was felt that other valuable criteria could be covered in bidder qualifications statements and in bidding screening procedures.

3.2.3 Accommodates Fast-Track Construction (P11)

One advantage of the D–B system was its inherent ability to use fast-track construction procedures. Because there was one contract for design and construction, construction could start prior to the completion of drawings and specifications. This permitted the later stages of design to proceed concurrent with the early stages of construction, the end result being compressed overall project time.

Many public owners were looking for ways to complete their projects as early as possible. Fast-tracking was an answer, but it was not compatible with the design-bid–build criteria they were essentially bound to.

3.2.4 Construction Expertise Availability During Design (P12)

The GC system precluded availability of the general contractor's construction expertise prior to construction. The D–B system provided for this facet, but in doing so, created a potential for conflict of interest resulting from the single responsibility contract format. It was considered essential that construction expertise be available to the owner and the A/E during design, while retaining the checks and balances not provided by the D–B system. This was the prime reason for creating a new system of contracting rather than trying in some way to change the GC system.

3.2.5 Continuity Between Design and Construction (P13)

The GC system precluded the reliability of continuity desired by owners. It was a two-leader process: the A/E during design and GC during construction. Although the A/E provides continuity for contract administration purposes, leadership during construction was vested in the general contractor. The continuity between design and construction provided by the design–build contracting system was ideal. Somehow it had to be emulated in a new system.

3.2.6 Uses Existing Industry Participants (P17)

To rapidly succeed, a new contracting system would have to use as many existing construction industry participants as possible and, as much as possible keep them in their traditional roles. The general contractor and design–build contractors were the hubs of their respective systems so their involvement would have to be drastically modified or eliminated.

It should be noted that general contractors and design–build contractors as business entities were not the problem; the problem stemmed from the contracting structure in which they functioned. Both had the opportunity to be part of the new system in a nontraditional role.

3.3 ELIMINATING THE GC SYSTEM NEGATIVES

3.3.1 Insufficient Construction Expertise Available During Design (N01)

Construction cost is established during the design phase. It is based on the scope of work included, the practicality of the construction details, and the quality and availability of the materials and equipment specified. Designability and constructability describe the cost factors.

A better system would provide experienced designability and constructability input during design. Because the cost benefit of both decreases as design develops (cost is incurred by changing what has already been finalized) designability must be provided early and constructability input must be provided as early as possible during the design phase.

3.3.2 Insufficient Contracting Expertise Available to the Owner (N02)

Contracting for services is a major facet of project delivery. Numerous contracts are necessarily involved and many contracting possibilities are available within the construction industry. Choosing the best way to contract for the most efficient engagement of services is called contractability.

A better system would provide owners with experienced based advice on the contracting possibilities available for a particular project, and provide the option to select the combination which provides the best end results.

3.3.3 Lack of Owner Decision-Making Involvement During Construction (N03)

The standard contract between general contractor and owner restricts the owner's interaction and involvement with the general contractor's decisions. During the course of construction, the general contractor's unilateral decisions sometimes create problems for the owner which could have been avoided through owner interaction and involvement.

A better system would position owners as close to the day-to-day construction decision-making process as they choose to be. Reducing unilateral decisions to a minimum decreases the possibility and consequences of surprise.

3.3.4 No Measure of a Bidder's Management Ability Prior to Award (N04)

The GC system provides no assurance that the low bid general contractor is a qualified manager of construction operations or that qualified personnel will be assigned to the project. The standard contract documents address the latter point, but not the former.

Owners have the right to ask the general contractor to replace personnel who are not performing to the owner's expectations. However, there is no recourse available to an owner, other than contract termination, if the general contractor is not performing at the level of management expected by the owner.

With the success of the project so dependent on well-managed performance during construction, a better system would allow the owner to separate the quality-based selection of a manager from the price-selection of a contractor.

3.3.5 Limits Owner Involvement in the Selection of Trade Contractors (N05)

The GC system requires the general contractor to select the trade contractors who will construct the project. Due to the intense GC system bidding competition, the selection criteria used by the general contractor may be price-oriented rather than performance-oriented. Their decisions may not be (and do not have to be) in the owner's best interest.

Especially on buildings, where trade contractors exclusively may provide hands-on construction, it would be appropriate and more effective for the owner to be involved in the selection of the trade contractors. A better system would involve the owner in the trade contractor selection process to a greater degree, directly or indirectly.

3.3.6 Insufficient Continuity Between Design and Construction (N06)

The responsibility of the A/E for contract administration notwithstanding, there is an obvious change in A/E project involvement once construction begins. This is much more evident on architectural projects than on engineering projects.

The change is most noticeable in A/E leadership roles. The project engineer or project architect who had a very active lead role during design does not always remain as active during construction. Responsibility for on-site liaison and contract administration is usually assigned to a field representative of the A/E firm, one who was probably not involved in the project during design.

Because many decisions made during the design, bidding, and award phases of the project are relative to problems during the construction phase of the project, a better system would improve the continuity of personnel through all phases of the project.

3.3.7 The Stressful Nature of the Proposal Assembly Period (N07)

The proposal assembly period or bidding period is unnecessarily demanding and prone to mistakes in judgment and calculations, especially in the public sector. Regardless of how long the period lasts, the current process concentrates the major effort by contractors in the final days and hours of the bidding period. It is common for a mistake in communication, arithmetic, or judgment to determine the low bidder. In spite of the fact that prime contractors and trade contractors are used to the GC bidding system and have shown reluctance to change, it is certainly not in the best interest of the owner when a mistake is the deciding factor in the selection of a contractor.

The conversion of complex contract documents into a firm dollar proposal, within a short span of time, has been a practice of the GC system for as long as it has existed. However, the increased number of specialized trades contractors and products involved on architectural projects has greatly increased the burden of the estimating effort.

While there have been attempts to relieve the stress connected with bidding, none has proved popular with all parties. The principal concern is the late receipt of offers from trade contractors due to pre-bid shopping and bid peddling.

Bid depositories, where trade contractors deposited written offers to individual prime contractors the day prior to the owner's receipt of proposals, eliminated bid shopping and peddling and the late rush of information, but was unpopular with those contractors who wanted to be involved in pre-bid shopping and peddling.

Bid depositories are used in some areas of the United States and Canada but have not become popular industry-wide. Where they are in use, care must be taken to comply with restraint of trade laws, which in a broad interpretation favor bid shopping activities.

A better contracting system would provide a bidding process less prone to mistakes—one that suitably recognizes the gravity of the financial commitment between an owner and a contractor and replaces chance with appropriate business acumen.

3.4 ELIMINATING THE D–B SYSTEM NEGATIVES

3.4.1 Competition is Provisional Without Complete Contract Documents (N08)

It is impossible for contractors to submit highly competitive proposals on a project without the benefit of a comprehensive work-scope to work from. The D–B system precludes this benefit. The bid–design–build sequence requires bidders to develop proposals from performance criteria, outline specifications and schematic drawings, the contents of which are open to broad interpretation and subjective decisions on quality, quantity, and in general, the credibility of the constructed project.

Price competition, based on complete drawings and specifications, is essential if not mandatory when conducting an equitable bidding process. A better contracting system would preclude subjective proposal interpretation in the bidding process.

3.4.2 Designed Quality Directly Affects the Contractor's Profit (N09)

When the responsibilities for design and construction are under a single contract, design decisions may lose owner objectivity beyond code requirements. As an independent contractor, with the right to place his own interests above the owner's, the design–build contractor can make biased cost-quality decisions without challenge from a design professional. A better system would separate cost decisions and quality decisions and allow the owner to make the final determination based on expert cost-quality information.

3.4.3 Owner Must Have High-Level Construction Industry Expertise (N10)

When using the D–B system, the single-responsibility contract places the owner in a prime decision-making position on all project delivery questions which arise. The owner must have considerable knowledge of all facets of design, contracting, construction, and the workings of the construction industry—total dependency on the independent design–build contractor can create misunderstandings and generate problems.

Many owners, especially in the public sector, do not have sufficient expertise to accept this role. Most are first-time users of the construction industry, proficient in their own endeavors but not capable of rationalizing the many demanding decisions they must make during the project. A better system would provide owners with reliable, impartial expertise in all areas where owner decisions must be made and where owner knowledge may be lacking.

3.4.4 Potential Conflict of Interest with Combined Services Contract (N11)

A potential for conflict of interest is present any time two parties enter into a contract, especially when one party has two or more contesting responsibilities to perform on the same undertaking for the other party. Can a party consistently make decisions to the advantage of the party he/she is serving, if the decisions would prove detrimental to the party making the decisions?

When legal or ethical consequences are considered, the potential for conflict of interest is minimized. When neither legal nor ethical consequences impede, however, the potential for conflict of interest is higher. A better contracting system would preclude dual, contesting responsibilities and rely on legal consequences to minimize potential conflicts of interest as much as possible.

3.4.5 No Decision-Making Checks and Balances (N12)

A single-responsibility contract suppresses debate and precludes the positive effects of checks and balances in decision making. Because of the independent contractor status, unilateral decisions often favor the contractor, not the owner.

Unilateral decisions should be avoided as much as possible. The many-faceted effects of project delivery decisions call for a synergistic approach to solutions, and owners, the ultimate decision makers, deserve the benefit of complete and diverse input for their decisions.

Checks and balances in decision making have become the essential elements of successful project delivery and should be part of an effective contracting system.

3.4.6 Cannot Be Used in the Public Sector without Modification (N13)

In the public sector, the major deterrent of the D–B system is its bid–design–build sequence. Most public statutes require a design–bid–build sequence by stipulating that competitive bidding is to take place based on a fully defined and documented scope of work. Each bidder must bid on the exact same work-scope, under uniform conditions.

To use the D–B system in the public sector would require owners to amend their statutes. This was considered a major imposition—one which would require either considerable owner education and time, or installation of a unique competitive D–B bidding system that respected existing statutes. This was considered too complicated to effectively install at the time. A better contracting system would be one that could serve both public and private sector owners without the need to amend existing statutes or provide selective modification.

3.4.7 Non-Synergistic Approach to Problem Solving (N14)

Vesting responsibility for project delivery in one contracting entity produces a provincial approach to decisions by limiting exposure to the project delivery experiences of other construction industry participants. While this may be an asset to provincial private sector owners, many owners wanted more diverse input and the choices which that input generates. A better contracting system would bring together parties with diverse experience to promote synergistic design, contracting, and construction solutions.

3.5 ELIMINATING THE COMMON NEGATIVES

3.5.1 Long Retention Time on Early Subcontractor Completions (N15)

The owner's retention on progress payments was a major concern of prime contractors and subcontractors during the late 1960s and early 1970s when interest rates soared along with inflation. Retainage is money already earned by the contractors which remains in the owner's custody, accruing interest for the owner's account instead of the contractor's. An example of retainage is when prime contractors hold back proportionate amounts from subcontractors under pay-if-paid clauses.

Retained amounts and the time of release by the owner are contractual and vary from project to project. A 10% retained amount based on the value of work in place is a common provision. The retainage is sometimes reduced when the project is half complete, but it often remains at a set level until the date of substantial completion or final completion of the project.

Subcontractors, who complete their work early in the project, are more affected by retainage because their monies are not freed up by the owner until months or years after they have completed their contractual obligations to the prime contractor. Subcontractors involved in a project during the late stages of a project are also denied full payment but for a shorter period of time.

Retained amounts increase the cost of doing business for contractors and consequently increase the owner's cost of construction. Although the owner earns interest on the retained balance, contractors add to their dollar proposals the cost of financing the retained money. A better contracting system would simplify the retainage provisions to make them easily administered and more equitable to trade contractors.

3.5.2 Covert Assembly of Contractor's Cost Proposal (N16)

The GC system typically requires prime contractors to submit cost proposals for the work-scope described in the contract documents. Proposals are usually in the form of a lump sum on architectural projects and unit prices on engineering projects.

The D–B system requires prime contractors to submit cost proposals as a lump sum or guaranteed maximum price based on preliminary work-scope documentation and performance specifications. The comparatively loose project definition lends itself to contract negotiation rather than to competitive bidding. In the private sector, most design–build contracts are arranged by negotiation.

Regardless of the type of proposal submitted, prime contractors accumulate costs and assemble their proposals in private, essentially in the same manner. The substantial quantity of figures and data involved and the pressure generated by time constraints make the process prone to error.

When made aware of prime contractor proposal assembly practices, owners became concerned that its informal, laborious, hurried, and tentative nature did not always produce a dollar proposal in their best interests. Looking to their own experience, owners sensed there was a better way to arrive at such an important cost decision. A better system of contracting would include a bidding process that is highly competitive but reduces the opportunity for bidder error. It would be one that provides more cost detail to the owner and essentially finalizes costs when bids are received.

3.5.3 Adversarial Relationship Exists Between Contractor and Owner (N17)

When any contract is signed, an adversarial relationship is created in a variable sense of the term. Both parties have a specific commitment to meet; a prescribed obligation to perform. However, as long as value and money are the basis of the contractual arrangement, contracted parties are inclined to act in their own best interests and extract as much money or value as they can from the arrangement.

Lump sum contracts commit the owner to pay the contractor, or the contractor to pay a subcontractor or supplier, to complete a defined work-scope or provide a product for a predetermined price, on time, and at specified quality. If these conditions require change during the course of construction, both parties react defensively to negotiate the change in their own best interest.

In the GC and D–B systems, the prime contractor must negotiate all contract changes because every change will affect either the owner and/or his subcontractors and suppliers. The size and nature of the change, and who is involved in it, will determine where the contractor's best interests and advocacy is positioned.

Changes that materially affect the contractor's financial outcome will get the contractor's attention more than those which do not, and the parties involved in the change will determine the contractor's advocacy position when negotiating. The contractor may be his own advocate, the advocate of the owner, or the advocate of his subcontractors or suppliers. A better system would minimize the creation of adversarial situations by limiting the parties involved in negotiating changes and eliminating the question of advocacy.

3.5.4 Pass-Through Progress Payments to Performing Contractors (N18)

Pass-through payment systems (where prime contractors are paid by the owner and subcontractors are paid by the prime contractor) create problems for the owner and for subcontractors. (See Appendix F.)

Owners do not know when, if, or how much subcontractors and suppliers are paid for services and products incorporated in their project. Subcontractors do not know when, if, or how much they will be paid. Prime contractors have their own contractual payment arrangements with subcontractors and suppliers, and the owner is not privy to them.

In the private sector, and in some public sector jurisdictions, unpaid suppliers and subcontractors have the legal right to file mechanic's liens against the owner's property. A lien gives the lienor a financial interest in the owner's property in the amount of the lien's face value, and payment of that amount to the lienor is the only solution available to the owner. While surety bonds which financially protect the owner are mandatory in the public sector and optional in the private sector, they cannot prevent liens from being filed against the owner or the surety and the inconvenience which accompanies them.

From the trade contractor's perspective, prompt progress payments from prime contractors are a positive influence on cash-flow. Payment delays, however, require additional financing and increase trade contractor costs. Trade contractors who subcontract to prime contractors known to be "slow pay" increase their dollar proposals to compensate for it, and owners suffer negative financial consequences from increased prime contractors dollar proposals.

Trade contractors have tried long and hard to eliminate pass-through payments. They have made some progress in the public sector but none in the private sector. Some federal contracting regulations require prime contractors to pay subcontractors within a stated time. Efforts by trade contractors to similarly change standard contract documents have met with refusal by prime contractor associations. A better system of contracting would eliminate the inconvenience and costly consequences of pass-through progress payments by requiring direct payment to trade contractors.

3.5.5 Questionable Competition Between Trade Contractors (N19)

Bid shopping, scope enhancement, and bid peddling have become an integral part of the GC and D–B contracting systems. All three detract from a logical competitive bidding process, produce questionable results, and can generate serious problems for the owner as well as trade contractors during the project.

Bid shopping is when a prime contractor bargains (truthfully or untruthfully) with trade contractors over a price for the work they intend to do. Scope enhancement is bid shopping where the amount of work, not the price, is the subject of the bargaining. Bid shopping and scope enhancement originate with the prime contractor. Bid peddling is bid shopping or scope enhancement that originates with the trade contractor.

When these practices occur before bids are received by the owner, they are termed *pre-bid*; if they occur after award of a contract to a prime contractor, they are termed *post-bid*.

Although pre-bid and post-bid shopping and scope enhancement intensify competition, the ultimate value of these practices to the owner is questionable. In both GC and D–B contracting systems, trade contractors depend entirely on prime contractors for their work load. Under pressure to obtain a subcontract, trade contractors sometimes cut their prices or expand their work scope to a point where their performance during construction is adversely influenced by their efforts to extract a profit that is not there or to mitigate losses from overzealous bidding.

Trade contractor (subcontractor) groups have had bid shopping elimination on their agenda for years but have made little progress. A/Es are in a position to reduce pre-bid shopping by having prime contractors list subcontractors and suppliers on proposals forms, but have not involved themselves in an effort to stop this practice. The A/Es claim that listing complicates the bid preparation process and is not in the owner's best interest. When listing is required, prime contractors often ignore it, knowing that the owner can waive proposal irregularities and are more interested in the dollar proposal anyway.

Prime contractors consider bid shopping their privilege and are openly adamant about retaining it. They insist that pre-bid shopping lowers their bids to owners—that it gives them an unbridled opportunity to find the lowest trade contractor offers and to assemble the lowest possible proposal to the owner.

Prime contractors claim that post-bid shopping gives them the opportunity to retrieve the profit they surrendered in their competitive bid. They readily admit that post-bid shopping (the more vigorous of the two) does not directly benefit owners but assert that it reduces all prime contractor proposals in the end.

To combat the negative effects of bid shopping and scope enhancement, trade contractors inflate their initial offers to allow room for shrinkage under shopping pressures. They also wait as long as possible in the bidding period to pass on offers to prime contractors. This shortens the time for shopping by reducing the number of bid shopping rounds that prime contractors can make.

Trade contractors do not make offers to all prime contractors bidding a project, nor do they give every contractor the same price. Offers are withheld or adjusted based on past experience with prime contractors. Seldom if ever does the low bid prime contractor have all of the available low trade contractor offers in the proposal he submits to the owner. A better contracting system would eliminate bid-shopping, ensure that all of the lowest bids are received, and retain a high level of competition. (See Appendix F.)

3.5.6 Prepackaged Risk Identification and Management (N20)

As owners became aware of the inherent risks of the project delivery process, they related them to risks they were exposed to in day-to-day operations of their own business. Whatever their endeavors, risk taking was part of it. They understood that risk, once identified and evaluated, could be managed, eliminated or assigned that the cost of assignment often exceeded the cost of management, and that elimination, when appropriate, was least costly of all.

There are two types of risks involved in any enterprise: static and dynamic. Static risks are those over which management decisions have no control. Accidents resulting in injury, loss of life or property damage, and situations leading to financial loss from contractor nonperformance are static risks which, on a construction project, can be assigned to others in the form of insurance or surety bonding.

Dynamic risks are the result of management decisions, and although the financial loss can be substantial, it is neither possible nor practical to insure or bond their financial consequences. These risks must be managed or eliminated by the party who is assigned the risks.

Both the GC and D–B systems remove the owner from risk involvement by essentially assigning all risks, static and dynamic, to the prime contractor, as part the dollar proposal. Protection against static loss is specified in contract documents as property and liability insurances, labor and material, and performance bonds. But protection against dynamic loss is covered by contingency amounts in contractor proposals and the day-to-day decisions made by the contractor.

Owners saw the prospect of cost savings and the benefit of synergistic decision making if they were involved in risk disposal and risk sharing. They realized many dynamic risk decisions directly or indirectly affected their best interests and wanted to be involved in those decisions. They also favored the self-funding of project delivery contingencies rather than paying contractors contingent amounts that may or may not be used. A better contracting system would permit risk sharing and involve owners in dynamic risk decision making to whatever extent they desire.

3.5.7 Questionable Supervision of Trade Contractors (N21)

Subcontractors rely on the prime contractor to schedule their work and coordinate contracting and construction interfaces on-site. A subcontractor's successful operation depends on getting on the project when enough work is available, getting off the project as quickly as possible, and experiencing few, if any, call-backs.

When scheduling and coordinating on-site activities, prime contractors are not always sensitive to the performance goals of their subcontractors. It is common for prime contractors to prematurely order subcontractors on-site and to prioritize their own work in a manner detrimental to the subcontractor's best interests.

Owners are not privy to the contracts between subcontractors and prime contractors. Subcontract forms are mostly nonstandard and often favor the prime contractor due to the relative bargaining positions during subcontract negotiations. Subcontractors resist one-sided agreements but face the reality that their access to work is through a prime contractor.

The importance of trade contractors as constructors, especially on an architectural project, suggests that their involvement in the project should have high priority with regard to scheduling work and coordinating contracting and construction interfaces. As well, their contractual responsibilities should not be encumbered by inconsistent and subjective contractual language. A better contracting system would recognize the operational goals of trade contractors, prioritize on-site trade contractor coordination, and eliminate individualized contracts and one-sided contract provisions.

3.5.8 Does Not Permit Direct Contact with Trade Contractors (N22)

Trade contractors are the constructors on the project. Consequently, performance and coordination communication either begins or ends with them. Both the GC and D–B systems preclude direct communication between the owner and subcontractors.

Standard contract documents interject the prime contractor as the liaison for communication in both systems. In the GC system, the A/E is also positioned between the prime contractor and the owner. As a result, problem solving is a pass-through activity prone to time delays and misunderstandings.

In both the GC and D–B systems, there is good reason for an intermediary. The insertion respects the contractual hierarchy designed into both contracting structures and protects the integrity of the design element in the GC system. Without intermediaries, construction decisions and contract changes could not be controlled. A better contracting system would bring the owner and performing contractors closer in the communication chain without weakening the design integrity link.

3.5.9 Delegates Authority for Determining Contract Change Costs (N23)

Few projects are constructed without making technical or administrative changes that increase or decrease the owner's cost after construction contracts are signed, and most changes involve more than one contractor or supplier and involve complex price negotiations.

On both GC and D–B projects, the prime contractor assembles the costs of a change from subcontractors and suppliers and submits the total for review and approval. In the GC system, a cost change is submitted to the A/E for review and passed on to the owner for approval. In the D–B system, the prime contractor submits a cost change directly to the owner for review and approval. The cost of a change is developed according to the provisions in the contract for costing changes.

The prime contractor in both systems stands to gain from a change in one of two ways: (1) indirectly through the mark-up on the accumulated costs of his suppliers and subcontractors, or (2) both indirectly and directly through the mark-up on suppliers and subcontractors and on the cost of his own work in the event he is also a performing contractor involved in the change.

The prime contractor, an independent contractor and a financial beneficiary of a change, has no obligation and little incentive to consider the owner's best interests when assessing and assembling the cost of changes. This is true in both the public and private sectors and for both competitively bid or negotiated contracts.

In the D–B system, the owner is expected to be the knowledgeable party when evaluating and negotiating changes. In the GC system, the A/E acts as an agent representing the owner's best interests, replacing the owner as the knowledgeable party in evaluating and negotiating changes. A better contracting system would put the owner in a better position to evaluate and negotiate changes and allow a qualified knowledgeable party obligated to look after the owner's best interests to closely scrutinize trade contractor and supplier costs.

3.6 CONTRACTING SYSTEM COMPARISONS

All three contracting systems—general, design–build, and construction management—have their own composite advantages and disadvantages that foster their use. Consequently, they can be rated on features that are of interest to owners.

Figure 3.1 provides a relative rating of the systems from the perspectives of project cost, project time, owner risk, and conflict of interest potential. CM is represented without consideration for its several forms and variations. The positive and negative influence of CM forms and variations on the four perspectives will be covered in later chapters.

In Figure 3.1, the systems are depicted under their best-use conditions from the owner's perspective:

Design–Build as a negotiated contract, fast-tracked, in the private sector;

General Contracting, competitively bid as a lump sum, in the public sector;

Construction Management as Agency CM, in the public sector.

Figure 3.1a reveals that multiple contracting CM provides the owner with the lowest project cost. General contracting is second best and design–build, third.

The reason for this difference is multiple contracting—the opportunity to obtain trade contractor proposals in direct competition without bid shopping or a second layer of profit added in. Multiple bidding and its affect on project cost is explained in Chapter 22, Multiple Bidding and Contracting.

Figure 3.1b reveals that fast-tracked design–build provides the owner with the fastest project delivery time. Fast-tracked construction management is second best and general contracting, third.

The basis for D–B's high rating is fast-tracking (starting construction before design is complete). The single point responsibility for design and construction permits this. General contracting requires complete design before bidding which precludes fast-tracking.

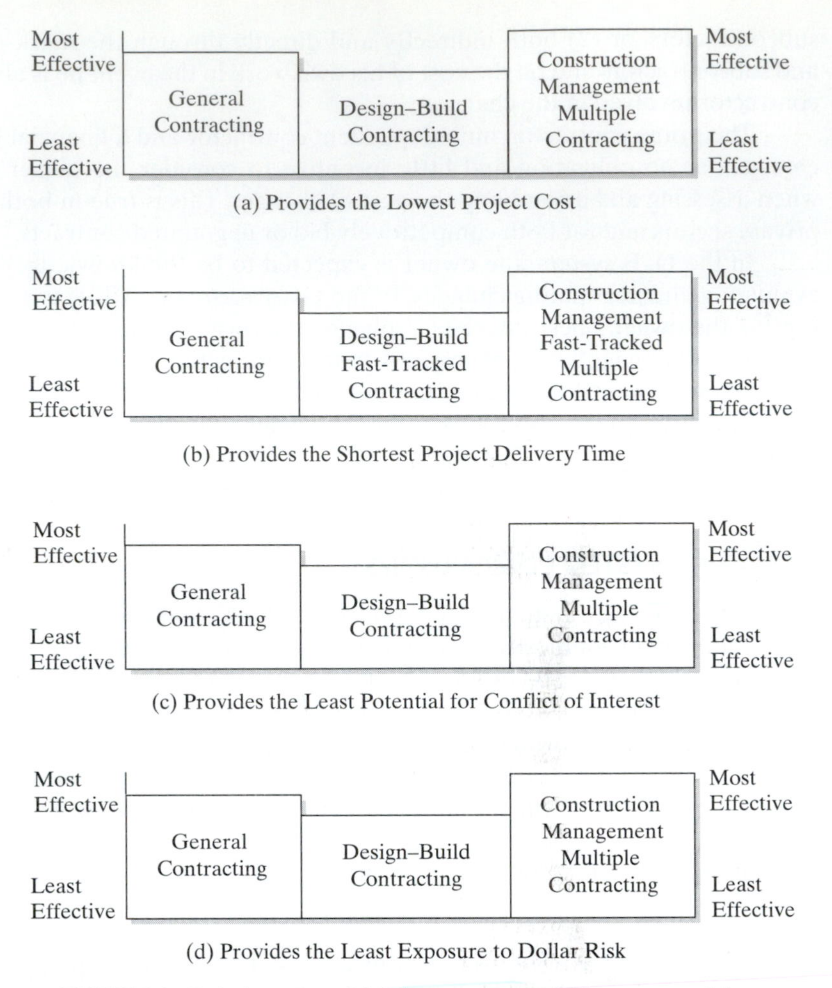

FIGURE 3.1 Relative rating of the three systems under best-use conditions.

The CM system facilitates fast-tracking with a series of design–bid–build sequences—completing the project almost as fast as D–B but with better cost results. CM fast-tracking is covered in Chapter 15, Value Management.

Figure 3.1c reveals that multiple-contracting CM has the best rating with regard to potential conflict of interest. General contracting ranks second and D–B, third. The construction manager's agency relationship with the owner and the team concept of checks and balances provides an almost conflict-free contracting structure.

General contracting generates a potential conflict situation between the owner, who believes the contract documents establish the project's minimum performance requirements, and the independent contractor who conversely believes that they establish the maximum requirements.

The combined design–construct responsibility of a D–B contractor vests responsibility for both developing and interpreting the contract documents without critical review or the benefits of checks and balances. The D–B contractor's reputation and professional acumen must be relied upon to protect the owner's best interests. The potential for conflict of interest is discussed in Chapter 5, CM System Forms and Variations, and Chapter 6, Construction Management Under Dual Services Agreements.

Figure 3.1d places the D–B system as the best for protecting owners against cost increases during design and construction. D–B has a bid–design–build sequence. General contracting and CM, with their design–bid–build sequence, rank second and third, respectively.

Assuming the D–B contractor provids a lump sum or guaranteed maximum price to the owner before design begins and there are no owner changes, the cost of the project should not increase. The D–B contractor should design to cost, build to cost, and do all possible to remain within the proposal amount.

The general contractor, on the other hand, has no input on the design and provides a price based exclusively on the contract documents produced by the A/E. There is no guarantee that GC proposals will be below the A/E's estimate or that the project will be completed without increased cost due to owner and A/E changes.

The CM system provides checks and balances during design which greatly reduce the incidence of errors and influence the A/E to design within budget. The likelihood that cost will not be controlled is offset by the diligence and expertise of the team's effort. However, multiple contracting does not provide a single lump-sum amount that limits the owner's cost by contract. For example, construction support items are part of AGC's and D–B's dollar proposal. In the CM system construction support items are budgeted for owner payment. They could inflate, however; budgeted items, such as construction support, only represent about 15% of total project cost. Inflation in construction support would be very small in terms of the total cost of the project.

3.7 CM AS A COMPOSITE SYSTEM

The ranking of CM in the four categories shown in Figure 3.1 indicates the direction construction managers took in the development of the ACM system. ACM received the highest ranking in two of the categories; it is a close second in one, and last in the other.

It should be understood that there are forms and variations that can change CM's ranking according to owner needs. They are discussed in Chapters 5, CM System Forms and Variations, and 6, CM Under Dual Services Agreements.

CHAPTER 4

The Agency Form of CM

Chapter 3 laid the groundwork that supported the need for beneficial change in the way construction projects were delivered. The option of changing either or both the GC and D–B systems was considered, but the changes needed by construction users and practitioners were too fundamental and extensive to fit within the basic framework of either system. The alternative was the development of a new contracting structure that could accommodate as many of the beneficial changes as possible. The new system was conveniently called Construction Management or CM after a similar contracting process being used by the General Services Administration at the time.

This chapter introduces the CM system by explaining ACM, the root form of CM, in sufficient detail to differentiate between the forms and variations of CM presented in Chapter 5.

4.1 THE ACM CONTRACTING STRUCTURE

One major difference between ACM and the traditional GC system is that one complete contracting tier has been eliminated. The general contractor, the traditional prime contractor, has been eliminated, and trade contractors, traditionally subcontractors to the prime contractor, become prime contractors.

Another major difference is that a new business entity, a new construction industry participant, is included in the structure: a construction manager; not in the role of an independent contractor but that of an agent with fiduciary responsibility to the owner.

The CM, whose services begin prior to the start of design, has the opportunity to add construction and contracting expertise to the common goals of the project. The availability of this expertise during design and construction favorably responds to many of the primary concerns expressed by construction industry users.

To provide a perspective, Figures 4.1, 4.2, 4.3, and 4.4 show the differences and the similarities of the GC and ACM systems of contracting. The distribution of responsibility for the six essential elements of project delivery, the contracting structure, and the project tenures are different. The design–bid–build sequence and the use of an A/E employed directly by the owner are common to both.

Both systems use the same construction industry resources. The trade contractors who are subcontractors and sub-subcontractors in the GC contracting structure are prime contractors and subcontractors in the ACM contracting structure.

For example, the Acme Masonry Co., employed by GC Construction Inc. on the local courthouse project, is employed by the Westport School District on the

	Project Delivery Element					
Contracting System	Design	Project Management	Contracting	Construction	Construction Coordinator	Construction Administrator
General Contracting	AE	AE	GC	GC	GC	AE
Agency CM (ACM)	AE	AE/CM	O	TC	CM	AE/CM

CM: Construction Manager
AE: Architect/Engineer
O: Owner
TC: Trade Contractor
GC: General Contractor

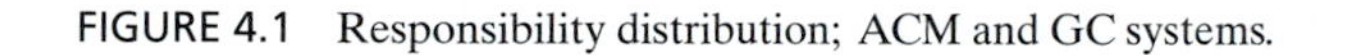

FIGURE 4.1 Responsibility distribution; ACM and GC systems.

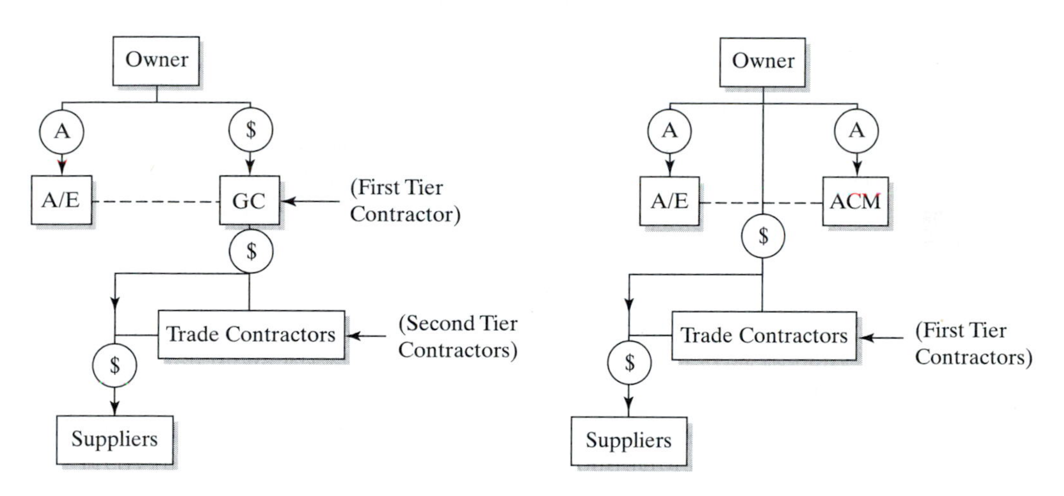

FIGURE 4.2 The GC and ACM contracting structures.

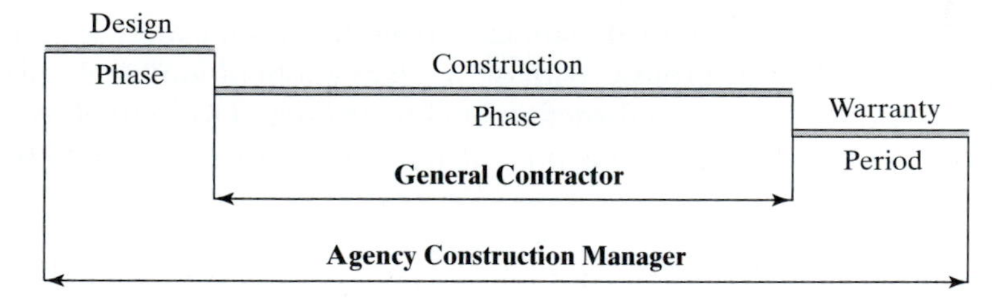

FIGURE 4.3 The project tenure of a GC and an ACM.

new middle school CM project. Acme is a subcontractor on the courthouse and a prime contractor on the middle school.

Keeping the obvious similarities and differences between the GC and ACM systems in mind will help the reader understand the numerous but less obvious differences between the two.

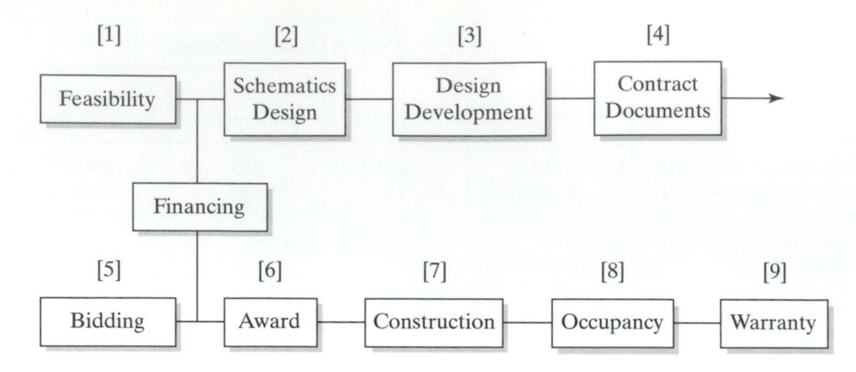

FIGURE 4.4 The nine phases of a GC or an ACM building project.

4.2 WHAT AN ACM IS AND IS NOT

A major misconception about CM is that the construction manager is the leader of a schedule-driven project team, an autonomous superior who makes unilateral decisions and issues orders for all others to follow. This is not so. Although the project is necessarily schedule driven, the decisions that guide the project are team decisions based on the unique expertise of each team member and propagated by the team's common commitment to the project. The CM is the team's facilitator.

4.2.1 Team Leadership

Team leadership agreeably shifts between team members according to the expertise that is relevant at that phase of the project or is germane to the topic being discussed. The overlapping knowledge of the A/E, CM, and owner is fully expected to provide checks and balances to team decision making. The unique knowledge of the owner is expected to provide limiting guidance, and as the employer of services, final decisions in the absence of team consensus.

This concept of mutual team leadership based on expertise, not status, is vital to the success of a CM project. Without it, team cooperation would disintegrate, synergism would be lost, and no benefits could be extracted from the CM system. The team concept and leadership sharing provide the foundation of the ACM form of CM.

4.2.2 Partnering

The common commitment of the ACM team to project success is the forerunner of the partnering concept used with some degree of success in the GC system. ACM is partnering in its fundamental and most workable form. The difference between GC and ACM partnering is that the commitment to synergize and cooperate is part of the ACM system rather than an add-on.

Partnering is an assured part of the project when the ACM form of CM is selected by the owner. It is put into effect when the CM and A/E are hired as team members by the owner.

GC partnering comes into play after an award is made to a general contractor and construction contracts are signed. It amounts to a congenial recommitment to the terms of the contracts; everybody agrees to do what they contracted for, but to do so in a more cooperative way. The beneficiary is now the project instead of the individual participants.

4.2.3 The ACM's Legal Status

In some political jurisdictions, construction managers are required by law to be licensed general contractors, registered architects, or professional engineers. While such requirements obviously must be observed, legal status in these three areas has very little to do with competency in providing construction management services.

Unfortunately, this requirement erroneously implies that general contractors, architects, and engineers have the expertise to be construction managers. This could not be further from the truth. A CM firm is a unique organization of all three.

A construction manager is a multidiscipline business organization with primary management expertise, fundamental construction and contracting expertise, and more than ancillary design expertise. The CM's expertise in all areas must be supported by related education and extensive experience.

When an owner hires a CM, he/she hires an organization, not a person. It would be physically impossible for one person to provide all of the CM requirements or render all of the decisions necessary to produce a project in a timely manner.

4.2.4 Performance Measurement

Unlike the A/E, who provides services according to a recognized format that has codified itself over the years, CM services are enumerated but not defined in the CM's contract with the owner and are subject to varying interpretations.

When performance is questioned, the services provided by an A/E or an ACM is measured by standards of care that prevail in the construction industry. An A/E's standard of care has been well established over a long period of experience. An ACM's standard of care is currently in the process of being established. State laws notwithstanding, it is inappropriate to consider ACM practitioners as architects or engineers or to measure their performance against an A/E's standard of care. It is equally inappropriate to compare ACM services to the construction services provided by a GC.

An ACM should be perceived as a firm that provides a professional service to the owner (professional in the context of the character of the services provided, not of the legal status of those providing services). Architects and engineers provide professional services which are performed by persons who are legal professionals by virtue of education, examination, and licensure.

Design is a professional service that must be provided by legal professionals. ACM is a professional service that may or may not be provided by professionals. Based on this difference alone, ACM performance should be judged on its own standard of care.

4.3 THE ACM FORM OF CM

As the CM system developed in various parts of the country, it became apparent that it had many variations. However, each variation had a block of core services that tended to bond them all together. This block of services became the basic form of CM, later called Agency CM or ACM.

ACM is a form of CM that has no variations. It is sometimes referred to as *pure CM* or *professional CM*. The other forms and their variations combine additional construction industry services with the ACM format. To fully understand the CM system, a clear understanding of the ACM form is necessary.

An ACM acts exclusively as a fiduciary agent of the owner throughout the course of the project; the limits of agency are specified in the Owner-CM agreement. The CM is expected to have sufficient, competent human resources available to perform ACM responsibilities in a timely manner.

4.3.1 ACM Mandates

The ACM form of CM mandates that:

1. the CM be a contractual agent of the owner
2. the owner hold all construction and design contracts, and
3. the CM has no contracted responsibilities to the owner as an independent contractor or as the A/E.

These three mandates establish a unique platform in the contracting structure from which the construction manager can effectively carry out his responsibilities as an agent of the owner and a peer of the A/E. The character of the services that support the ACM form can only be provided from such a platform.

4.4 THE CM'S PLACE IN THE ACM CONTRACTING STRUCTURE

The CM's knowledge and experience, agency relationship with the owner, and strategic position in the contracting structure facilitates timely and useful construction and contracting expertise during all phases of the project. Unlike the GC system, where pragmatic construction and contracting expertise is lacking prior to construction, the CM makes it available during the design phase.

4.4.1 The Team Relationship

The first mandate of the CM's role, that the CM must be a contractual agent of the owner, places the CM in the same legal relationship with the owner as the A/E. They are team peers along with the owner. Both are selected on their qualifications through an interview process. Their involvement is collaborative, neither have a financial stake in the project, and both are working for a professional fee. They have one goal in common with the owner—the success of the project—and both are aware that their future depends on the success of the project from the owner's perspective.

4.4.2 Conflicts of Interest

The second and third mandates are necessary to retain the CM's objective expertise and sustain the subjectivity of the ACM's relationship with the owner, A/E, and the goals of the project. Both mandates preclude the possibility of a conflict of interest in the performance of the CM's responsibilities and the creation of an independent contractor relationship between the CM and the owner. Chapter 6, Construction Management Under Dual Services Agreements, addresses the relationships when providing other forms and variations of CM.

4.5 THE CM AS A TEAM MEMBER

There should be no doubt that the ACM system is based on the team concept in the most literal sense of the word. Perhaps the best analogy would be a design team where contract documents are developed cooperatively by an architect and electrical and mechanical engineers. In the case of ACM, the project is delivered through the same cooperative efforts of the owner, A/E, and CM.

Design team members pursue their specialties in developing the unique components of design, and the architect coordinates their work into a total design presentation. This dovetailing of all design disciplines into a set of contract documents is referred to as design coordination. In the case of ACM, the unique expertise of each project team member is drawn upon and all efforts are coordinated by the CM. The CM's effort could be called project delivery coordination.

4.5.1 CM-A/E Peer Relationship

To be an effective team member and facilitator, the CM must be a respected peer of the A/E and a sanctioned ally of the owner. This is a unique position in contracting structures that only exists in the ACM form of CM.

In the GC system, the A/E has no peers to work with, and the independent contractor status of the general contractor creates competition rather than cooperation. It has been often said that the GC system creates an adversarial relationship between the A/E-owner and the GC.

In the D–B system, a quasi-peer relationship exists between the A/E and contracting entities. A degree of autonomy will exist because design is a profession from a legal perspective and contracting is a profession from a service perspective. The situation is not as adversarial as in the case of GC, but friction exists nonetheless.

4.5.2 The CM's Acceptance

Although mitigated by good experiences over the years, the attitude of many A/Es towards construction managers still varies from one of high regard to virtual nonacceptance. There are several reasons behind this broad position, foremost of which is the intrusion of the CM into what was an exclusive A/E-owner client relationship.

Another reason is the A/E's forced allegiance with a construction/contracting type organization, which is certainly nontraditional for A/Es. A third reason is the A/E's reluctance to accept the premise that a construction manager can provide expertise to the project that an A/E cannot.

Owners did not share the A/E's attitude toward CMs to the same degree. Many were occasional users of the construction industry and unfamiliar with its hierarchy and participants. They formed their opinions from their experiences when building projects and from the comments made by A/Es and fellow construction industry users.

Prior to the advent of CM, their opinion was that the construction industry consisted of A/Es and general contractors; A/Es were their only available advocates and general contractors were contentious opportunists that they had to do business with if they wanted to build.

Occasional construction industry users unwittingly connected CMs with general contractors and had difficulty understanding and accepting the allegiance offered by CMs in the new system. The more knowledgeable users of the construction industry welcomed the support provided by CMs and had no difficulty understanding who and what a construction manager was. The inconsistent uncertainty of owners and A/Es compels construction managers to make an extra effort to become and remain effectively involved in their projects. They were and often still are constantly required to prove their value and maintain their status as a competent team member with the owner's best interests foremost in their minds—something that the CM contracting structure cannot automatically provide.

Although the ACM's contractual status and team involvement creates a positive image and facilitates interaction, acceptance as a peer of the A/E and an ally of the owner must be earned through exceptional performance on every project.

4.6 THE CONSTRUCTION MANAGER'S EXPERTISE

To the same degree that we understand that the management of construction (a task which must be performed on every project) and construction management (a unique contracting system) are not synonymous, we must also understand that a CM is an organization and not an individual.

While it is possible for an individual to be qualified to perform all of the tasks required of a CM on an ACM project, it is physically impossible for one individual to accomplish the tasks in a timely manner, except perhaps on a very small project. To fully understand CM, we must acknowledge that a CM is an organization comprised of individuals qualified in one or more of the several disciplines required to perform the ACM's block of core services.

As we will see in Chapter 5, CM System Forms and Variations, all other forms and variations of CM incorporate design, construction, and contracting services in different combinations with the ACM services covered in this chapter. Consequently, a CM must be expert and experienced in ACM services to properly function in the other forms and variation of CM.

The expertise required to perform ACM services is vested in twelve areas of knowledge:

Budget Management	Project Management
Contract Management	Quality Management
Decision Management	Resource Management
Information Management	Risk Management
Material/Equipment Management	Schedule Management
Value Management	Safety Management

Each area will be discussed in depth in later chapters; they are briefly covered here to provide preliminary insight to the CM's required ability.

4.6.1 Budget Management

Budget Management encompasses all cost-related aspects of the project and focuses on the ability to complete the project within the budget established by the owner.

The construction manager should formulate and install a budget-control system which provides for the accurate and timely estimating/tracking of project costs from the conceptual estimate to the final budget accounting. The system must be consistent with the owner's budget requirements and format, and all aspects of project cost should be subject to team review.

4.6.2 Contract Management

Contract Management encompasses the content, execution, and administration of the contracts in force on the project. The CM system realigns traditional contracting roles and responsibilities and has unique requirements for each CM form and variation used.

The CM should perform a contractability review, recommend an appropriate contracting format, and coordinate its requirements. Although not responsible for drafting contracts, the CM should provide guidance to the owner to ensure that all unique requirements are included in the contracts issued by the owner.

4.6.3 Decision Management

Decision Management encompasses the decision-making interrelationship of the Program Team (owner, A/E, and CM) and the Project Team (the Program Team plus contractors) during the course of the project. The team makes collective decisions, recognizing each other's expertise and contractual obligations. Decisions should be oriented toward synergistic results and include checks and balances.

The CM should formulate and propagate a decision-making process that extracts the best input from each team member and favors the owner's best interests. The specific responsibilities of team members must be agreed to, defined within the context of their contracts with the owner, recorded, and published as a reference.

4.6.4 Information Management

Information Management encompasses the issuing, collection, documentation, dissemination, safe keeping, and disposal of written, verbal, and graphic project information. The volume of information made available by the CM system, which is freely shared with team members, requires a comprehensive, well-organized gathering and reporting structure based on a "need-to-know" philosophy.

The construction manager should install an effective information management system, the goal of which is to provide an impeachable information paper trail for the duration of the project and for future recall.

4.6.5 Material/Equipment Management

Material/Equipment Management encompasses activities relating to construction materials and equipment from specifications to installation. The execution of direct purchasing must be efficient and definitive to extract maximum benefits and conform to client's acquisition policies. The CM system facilitates direct purchases of material and equipment, the impact of which can be significant.

The construction manager should identify direct purchase items as soon as possible, evaluate them with regard to availability and cost, competitively bid and appropriately purchase them, and arrange transportation, safe storage, and proper installation.

4.6.6 Value Management

Value Management encompasses the interfacing of cost and value on the project. There are three basic areas of value: designability, constructability and contractibility. Designability and constructability optimize project value through the efficient integration of design and construction. Contractibility is the optimization of value through the efficient integration of owner requirements and available contracting options.

The construction manager should extract maximum value from all options available to the owner through the decision-making process. A clear definition of "value" from the owner's perspective must be established and consistently used in team deliberations. The CM should point out alternatives and assign cost to those selected as practical by the A/E and owner.

4.6.7 Project Management

Project Management encompasses the operational aspects of the project from predesign to final acceptance. It is the most sensitive area in which the construction manager must function. Success depends on how well the CM has postured itself within the team. The CM's contractual platform provides authority, but only a CM representative's demeanor as a team member can produce the required results.

The CM should decide direction, determine emphasis, motivate participants, guide activities, coordinate efforts, and document the performance of itself, the owner, the A/E, and the performing contractors in pursuit of completion requirements.

4.6.8 Quality Management

Quality Management encompasses all activities which contribute to the quality of the constructed project. Quality is identified by the owner, designed into the project by the A/E, reviewed for adequacy by the team, and conformed to by the constructors. Quality only has a range during design. Once specified by the A/E, quality from the team's and contractor's perspective is conformance.

The CM is not an inspector. The CM's responsibility should be to install and execute a quality plan for the project which will ensure that the definitive needs of the owner are properly incorporated into the project.

4.6.9 Resource Management

Resource Management encompasses the selection, organization and utilization of the human, material, equipment and service resources for the project. The CM contracting structure positions consulting, design, contracting and construction functions at the project team level and assigns coordination of these services to the construction manager.

In addition to managing project team resources, the construction manager should guide contractors in using their resources and strategically integrate his own multidiscipline resources into the project effort.

4.6.10 Risk Management

Risk Management encompasses the dynamic and static risks which are inherent to all business ventures and to construction project delivery in particular. Dynamic risks (avoidable occurrences resulting from operational decisions) and static risks (unavoidable occurrences related to uncontrollable forces) must be identified, evaluated, and disposed of with minimum consequences accruing to the owner.

Occurrences resulting from risk taking affect project time, cost, and quality and create costly business interruptions for the owner. The construction manager must design, install, and operate risk-management procedures that minimizes owner exposure to occurrences resulting from static and dynamic risks.

4.6.11 Schedule Management

Schedule Management encompasses all scheduling requirements during the course of the project. The construction manager is responsible for the synthesis of project time and project resources at the team level and at the constructor level of the contract structure.

The goal of CM scheduling is the mitigation or elimination of resource crisis and the practical prediction of project completion. Scheduling is considered a prime management tool of the construction manager. It is used sparingly as a means to an end and best exemplifies the CM philosophy which drives the project-delivery process.

4.6.12 Safety Management

Safety Management encompasses the procedures necessary to promote safe practices at the construction site, in accordance with prevailing construction industry practices, federal and state laws, and local ordinances.

The construction manager is not in charge of safety at the construction site but must accept an appropriate obligation to protect the owner's best interests with regard to contractor safety practices.

The construction manager's responsibility should be to verify that every contractor operating on the site has a written safety program, that they hold regular safety meetings, that they carry out site posting requirements, and file required communications promptly.

4.7 EARLY SERVICES

Prior to the recognition and establishment of the block of core CM services, the disconnected development of the CM system in various parts of the United States produced programs with varying emphasis. Firms entering the CM field stressed their stronger abilities when explaining CM to prospective clients.

General contractors stressed their expertise in construction and contracting. A/Es emphasized their knowledge of design and contract administration. Scheduling consultants pointed to project and construction scheduling as the important element in successful CM. Estimating consultants put cost expertise above all else. Unaware of the scope of services really required, owners selected a construction manager on their own impression of what effective CM services should encompass.

It took several years before owners and pioneer construction managers recognized or admitted that CM was a multidiscipline service, one that required the combined strengths of all early CM practitioners who believed themselves adequate. It took even longer for CM practitioners to realize that multidiscipline ability had to be supported by managerial expertise in order to successfully execute CM services.

By the mid-1980s, CM had matured enough to define the resources required to provide successful CM services to owners. The resources had to be able to provide the core services and any other services the CM practitioner wanted to include. However, the availability and degree of competency of the resources remains unresolved.

4.8 CM SERVICES PERSONNEL

The personnel in a CM firm can be classified in five groups: resource, operations, administration, company officers, and support persons. Each group collectively contributes to the success of a CM project.

Resource persons are those with special expertise, such as value engineers, schedulers, estimators, accountants, planners; contract, construction and risk specialists, and so on—people who can be depended upon to provide competent input to a project. The level and the range of expertise within the resource group are fundamental to the successful execution of CM's core services. Expertise must be available when needed and competently disseminated as needed. Resources must be well managed within the CM firm, at least as well as when being used in behalf of a client.

Operations personnel are the managers. They represent the CM on the project team, formulate/administer the CM format, coordinate performing contractors, and utilize the services of the resource group. To facilitate CM operations there are three distinct levels of management authority; executive, management, and administrative. Operations personnel function at two levels: management, and administrative; the third level is reserved for the company officers, the executives. The management hierarchy is explained in Chapter 8, The CM Organization.

The administration group handles the day-to-day operation of the CM firm (personnel, payroll, accounts payable/receivable, finances, etc.) and also contributes to CM projects when appropriate. The responsibilities of the administration group are similar to those found in any business enterprise except when called upon as a resource by operations.

The company officers function as all company officers do; they establish company policy and CM policy and overview the operations of the firm. An added responsibility is that of interfacing with owner and A/E team counterparts on CM projects, establishing team policies, and responding to situations which cannot be handled at the administrative or management levels.

Support personnel provide service as the title implies: support to the operations, resource, administration, and company management groups in the reception, clerical, computer, travel, and secretarial areas. Support personnel provide the glue that keeps company operations together.

4.9 THE PROJECT TEAM STRUCTURE

The CM team structure should have at least three levels of matched authority. The owner, A/E, and CM each provide representatives with specific authority to act in their behalf, at each level. Each representative's authority should be understood by all team members. Decisions should be made at the lowest practical level of authority. When decisions falter collectively at one level, they should be referred to the next highest level. Issues which are not collectively resolved at the top level of the decision structure should be decided upon by the owner's team representative at that level.

In practice, the team hierarchy should function consistently but without rigid formality, especially at the two lowest levels. The owner, CM, and A/E team representatives are matched at each level based on their authority to act for their employer. Each is represented at the third or top level of authority by a principal or executive who can obligate her/his firm by contract or agreement.

The second or middle level should be team members who are managerially involved with their own firm. The A/E's project engineer or architect and the CM's project manager match up with a competent construction/contracting person who represents the owner. Project direction, design, construction, and contracting issues are decided at this level. In the case of the A/E and CM, the representative at this level should be the managing liaison between home office resources and field operations.

The third level is field operations. The CM's chief on-site person is the CM's representative at this decision-making level. The owner and A/E may or may not have persons assigned full-time on-site. However, it is critical to team synergism that representation be assigned at this level to persons readily available and qualified to interact with the CM's on-site person, day or night.

The size of the project dictates the size of the team, the number of authority levels, and the ability of the representatives. Large and complex projects require large site organizations, suborganizations, and multiple decision-making levels. Lesser projects can be handled adequately by one full-time, on-site CM person and served adequately by A/E and owner representatives whose decision making authority spans two or all levels.

4.10 TEAM DECISIONS

On complex projects, especially those that are fast-tracked and those involving additions and renovations, it is important that the authority of representatives of the owner, CM, and A/E is specifically and accurately defined. Projects in these categories usually have numerous technical and physical interfaces that can be seriously problematic unless resolved expeditiously and properly.

Expedience is necessary to mitigate the effects on an owner's business operations and avoid any negative impact on the momentum of the contractors involved. Many interface issues rely almost entirely on the cause conditions which, if not interpreted correctly, could have serious effects on the solution. The project team should approach interface issues collectively and use the checks and balances of the team decision-making process toward proper resolution.

The CM's concern for interface issues relates to the combined effect of the problem and its solution on the cost and progress of the project. Delay in decision making or as a consequence of the solution could mean both increased cost and additional contract time. The CM's team representative should approach the decision-making effort from this perspective.

The A/E's responsibility when facing interface issues is to find the technical solution that conforms to the owner's established needs and standards and meets the A/E's professional responsibility. While the decision-making process should be handled in an expeditious manner, a fast solution may not be the best answer. Alternatives should be considered if time allows. The A/E's team representative should approach the decision-making effort from this perspective.

The owner's involvement in interface issues is to evaluate the input of the CM and A/E from the perspective of the problem's affect on current and future consequences. This is value judgment ("pay now or pay later") time for the owner. The owner is the only team member who is qualified to make the right decision when faced with the facts presented by the A/E and CM.

4.11 QUALITY OF PERFORMANCE

The CM is a contracting/construction resource and a responsible manager who, in conjunction with the A/E, owner, and a complement of capable constructors, has a responsibility to deliver a project on time, within budget, at specified quality, and with minimum inconvenience to the owner.

The role of the ACM has become ubiquitous and very consequential, to the point where project success is more dependent on the performance of the construction manager in the ACM system than on the performance of the general contractor in the GC system.

From the start of design to occupancy, the CM is functioning in behalf of the project in some way. As an agent of the owner, with legal and ethical obligations to act in the best interests of the owner, there is constant pressure on the CM to do the right thing, do it once and do it now. Procrastination has no place in the CM's vocabulary when performing ACM services.

In the GC system, the day-to-day decisions of the general contractor do not directly affect the owner. They may ultimately be consequential to the owner in terms of the owner-GC contract, but on a day-to-day basis, GC decisions are consequential to the GC. Delaying a decision, closing the project down for a day, trying a certain construction technique: these actions are all within the authority of the GC because they affect no other.

Each time the CM is faced with a decision on an ACM project the first question to be asked is which solution will be in the best interest of the owner. The CM must place himself in the position of the owner at every juncture in the project.

The quality of the CM's performance on an ACM project simply cannot be compromised. Every decision made, every action taken, and every piece of advice given or information disseminated must pass the test of agency representation. Performance quality is driven by the agency agreement between the owner and CM.

4.11.1 The ACM Standard of Care

Although ACM is established and in wide use today, universally accepted standards of practice that stipulate acceptable performance do not exist and probably never will. ACM performance is not susceptible to published performance standards and will eventually be evaluated on its own standard of care. However, establishing a "standard" of care will not be easy because ACM practices and procedures differ substantially from one CM to another on an arbitrary basis. Two reasons can be cited:

1. all CMs do not provide all available ACM services to clients, and
2. all CMs do not provide ACM services to the services' full potential; and there are two reasons for this:
   a. all CMs are not staffed to an adequate level either in numbers or talent, and
   b. price-based competition for ACM services is conducive to minimizing the services provided.

4.11.2 Measuring CM Performance

ACM services only have four culmination points where success can be objectively measured:

1. when the bidding process produces costs that are within the budget
2. when owner occupancy occurs on schedule
3. when an owner walk-through reveals expected quality, and
4. when no claims are brought against the owner by contractors.

If the project is completed within budget, on time, at specified quality, and does not cause undue business interruption for the owner, did the owner get a fair return on the CM's fee?

Did the CM extricate maximum value for the owner? Could the cost have been reduced? Was it possible to complete the project sooner? Could quality have been upgraded? In other words, did the CM provide a complete and maximum effort in the owner's best interests, or did the CM simply control a controllable set of circumstances?

Owners have no way of telling if a CM is providing ACM services at an acceptable standard of care during the course of the project unless they had ACM experience on previous projects or unless the CM makes it a point to inform the owner of progress and accomplishments. Subsequent chapters of this book cover the process of keeping the owner and the A/E informed so performance can be evaluated at other than the four culmination points.

4.11.3 The Standard of Care Quandary

The standard of care for ACM services has become a very important issue for construction management advocates. If CM is to someday be the preferred contracting system of owners, the standard of care must be set at its highest level. To achieve this, both owners and construction managers will have to change their current practices.

Owners will have to place more emphasis on a CM's qualifications than on the amount of the CM's fee. Quality-based selection (QBS) must completely replace price-based selection (PBS) when owners acquire ACM services. As long as owners visualize CMs in the context of a contractor and buy services on price, this will be hard to accomplish. As long as CMs willfully abbreviate the scope of their services and are satisfied with half an effort, it will be even more difficult.

The same argument that supports QBS when acquiring design services can be used when acquiring ACM services. However, the argument for QBS in the case of ACM is even more valid, if that is possible. ACM services are much more diversified and subject to wider latitudes than design services. It seems ludicrous for owners to choose a contracting system specifically designed to protect their interests and then price shop for a CM to operate the system.

Appendix B, A Model Program for the Certification of Construction Managers, describes a way to assure owners that full service CMs are included in the selection process described in Chapter 24, Acquiring CM Services.

4.12 EARLY RESPONSIBILITIES

The comparatively short history of CM and the complexities of its many innovations imposes a professional responsibility on the construction manager to inform the owner of its features, functions, and use before an agreement for services is entered into. A determination of the form and variation of the CM system to be employed was the initial and most consequential decision made.

The CM should candidly guide the owner to a form and variation selection with the professional integrity inherent to agency and apprise the owner of potential problems and benefits inherent to that selection.

The broad areas of CM responsibility, established in most standard CM contract documents, are management oriented and cover the twelve areas of CM knowledge. Responsibility in all of these areas are generously implied but there are few specifics regarding the procedures that should be followed. The CM covenants to provide these functions using its best skills and judgment.

The CM is obligated to explain to the owner at an understandable level of detail how responsibilities in these areas will be carried out. The CM must take into consideration the owner's familiarity with the construction industry and the process of project delivery. The goal of the CM should be to preclude service-related surprises once the project begins.

4.13 CHECKS AND BALANCES

All forms/variations of the CM system depend on team interaction and a synergistic approach to decision making. The program team—the owner, A/E and CM—is structured to act in concert when making decisions and determining the course of project delivery.

Unilateral decisions on CM projects are precluded by the team concept and by a built-in system of checks and balances. In effect, the CM's expertise in contracting and construction is monitored and tested by the A/E's knowledge of those same areas. Conversely, the A/E's expertise in design is similarly tested by the CM's experience and knowledge of the design process.

The team's system of checks and balances precludes autonomous actions and intensifies deliberation but does not inhibit decision making in any way. Experience has shown that self-imposed preparation by team members actually speeds the decision-making process up rather than slowing it down and provides a high level of comfort with the decisions made.

4.14 THE ACM PROCESS IN BRIEF

ACM is a design–bid–build procedure that facilitates competitively bid single-phase, multiple-phase and fast-track projects. The ACM process uses traditional trade contractors as prime contractors by dividing the project into strategically defined individual work-scopes with the owner holding all contracts. The CM is exclusively an agent of the owner and does not provide either design services, contracting services, or construction services on the project.

A broad-scope list of typical ACM activities includes, but is not restricted to, the following:

1. Develop the project budget from information provided by the owner and A/E
2. Develop the management strategy and management plan based on the owner's parameters
3. Assemble and maintain the project procedures manual
4. Schedule the project delivery process from design through construction to occupancy

5. Apply value management procedures to ensure the designability, constructability, and contractability integrity of the project
6. Formulate contract provisions to facilitate the use of the ACM project delivery system
7. Review the bidding documents from the perspective of quality prior to issuance to bidders for proposals
8. Determine a work-scope list that facilitates the multiple bidding process
9. Write individual descriptions for all work-scope divisions
10. Qualify all trade contractors interested in bidding work-scopes
11. Identify long-lead items and other beneficial direct owner purchases
12. Assist the owner in obtaining proposals for materials, equipment, and contracts
13. Evaluate the labor pool and the contracting practices in the area of the project
14. Provide team leadership during the time that the expertise of the construction manager is germane
15. Develop bidding competition to generate the most favorable pricing conditions
16. Communicate with bidding contractors to clarify conditions and resolve discrepancies in bidding documents
17. Assist the owner during the bidding process to ensure the proper receipt of proposals
18. Review trade contractor proposals to determine if those being considered are complete and in the owner's best interests
19. Lead necessary negotiations with contractors on behalf of the owner
20. Provide assistance in the signing of contracts and the accumulation of required contractor documentation
21. Organize and chair preconstruction meetings with trade contractors
22. Develop and implement the on-site construction schedule
23. Coordinate contractors at the site on a full-time basis
24. Chair project and progress meetings during construction
25. Organize and administer a system for expediting owner-purchased and contractor-purchased material and equipment
26. Develop and administer financial and management information systems
27. Develop and coordinate a progress payment procedure for contractors
28. Procure and administer the construction support requirements
29. Assist the owner and contractors with labor relations connected with the project
30. Develop, implement, and maintain a project quality-management program
31. Administer contract changes and the project's change order procedure
32. Track project costs and administer the owner's project cost accounting program
33. Assist in the resolution of disputes arising from the performance of contracts

(The terms used in the above list and throughout the chapters may not be those used by all readers. A glossary of terms appears at the back of the book.)

4.15 WHAT THE ACM FORM OF CM IS DESIGNED TO DO

The ACM form of CM was designed to improve the delivery of architectural (building) projects in the public sector from the owner's perspective. Similar improvements will occur on engineering projects or on private sector projects if the procedures used in the private sector on architectural projects are emulated.

The following is a list of the things the ACM form of CM was specifically designed to do on public building projects:

1. Allow the owner to select the manager of the project strictly on an ability basis
2. Allow the owner to be more involved in the selection of trade contractors
3. Allow trade contractors to compete for work on a fair competitive bidding basis
4. Pay earned progress payments directly to the contractors who did the work
5. Eliminate pre-bid and post-bid shopping of trade contractors' prices
6. Introduce pragmatic designability and constructability values to the design phase
7. Inject contractability opportunities into the preconstruction phase of the project
8. Use synergistic team decision making throughout the project
9. Provide checks and balances to the decision-making process during the course of the project
10. Provide positive continuity between the design and the construction phases of the project
11. Emphasize project management as a planning tool as well as an administrative tool
12. Provide a practical platform for the use of modern management tools and methods
13. Ensure that all of the lowest contractor proposals were available for award
14. Eliminate the adversarial relationships that develop in the traditional system
15. Reduce the opportunity for conflicts of interest to surface in project participants
16. Permit the use of competitive bidding on projects that require the use of fast-tracking
17. Economically deal with the static and dynamic risks inherent to a construction project
18. Provide a new level of owner awareness and optional owner participation in the project-delivery process
19. Gain insight from a construction and contracting expert who is the owner's ally
20. Provide a way to use the services of both union-affiliated and non-affiliated contractors on a competitive basis

It should be remembered that as many ancillary benefits surfaced during the early use of the ACM form of CM as were originally and purposely built into the system. Many of the above goals were developed during the use of CM between 1975 and 1990.

CHAPTER 5

CM System Forms and Variations

Chapter 4 introduced the CM system by describing its root form, ACM, in some detail. This chapter confirms the ACM system and extends CM understanding by presenting the variations that can be derived from the ACM form.

The CM system slowly acquired identity after its spontaneous and controversial start in the late 1960s. ACM, had to be defined first, so that variations of the system could be explained from a common perspective.

CM firms generated these forms and their variations to accommodate the specific needs of owners, and many owner needs were unique. Consequently, the contract structures which satisfied their needs had to be unique.

It should be understood that when CM contracting began, it only had a name, not a definition. The practices that developed during its brief history eventually provided definition. It certainly was not a case of CM firms understanding ACM and developing mutations that became forms and variations.

Many of the forms and variations of CM were the original programs developed by CM firms in their effort to satisfy owner needs. Many providers of CM services have never practiced ACM. In fact, many current CM practitioners will gain their first comprehensive understanding of ACM in the pages of this book.

CM and its variations has sufficiently stabilized to permit bounded definitions and are diagrammed from a contract perspective. The inherent relationships between each are explained.

5.1 REVIEWING SYSTEMS INFORMATION

CM contracting is a unique alternative to GC contracting and D–B contracting. The signal difference between the systems is the philosophical approach to their execution, fostered by the configurations of the contracts between the parties in the project delivery process.

In the CM system, the independent contractor relationship inherent to the GC system (and to a greater extent, the D–B system) is altered by eliminating the general contractor or design–build contractor and adding an agent/fiduciary construction manager.

The intent of CM is to provide a more synergistic approach to the delivery of the project; to mitigate adversarial relationships common to prime independent contractor configurations, and enhance the contracting and construction expertise available to the owner during the project. The goal is to improve the quality, time, and cost parameters of the project and to eliminate problems attributed to traditional construction such as litigation stemming from claims against the owner after the project is completed.

The construction manager, a multi-discipline organization specifically created for the role, has the responsibility of guiding the owner through all phases of the project. In essence, the owner becomes his own general contractor who, on the advice of the CM, assigns the responsibilities and eliminates or assigns the inherent risks involved in the delivery process. The CM's competence, expertise, and initiative keep the project on course from feasibility to occupancy.

5.1.1 General Contracting

The parties involved in a general contracting project are the owner, GC, A/E, trade contractors, and suppliers. The A/E provides design and guidance to the owner. The GC provides contracting and construction services, including the management of construction. The A/E designs the project, produces drawings and specifications for bidding and construction purposes, and administers the project overall in the owner's behalf.

The GC, an independent contractor, privately develops a proposal for the project's construction cost, uses his own resources to purchase materials, subcontracts work, and constructs work in accordance with the terms of his contract. On buildings, 80 to 100% of the construction is accomplished by subcontractors. Extensive subcontracting postures the general contractor as a manager of subcontractors rather than a constructor. (See Appendix E.)

5.1.2 Design–Build Contracting

The parties involved in a design–build project are the owner, D–B contractor, trade contractors, and suppliers. The D–B contractor provides all required services under a single contract with the owner. Design services are accomplished by the D–B contractor's own forces or by an A/E firm hired directly by the D–B contractor. Construction services are provided as in the GC system; D–B contractors are sometimes GC contractors functioning in a design–build capacity on a project basis. (See Appendix E.)

5.1.3 CM Contracting

The parties involved in a CM project are the owner, A/E, CM, trade contractors, and suppliers. The involvement of the A/E and CM can vary on a project basis.

The service assignment options in the CM system are numerous, and a variety of contract configurations are produced as a result. This flexibility captured the imagination of the industry and contributed significantly to the popularity of the system. Its versatility facilitates its use. Owners can customize a CM form and variation to suit a specific need. (See Appendix E.)

CM has often been referred to simply as Innovative Contracting.

5.2 COMPARATIVE DEFINITIONS

General Contracting (Public Sector): The use of a single prime contractor, hired as an independent contractor, who has a financial stake in the construction phase of a design–bid–build sequence project. Project design, management, and administration responsibilities are assigned to an A/E by the owner. The owner is not privy to the development of the project's construction costs.

Design–Build Contracting (Private Sector): The use of a single prime contractor, hired as an independent contractor, who has a financial stake in both the design and construction phases of the bid–design–build sequence project. Project management and administration is the responsibility of the owner. Usually the owner is not privy to the development of the project's construction costs.

Construction Management Contracting (Public and Private Sectors): The use of multiple prime contractors, hired as independent contractors, who have a financial stake in the construction phase of a design–bid–build sequence project. Project design, management, and administration responsibility is assigned to a team, consisting of the owner, A/E, and a CM who provides construction and contracting expertise and advice. Some forms of construction management assign other responsibilities to the CM. The owner always is always privy to the development of the project's construction costs.

The three basic contracting systems are diagramed in Figure 5.1. Note that the CM system shown in the figure is the ACM form, the root of all CM forms. Review Chapters 1 and 4 if the particulars shown in the diagrams are not clearly understood.

5.2.1 Diagram Conventions

The solid lines connecting the parties denotes privity of contract by written agreement. The dashed lines indicate a cooperative relationship between parties established by provisions in each contract held by the owner. As previously established, there is no contractual relationship created between the parties connected by a dashed line. The dotted lines in the ACM diagram enclose the team members.

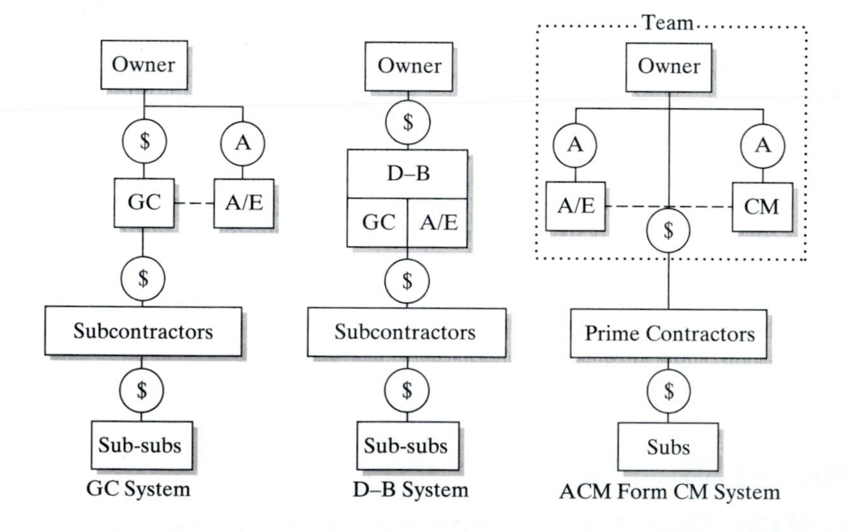

FIGURE 5.1 The three basic contracting system structures.

As in previous diagrams, Ⓢ denotes an independent contractor relationship between contracted parties. Ⓐ indicates a legal agency relationship between contracted parties that is stipulated and limited by the terms of the agreement entered into by the parties.

5.2.2 ACM Differences Restated

Viewing the GC, D–B, and CM contracting structures side-by-side clearly shows the differences between them. The ACM form:

1. eliminates one complete contracting tier
2. adds a construction manager to the structure
3. substitutes multiple contractors for the single prime conractor
4. elevates trade contractors to prime contractors
5. allows the owner to contract directly with trade contractors
6. increases agency relationships, and
7. creates a team to design, manage, and administrate the project.

5.2.3 Tenure of Services

Another obvious difference between the three systems is the involvement of the principal participants with the owner and the project. In all three systems, the A/E is involved from the start of design until the end of construction. Design services may begin as early as the planning or feasibility phase, depending on the needs of the owner, and may extend into the occupancy or startup phase, depending on the type of project.

The GC system excludes general contractors from the design phase and leaves construction/contracting exclusively in the hands of the A/E. In the D–B and CM systems, construction/contracting input during design is provided by the D–B contractor and the ACM, respectively. In essence, the D–B and CM systems provide a team approach to design that is not available in the GC system, unless competitive bidding on completed plans and specifications is replaced by predesign negotiations.

The service tenures of the GC, A/E, and CM are compared in Figure 5.2.

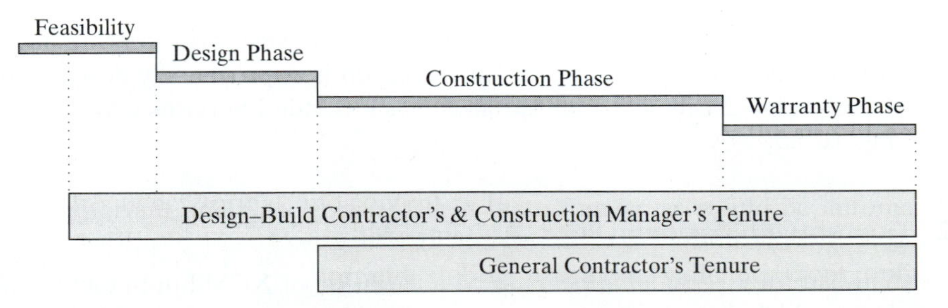

FIGURE 5.2 Comparative tenures of service.

5.3 CM FORMS AND VARIATIONS

All forms and variations of the CM system, *except the ACM form*, have a second, third, or fourth contractual relationship added to the primary ACM relationship between the CM and owner. Added relationships compromise the "pure" advantages of ACM, but compensate by packaging project-delivery responsibilities according to the owner's preferred requirements.

Unlike the GC and the D–B systems which have only one contracting structure, the contracting flexibility of CM permits a variety of contracting structures, each of which has use-benefits to owners. However, to derive benefits, owners must accurately assess their project requirements and properly match a CM form/variation to them. An obvious key to an owner's success with CM is the proper match-up of CM form/variation with his/her requirements.

Referring to ACM as the root form, three distinct subforms evolve by changing the vested responsibilities of the ACM team's members: Extended Services CM (XCM), Guaranteed Maximum Price CM (GMPCM), and Owner–CM (OCM). In each case, responsibility for more than one of the six essential elements of service, in whole or part, must be assigned to team members.

The segregation of services that identifies the ACM form becomes an integration of services in the subforms. Additionally, variations of the three subforms can be created by merging subforms and further combining service responsibilities. Subforms and variations of subforms have a practical limit based on the number of participants involved and services required.

5.3.1 Extended Services CM (XCM)

The XCM subform has three variations, each related to the types of organizations providing services to the owner. *Design–XCM*, the combining of design service with CM services, can only be accomplished with the A/E firm providing design services for the project. The team's A/E extends design responsibilities to include CM responsibilities or enters into a second contract with the owner to provide CM services.

Constructor–XCM and *Contractor–XCM* are variations which can only be provided by the team's CM, and then only if the CM has the capability to construct with their own forces or to enter into and bonded contracts for construction. In these variations, the CM's responsibility is extended to include construction services and contracting services, respectively.

The ultimate variation of the XCM subform vests both contracting and construction responsibilities in the CM. This variation is appropriately designated *Contractor/Constructor–XCM*. The four variations of Extended Services CM are diagrammed in Figure 5.3.

5.3.2 Guaranteed Maximum Price CM (GMPCM)

This subform of ACM appears to be a variation of XCM but in fact is a separate subform which has its own variations in combination with XCM forms. *Guaranteed Maximum Price CM* can only be provided by a CM with an appropriate contracting capability who can provide the owner with a surety bond covering the maximum price

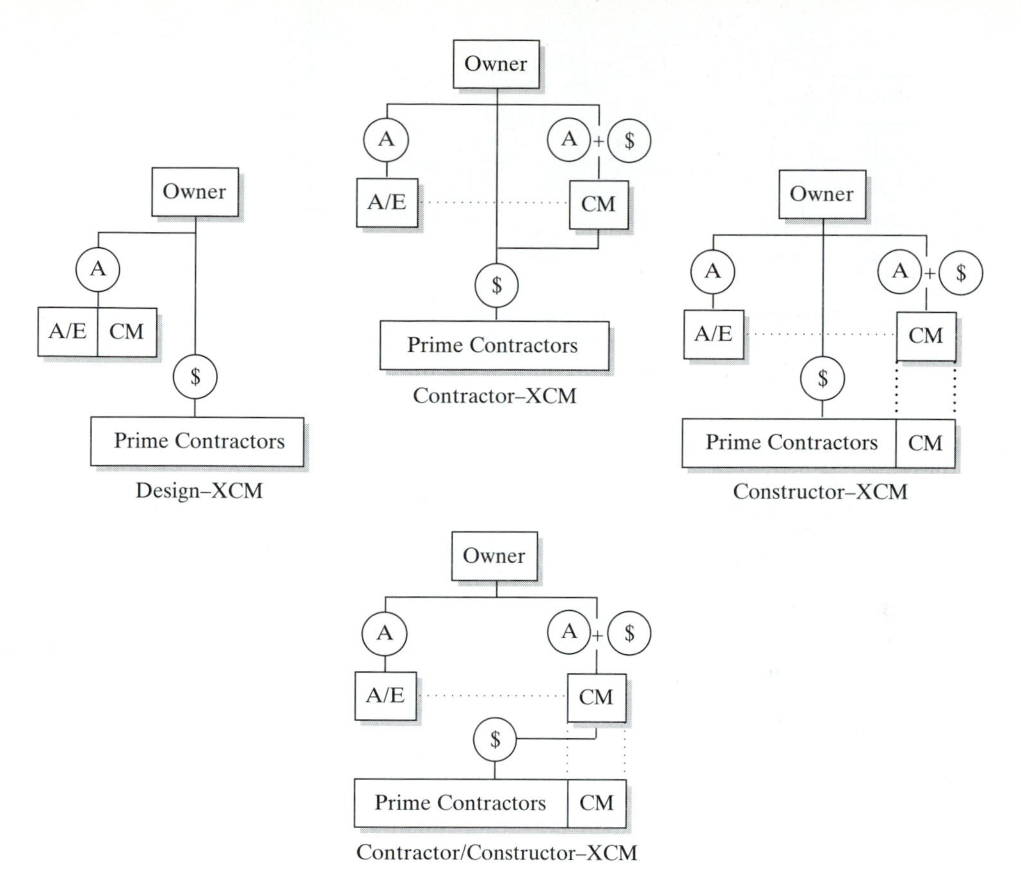

FIGURE 5.3 The four variations of Extended Services CM.

for the complete construction of the project. This subform, Guaranteed Maximum Price CM, is diagrammed in Figure 5.4.

The basic GMP subform referred to as GMPCM is essentially an Agency CM arrangement in which the CM provides a guaranteed maximum price (GMP) to the owner. The GMP covers the cost of the completed construction of the project before design is complete.

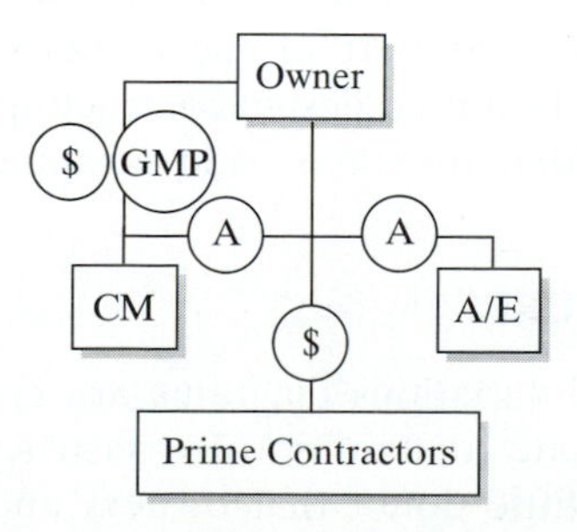

FIGURE 5.4 The basic subform of GMPCM.

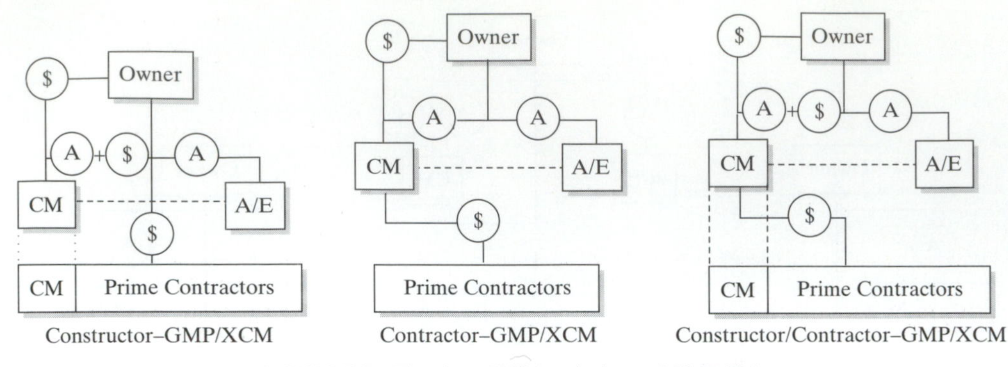

FIGURE 5.5 The three XCM variations of GMPCM.

The three XCM combination forms of GMPCM are diagrammed in Figure 5.5.

The basic form of GMPCM combined with Constructor–XCM, Contractor–XCM, or both produces variations that stretch the CM system to its definable limits. The contract structure of the *Constructor/Contractor–GMPCM* variation very closely resembles the contract structure of the GC system but remains a CM form and variation. The credibility of Constructor/Contractor GMPCM as CM depends on the final requirement of the comparative definition of CM, which is that *the owner must be privy to the development of the project's construction costs*. If this is so, Constructor/Contractor–GMPCM is a CM form and not general contracting.

5.4 THE OWNER FORM OF CM

As the contracting structures of the CM system were owner inspired, an owner subform probably existed before CM firms came into existence. Referred to as *Owner–CM or OCM*, this form depends on owner in-house staffing to provide some or all of the required CM services.

There are three variations of OCM: *Owner-Manage–CM*, *Owner-Design–CM*, and *Owner-Design/Manage–CM*. As the names imply, Owner-Manage–CM uses outside design services and internal CM services. Owner-Design–CM uses inside design services and outside CM services. Owner-Design/Manage–CM uses internal design and CM services. (See Figure 5.6.)

The owner form of CM requires personnel who are qualified in the disciplines of the construction industry as part of the owner's organization. The owner has to develop an appropriately staffed design/contracting department that can undertake a construction project with minimum outside assistance.

5.5 CM'S IDENTIFYING FEATURE

As the CM forms and variations diagrams are compared, it becomes obvious that perhaps the appropriate definition of construction management is Innovative Contracting. There is little doubt that owners and participants in the construction

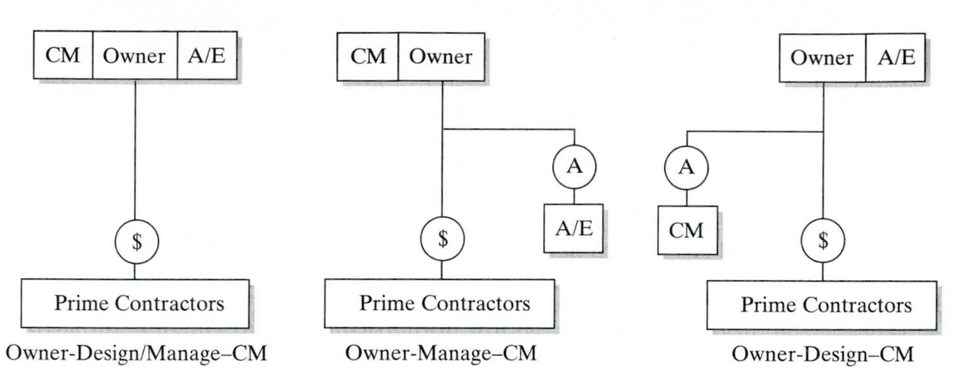

FIGURE 5.6 OCM variations.

industry have pushed their contracting structure imagination to its limits once they had the opportunity and confidence to do so.

Without actually diagramming the contracting structure and designating the legal relationships between parties, it is difficult to determine where CM starts and ends. Without careful examination, CM could easily be confused with the GC or D–B system. For this reason, careful study of this chapter is recommended. Construction industry users, participants, and students should have solid knowledge of CM's parameters to round out their expertise.

Figure 5.7 shows the similarities between GC contracting and Constructor/Contractor–GMPCM. Figure 5.8 shows the resemblance of D–B to Design–XCM, or Design/Manage–CM. The presence of a CM in a contracting structure *and* the agency relationship between the CM and owner determines if the structure is or is not a form of the CM system.

Readers may ask, "Why bother to identify CM where such a fine line must be drawn? Does it really matter that we know if the contract structure is CM or not?"

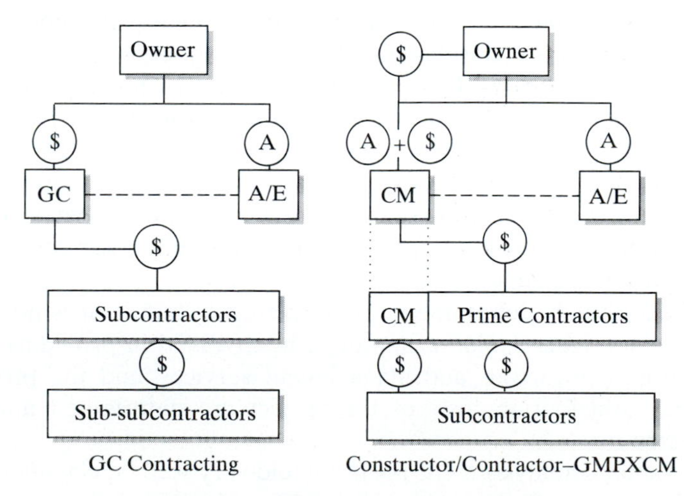

FIGURE 5.7 GC compared to CCGMPXCM.

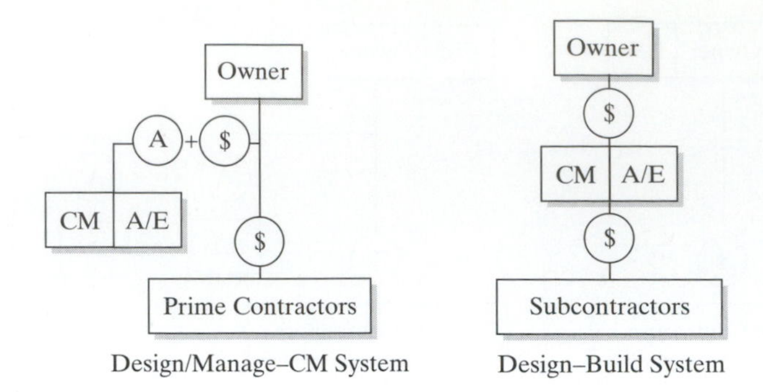

FIGURE 5.8 Design–XCM compared to D–B.

The answer does matter, because a CM contracting structure acknowledges an existence of agency not found in either GC and D–B—contractually and more importantly, philosophically. The philosophy of CM is *different* from either GC or D–B contracting.

It should be remembered that the CM system emerged only because traditional systems were not providing owners with satisfactory results. The success of GC and D–B contracting is heavily dependent on the performance of a prime independent contractor. CM contracting changes this situation by placing success challenges on the performance of an agent of the owner. This is a significant change considering the legal responsibilities of independent contractors and agents.

5.5.1 Project Cost Determination

The one unique feature of the CM system, that survives all attempts to confuse it with D–B contracting and GC contracting, is the overt manner in which the costs of the project are obtained and assembled. In both the GC and D–B system, construction needs (services, subcontracts, materials, equipment, etc.) are *unilaterally* determined, priced, and presented to the owner by the contractor. Owners are not part of the proposal process or the selection of subcontractors and suppliers.

At the heart of the CM system, regardless of form or variation, is the overt process by which project costs are determined. Project needs are established by the team, and all decisions are exposed to checks and balances. The project is selectively divided into numerous work-scopes, allowing the owner to be exposed to project costs at the trade contractor level. Cost exposure is exemplified when using ACM in the public sector, where each work-scope is individually bid by trade contractors in open competition with their peers.

To definitively determine if a contracting system is or is not CM, the following will suffice: an agency relationship must be involved in providing some or all of the construction, contracting, and management services, and the process by which the needs and contributing costs of the project are determined must be open to the owner's scrutiny.

In the final analysis, a GC project could very well be classified as a CM project if the owner has complete access to the GC's estimating information and has the final word in the selection and awarding of subcontracts. However, this is not typical of GC projects.

5.6 MAINTAINING AN OWNER ORIENTATION

The contracting structure of the CM system was strategically designed to be owner-oriented. The contractual arrangements and processes were designed to accrue benefits to the owner—benefits that either could not or did not consistently accrue from the structures of general contracting or design–build contracting. CM accomplished this by substituting an agency relationship for the major independent contractor relationships that drives the GC and D–B contracting structures. (The ACM form of the CM system demonstrates this very clearly.)

ACM is totally dependent on agency performance until the start of construction when trade contractors are hired. From then on, although independent contractors are performing construction, the day-to-day management and administration of the project remains dependent on agency performance. Recognizing that the decisions of an agent must always be made in the best interest of the owner (but decisions by independent contractors need not be), it is apparent that ACM provides a better opportunity for owners to benefit from the decisions made.

Agency performance throughout the project facilitates the pursuit of a common goal by the owner, A/E, and ACM and encourages synergistic decision making. An appropriate overlap of team member expertise introduces critical and impartial checks and balances to the decision-making process, and the agency status of both A/E and ACM essentially eliminates a potential for conflict of interest in the decision-making process.

5.7 OTHER FORMS AND VARIATIONS OF CM

Relaxing the optimum agency relationships in ACM, which occurs in all other forms and variations of CM, dilutes the intrinsic benefits of an agency relationship. However, other forms and variations of CM may be more appealing or may more appropriately satisfy the owner's requirements. Using forms other than ACM could produce tradeoffs which compensate for a less-than-optimum agency presence.

Project delivery is dependent on the collective results of individual performances. Exercising competent management achieves collective results, and involving credible individuals results in individual performance. Additionally, the owner must feel comfortable with the contracting strategy used.

In construction, the "individuals" are groups of individuals; business firms, that collectively provide the six required elements of service on which project delivery depends. Although ACM theoretically represents an ideal owner-oriented contracting structure, construction project delivery has too many primary and ancillary facets to blindly recommend its use on every project.

Contractability, explained in Chapter 2, The Reasons for a Third System, is the process of determining the best contracting arrangement for a particular project. A contractability review should be undertaken to determine which form of CM is best suited to the project. The review should take into consideration project type, size, complexity, location, schedule, financing, owner involvement and restrictions, and any other potential influences.

If all things are considered and prioritized, an appropriate contracting structure will be identified at the end of the study. The initial choice will be either general contracting, D–B contracting, or CM contracting. If the choice is CM contracting, further consideration must be given to select a suitable form and variation.

Initial consideration should be given to ACM, unless owner requirements or preferences point otherwise. The most suitable form CM can be selected based on how and to whom the owner prefers to allocate project delivery responsibilities. It is suggested that the form selected be the most conservative of those under consideration (i.e., if Construct–XCM meets all the owner's criteria, Construct/Contract–XCM should not be considered).

5.8 POTENTIAL FOR CONFLICT OF INTEREST

When responsibilities for the six essential elements of project delivery are combined—such as when the A/E also provides CM service or the CM also performs as a contractor or constructor—the credibility of the team's actions is subject to question. A potential for conflict of interest is created which must be dealt with in a productive manner if the project is to extract maximum performance from the team.

A potential for conflict of interest exists in every contract, regardless of whether it is between an independent contractor and an owner or an agent and an owner. This potential is independent of a party's technical abilities, and it must be accepted that an assertion of professional ethics provides no guarantee that a conflict will not arise or that the conflict will not adversely affect the owner. Unfortunately, we are exposed to examples of conflict of interest on a daily basis through the media. Ethical behavior whether in society, business or the professions is ebbing world-wide. The United States and its construction industry is certainly no exception to this trend.

5.8.1 Combining Contracts

If one party simultaneously has an agency contract and an independent contractor contract with another party, and the contracted responsibilities of each contract interact, it is realistic to assume that the agency relationship would at some point, consciously or unconsciously, defer in favor of the independent contractor's position. The only mitigating influence in such a situation would be professional ethics.

The same issues will arise if one party simultaneously has two agency contracts or has a combined services agency contract with another party, and the contracted responsibilities of each contract or services interact.

5.8.2 Potential as a Qualifier

The word "potential" is used to modify the expression "conflict of interest" out of respect for the construction industry participants who provide services on CM projects. It is important to remember that it is the contracting structure that creates a potential for conflict of interest, not the people. When a conflict of interest situation presents itself, the common goal of the project will be the beneficiary if all project participants rigidly adhere to their contractual commitments to the owner.

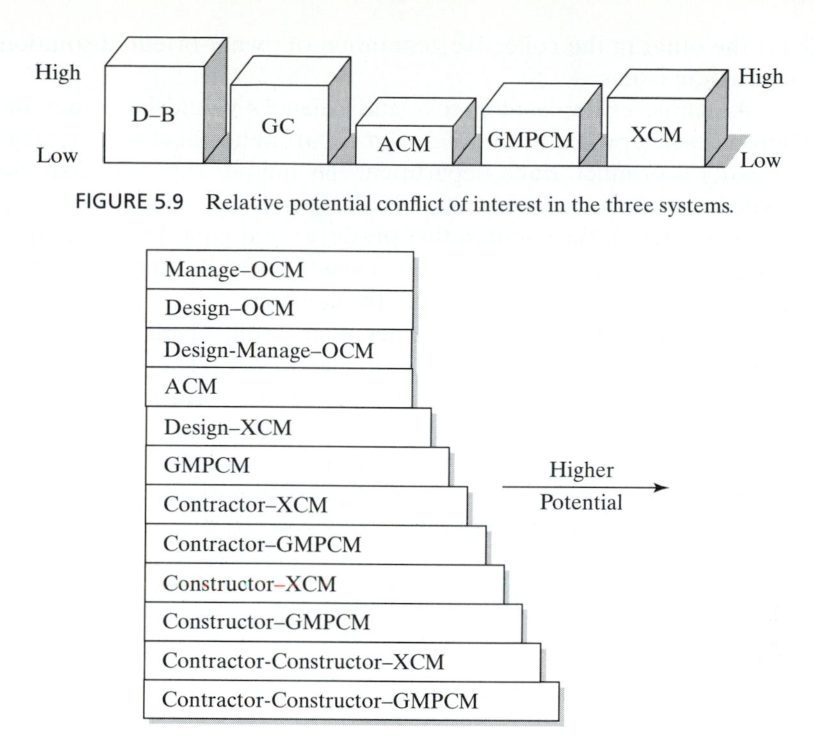

FIGURE 5.9 Relative potential conflict of interest in the three systems.

FIGURE 5.10 Relative potential conflict of interest in the CM system.

A comparison of the forms and variations of CM reveals varying exposure to potential conflict of interests. Figure 5.9 shows the *relative* potential for conflict of interest of the GC and D–B systems and the three CM forms. Figure 5.10 shows the *relative* potential for conflict of interest of the forms and variations of CM. Relativity is nominally based on the opportunity for financial gain by persons other than the owner in the team decision making process.

It should be noted that the three OCM structures and the ACM structure are similar with regard to potential for conflict of interest. However, the three OCM forms do have a potential to not serve the owner to the level of ACM. OCM is a cloistered form of CM—it excludes diversified cutting-edge management, construction, contracting, and design expertise from the project, expertise that expands with exposure to a variety of projects in the CM and A/E disciplines. To compensate for this exclusion, owners who use OCM should periodically use ACM as a means of updating their knowledge and abilities in these areas.

5.9 CHECKS AND BALANCES REVISITED

It is assumed that each team member, owner, A/E, and CM has unique ability in his/her field of expertise. It must also be assumed that each has value to contribute to the other's performance, based on their individual construction industry knowledge and experience. As peers on the project team, it is expected that each will appropriately assist the other in the collective generation of owner-oriented solutions to design and construction issues.

A simple example of checks and balances would be a manufacturing process where the sales, production, and design departments meet with management to design or modify a product. Each department has unique expertise, experience, and fringe knowledge of the other's expertise. If all four have a common goal—profit—they can collectively design the solution that produces that goal. There will be give and take in the discussions, and management may have to assert its final authority, but the eventual solution will be in the best interest of the company.

By analogy, CM puts the owner in the role of management; A/E, the design department; and the CM, the production department. The goal is performance in terms of time, cost, and quality, with minimum business interruption to the owner. In the CM system, the performance goal is not an event but rather a series of events which continue throughout the project. The CM system continuously pursues its goal through a comprehensive, ongoing management process that draws upon the 12 areas of knowledge covered in Chapters 10 through 21.

CHAPTER 6

Construction Management Under Dual Services Agreements

Chapter 5 explained the various CM forms and variations which emanate from the ACM form of CM. Every form and variation has its own potential for conflict of interest, as shown in Figure 5.10.

This chapter discusses this conflict of interest potential when using the CM system and offers suggestions for mitigating any perceived consequences.

From its beginning, construction management has been provided to owners under several unique contractual configurations, each requiring different combinations of construction/contracting services. Each form and variation of the system relies on appropriate expertise and operational procedures for success.

6.1 CONCERN FOR THE OWNER

Unlike the rigid GC and D–B contracting structures, the CM contracting structure has alternatives worthy of consideration by owners. Owners who use CM should understand the alternatives and all the subtleties involved before entering into a CM agreement. The individuals providing construction management services are responsible for making certain that the owner is properly informed.

Dual services are vested in a single organization in all CM forms except ACM. Consequently, the potential for conflict of interest is an accepted part of the other forms of CM contracting, just as it is in GC and D–B contracting.

As stated previously, the way to reduce or control this potential is to maintain effective checks and balances in team decision-making. It is essential that contractual and organizational separations be installed to prevent the overt consolidation of two services. Failure to do so could negate the inherent benefits of CM performance.

It is the CM's responsibility to: (1) alert owners to the fact that several forms and variations of CM are available for use, (2) explain how each form and variation works, (3) pinpoint areas where potential conflicts of interest exist, and (4) suggest ways to mitigate the potential to the owner's satisfaction.

Items 1 and 2 are academic; they can be demonstrated by diagrams such as those in Chapter 5. Item 3 can be a problem because there are some important but subtle differences that require considerable comprehension of the three systems. Item 4 usually can be accomplished by finding ways to maintain the checks and balances that will make the form and variation work to the owner's best advantage.

It is important to note that none of the standard contract forms for CM services should be used without modification through supplemental and special conditions. Modification is not necessarily the best approach to contracting for certain forms and variations of CM. Separate contracts for each service is a much better alternative when contracting for dual services.

6.2 THE CM/OWNER RELATIONSHIP

In the agency form of CM, the construction manager acts as an agent of the owner and, among other things, provides the construction and contracting expertise generally characteristic of a general contractor in the GC system. ACM permits the construction manager's expertise to be beneficially used by the design professional during the project's design phase, the phase that is inaccessible to a general contractor when using the GC system.

As agents, construction managers are consultants rather than contractors and are hired to perform a specific task that can be evaluated by a standard of care. This relationship holds true until the construction manager's contracted responsibilities extend beyond those usually assigned to an agent (which occurs in all forms and variations of the construction management system except ACM, OCM, and Design–XCM).

6.3 EXTENDED SERVICES CM

When construction and contracting responsibilities are assigned to the construction manager, a dual role of agent/independent contractor is created. The construction manager holds the status of an agent during the preconstruction phase but forfeits that status when performing construction or when assuming contracting responsibilities after the design phase has been completed.

The dual agent/independent contractor status creates a potential for conflict of interest during the feasibility and design phases, when agency services are being provided, as well as during the contracting and construction phases when independent contractor services are being provided.

It is understandably impossible for a CM to represent the owner's best interests when making preconstruction decisions, when the CM is aware that the decisions will impact the CM's financial stake in the project during construction. Dual agent/independent contractor status interferes with, if not eliminates, the checks and balances installed in the "agents only" environment of the ACM form of CM.

A similar situation develops when the project's A/E accepts the added responsibility of construction manager. In this case, although both involvements comply with agent and consultant criteria, the unique areas of expertise provided by the same party obscures the checks and balances inherent to ACM.

It is difficult to extract a level of designability and constructability that reflects the owner's best interests without the presence of professional contention. Deliberations in areas that rely heavily on checks and balances are subject to potential conflict of interest when design and CM services are combined in the A/E's contract with the owner.

6.4 THE POTENTIAL FOR CONFLICT OF INTEREST

As stated in the previous chapter, it is important that the adjective "potential" precede the phrase "conflict of interest" whenever it is used to describe a contracting structure.

The only deterrents to self-serving decisions made in a conflict of interest situation without checks and balances are business and professional ethics. There is a significant difference between the two. Business ethics are accepted rules of conducting business that have evolved through years of commerce. Business ethics are not supervised by a higher authority; professional ethics are.

It could be assumed that parties with dual responsibilities will act ethically in every decision that affects the owner; however, in this day and age every effort should be made to eliminate a sole dependency on ethics. More positive steps should be taken.

6.5 ETHICS

6.5.1 Ethics in the GC System

The GC system of contracting is a good example of a system that mitigates reliance on ethics for owner-oriented performance. The GC has a contract with a measurable end result and limited options during performance. The end result can be provided any way the GC sees fit, so long as the terms of the contract are fulfilled.

The performance options available to the GC are specifically spelled out in the contract for construction. The business ethics of the GC affect the GC's subcontractors and suppliers and have few opportunities to be tested in the GC's relationship with the owner.

If the GC's business ethics adversely affect the owner, they do so most often indirectly through the GC's dealings with subcontractors or suppliers. An example is nonpayment of obligations to suppliers that prompt the filing of mechanic's liens against the owner's property.

An example of a direct adverse effect of GC business ethics is front-loading the schedule of values on a lump sum construction contract. (*Front-loading* is when the owner is paying more than the value of the work done by the contractor during the early part of the project.)

In the GC system, the A/E's performance is highly dependent on professional ethics. When providing design services, A/E business ethics usually fall under the heading of professional ethics.

Professional ethics are subscribed to by the A/E through design profession licensing and A/E societies and associations. They are codified and monitored internally. Breaches of professional ethics are responded to by a hearing and a penalty such as suspension or revocation of the A/E's license to practice in a geographic jurisdiction.

The GC contract structure precludes interaction between the contractor and A/E during design; therefore, no interrelated decisions can be made. During construction, the hard money contract excludes A/E involvement except in the approval processes where the A/E represents the owner's best interests on the basis of A/E professional ethics. There are no checks and balances to provide accountability to A/E decision making.

6.5.2 Ethics in the D–B System

The D–B contracting structure, where A/E and GC services are provided by a single business entity, is very dependent on ethical performance for success, because a productive level of objective checks and balances are not part of the system.

The D–B contract's single point responsibility vests design, contracting, and construction accountability in one party. Oversights must be spotted by the owner or by a consultant hired by the owner. It is common for users of the D–B system to rely on the independent consultant who established the bidding parameters for the project to provide oversight and assist in periodic reviews during design and construction.

Other than specifying that the D–B contractor must employ an external A/E firm to provide design services, the owner has no way to install persistent checks and balances in the D–B system (as they exist in the ACM system) or to reduce the influence of ethics (as in the GC system). Consequently, the success of a D–B project is dependent upon the ethical performance of the D–B contractor and the construction and contracting expertise available to the owner.

6.5.3 Ethics in the ACM System

The ACM contracting structure uses both business and professional ethics to their maximum advantage by establishing a system of checks and balances that precludes sole dependence on ethics for project success.

The A/E is bound by professional ethics and by the agency requirement to act in the owner's best interest in all decisions. The CM is bound by an agency agreement to do the same. All of the six necessary elements, except construction, are vested in the A/E and CM. Consequently, the collective performance of the A/E and CM is a major key to project success.

Although CM firms are not professional firms in the same sense of the word as A/E firms, many CM firms employ professional architects and engineers (A/E peers who adhere to professional ethics commitments) to perform CM services.

Ethical performance in the owner's behalf and an agent's commitment to act in the owner's best interests are analogous. Ethical performance is inspired by professionalism and agency performance is dictated by a standard of care, and the goals of both are the same. Project team checks and balances simply reassure the owner that all decisions made by the A/E and CM will serve the owner's best interests.

6.6 DEALING WITH CONFLICT OF INTEREST

Each form and variation of the CM system has unique performance characteristics that accommodate specific owner needs. All can be accommodated by a broad spectrum of services provided by construction managers. The forms and variations of CM are contractual departures from the idealistically-conceived, owner-oriented ACM form.

One problem owners have with the CM system is a lack of knowledge of how the forms and variations of CM function. Preconceived ideas and unwitting comparisons with general contracting preclude a clear understanding of the CM system and obscure the precise role of the construction manager within each form and variation. As a result, owners often lack the ability to reconcile the CM services they receive with the services they anticipated.

To avoid such problems, construction managers and owners should take sufficient time to fully understand the pros and cons of a CM form and variation before a contractual arrangement is consummated. The potential for conflict of interest generated by dual responsibilities should be understood. An evaluation of the resultant trade-offs should be made, and then action should be taken to deal with the trade-offs during the project.

Even with due consideration, a potential for conflict of interest will persist but can be prevented if sufficient steps are taken. Owners can accrue the unique benefits offered by all forms of CM when they are thoughtfully structured and understood.

6.7 TEAM ACTIONS AND DECISIONS

The six service elements essential for project delivery—design, project management, contracting, construction, construction coordination, and contract administration—were discussed in Chapter 1. A review of that chapter would be appropriate here.

In the ACM form, the A/E and CM share responsibility for project and contract administration; design responsibilities are not shared. The owner does the contracting with guidance from the construction manager and technical input from the A/E, and the construction manager provides coordination of contractors during construction. This alignment of responsibilities provides maximum owner protection by eliminating dual services as well as providing checks and balances on team decisions.

All other forms and variations of CM shift responsibility for service elements, in part or in whole, to either the A/E or the CM. Shifting responsibilities introduces the potential for conflict of interest by influencing team member decisions and diluting the synergistic team effort inherent to ACM. To compensate for this, the checks and balances that produce owner-oriented decisions must somehow be preserved and more closely monitored by the owner.

6.8 DUAL CONTRACT SOLUTIONS

When CM forms and variations are used that assign dual responsibilities to either the CM or A/E, steps should be taken to preserve the synergistic composition of the team. The fact that a team member is providing combined services does not mean that member's representation on the team can also be combined. To preserve the checks and balances under dual service arrangements, separate representation for each service being provided should be a contractual condition.

The best approach to separate representation is to have separate contracts for each service being provided. Rather than extending the A/E's design services contract to include CM services, the A/E and owner should enter into two separate contracts, one for design services and one for CM services.

Instead of extending the CM's ACM contract with provisions to include construction or the holding of construction contracts, owners should enter into a separate contract for each service provided by the CM.

Separation permits each contract to speak for itself with regard to provided services, prevents the use of crossover procedures, and precludes dual personnel assignments which might be inferred by a single extended-services agreement.

6.8.1 The Design–Build Example

The D–B system combines design services and contracting/construction services in one contract. This combination permits the D–B contractor to intimately connect design decisions with contracting/construction decisions. The question of who benefits from decisions made in a dual responsibility structure, especially one without positive checks and balances, always exists: who is the ultimate beneficiary, the owner or the D–B contractor?

One way to strengthen results from ethical dependency in the D–B system is to structure the contracts so as to strengthen ethical responsibility. Figures 6.1 and 6.2 show contract structures that provide the same advantages of the D–B system to the owner. However, one favors the owner more than the other with regard to the potential for conflict of interest.

The best owner-oriented contractability decision would be to use the structure shown in Figure 6.1a. Note that it contractually separates the design entity from the construction entity but still permits single-point responsibility between the owner and D–B contractor.

Separating the design entity from the contracting/construction entity offers a better opportunity to obtain the checks and balances provided by a physical separation

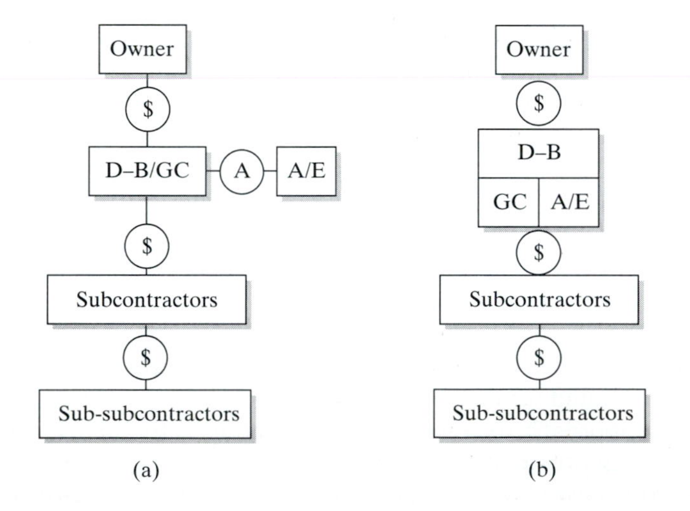

FIGURE 6.1 Optional contract structures for the D–B system.

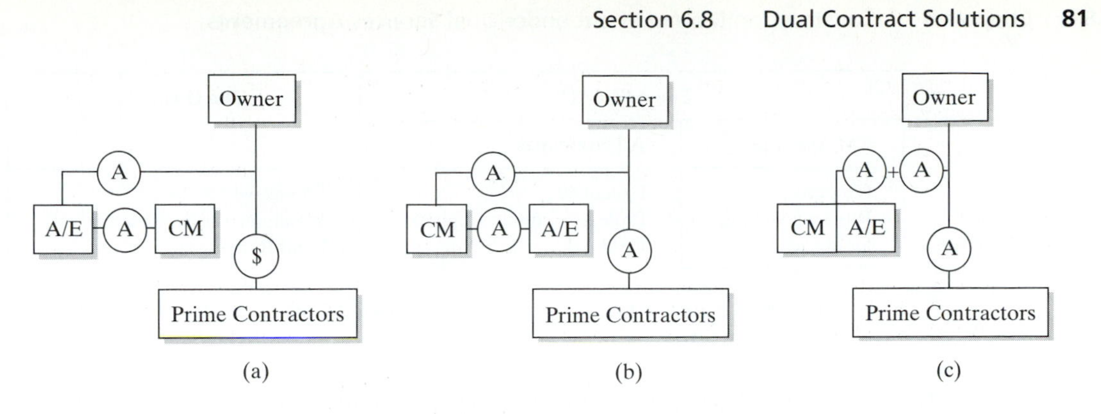

FIGURE 6.2 Optional contract structures for Design–XCM.

between the independent D–B contractor and the agent design professional. The professional ethics of an independent design firm have a better chance of influencing decisions than a design entity that is an integral part of the D–B contractor's organization.

6.8.2 Design–XCM Services

There will be times when an owner prefers a single-responsibility contract for A/E and CM services rather than working in a team situation with a separate A/E and a CM. The D–B system satisfies this requirement, providing the owner wants to hold the construction contracts. If the owner wants to use CM multiple bidding and hold the construction contracts, Design–XCM, diagrammed in Figure 5.4, is a CM alternative.

If the owner is concerned about potential conflicts of interest inherent in dual responsibility contracts (even though both are agency contracts), solutions similar to the one suggested for D–B in Figure 6.1 are available for Design–XCM as shown in Figure 6.2. The owner has three Design–XCM options available.

The Design–XCM structure (Figure 6.2c), where the owner holds an A/E contract and a CM contract with a single firm, provides less potential for conflict of interest than the structures in Figures 6.2a and 6.2b. The team concept is preserved in Figure 6.2c because the owner holds two agency contracts, although with one party.

6.8.3 Design–XCM Solutions

When the A/E also serves as the CM, it is prudent to contractually separate the services—one contract for A/E services and another for CM services. The contracts should stipulate that personnel assigned to one service not be involved in the other.

Dual services are usually suggested as an economic advantage or a beneficial expedient from which the owner will gain. However, the potential loss resulting from diluted checks and balances heavily outweighs the perceived savings in both time and in the cost of services. Dual personnel assignments are a blatant invitation to convert potential conflict of interest into reality.

	The A/E-CM		The Owner
Level	CM Activities	A/E Activities	
Executive Management Administrative	Executive Project Manager Field CM	Executive Project Architect/Engineer Field A/E	Executive Management Representative Field Representative

FIGURE 6.3 Project team representation (Design–XCM).

Every precaution should be taken to contractually prevent this compromise. Each service should be required contractually to provide an exclusive cadre to function on the project team at the executive, management, and administration levels in order to maintain the team structure and preserve the benefits of checks and balances. Figure 6.3 shows the composition of the project team when Design–XCM is used. It can be compared to Figure 13.1 (Chapter 13) for clarification of the minimum cadre.

It would be advisable, and certainly within precedent, for the names of the three project team representatives and their back-ups to be written into the appropriate contracts to assure the mechanics of separation as a minimum.

6.8.4 Contractor–XCM Services

There will be occasions when the owner prefers to use ACM but does not want to hold the contracts for construction. An obvious solution is to assign the awarded contracts to the ACM. When this occurs, the ACM becomes a contractor with the accompanying legal responsibilities to the trade contractors who are now subcontractors.

If the owner's reason for assigning the contracts to the CM is to shed responsibility for the contractor's performance, the ACM becomes a CM providing contractor XCM services. She/he also enters into an independent contractor relationship with the owner that eclipses the agency responsibility of an ACM.

When progress payments are due, the owner issues a single check to the CM to cover each total progress payment. The funds are deposited into the CM's account and then redistributed to the individual subcontractors in separate checks.

The CM has control over the disbursement of the funds because it holds the contracts with the subcontractors. The payment situation is the same as if the CM were a general contractor in the GC system and exposed to the abuses pointed out in Chapters 2 and 3.

The CM represents the interests of the subcontractor in any claims or disputes that arise under their contract with the CM, but the CM must also pursue their interests to the owner. It is difficult, if not impossible to serve both the owner and the subcontractors under this contractual relationship. However, if the owner's concern is simply the inconvenience of tracking so many contracts and purchase orders and having the accounts payable department issue multiple checks on each progress payment date, a simple solution is available. The owner can open an escrow account at a local bank. The amount of the progress payments due to each work-scope contractor and supplier are listed and supported in the monthly financial report assembled by the ACM as described in Chapter 13, Information Management.

The owner deposits the amount of the total progress payments to the escrow account, enclosing a copy of the application summary found in the monthly financial report. The bank issues checks for each applicant in the amount listed and forwards them to the CM for distribution to contractors and suppliers. The bank is entitled to a minor service fee for administering the escrow account.

An alternate solution that is available (but not recommended) is to have the ACM act as an escrow agent instead of setting up an escrow account in a bank. The ACM deposits the total payment check in its own account and disburses payments to contractors and suppliers according to the application summary list. The ACM is entitled to a small service fee for making the transactions.

One problem with using the ACM as the escrow agent is that it involves the ACM as a pass-through payment agent when it is not necessary. An ACM should not be put in a position where it receives and disburses project funds. Although an ACM is a fiduciary agent of the owner, it should not be put in the position of financial agent of the owner.

An element that should be considered is that by using the ACM as an escrow agent, the ACM's name will be obvious on the progress payment disbursement checks. Some believe that contractors and suppliers will cooperate better with the ACM if they think that payment is coming from the ACM.

Others believe the opposite. Trade contractors should not be given the opportunity to relate the construction manager to a general contractor in any way. Payments made on the owner's checks remind them that they are prime contractors with direct access to the owner, completely unlike their status in the GC system.

6.8.5 Constructor–XCM Services

There will be occasions when a general contractor-based CM who has an owner-ACM contract will convince the owner that he could construct one or more work-scopes more economically, or better in some way, than another contractor. An owner may have a close relationship with a GC-based CM and prefer to use the CM as a constructor on certain parts of the project.

When an ACM contracts with the owner to construct work-scopes, the ACM becomes an independent contractor which eclipses the agency status and turns ACM into Constructor–XCM. This dual status of the CM should be handled discretely to retain the confidence of trade contractors who are planning to submit work-scope bids on the project.

There are several ways to handle this situation; none have particular merit from the best-use perspective of the CM contracting system. All require open communication with potential work-scope bidders to maintain as much of the credibility of the CM system and multiple bidding process as possible.

One option that can be used on private sector work but not on public sector work is to inform prospective bidders that specific work-scopes will be negotiated between the owner and CM and will not be bid competitively. Another is to inform prospective bidders that the CM will submit prices for specific work-scopes in an overt competition and will be considered for award on the same basis as the other bidders.

Still another option is to inform bidders that the CM will review competitive bids when they are received by the owner, and if the CM decides it can do the work for a lesser price, then the CM will be awarded a contract by the owner for the work covered in the selected work-scope.

One final approach is for the owner to award the CM services contract using a price-based selection process and parametrically include the cost of constructing specific work-scopes in the fee structure for Contractor–XCM services.

As noted, none of these have merit. It is nearly impossible to cite a situation where the construction capability of a CM is so superior to that of the available competition that it would be to the owner's advantage to award or assign a specific work-scope(s) to the project's CM. It is equally impossible to comprehend how owners might believe that a CM will criticize construction work done by the CM's employees as impartially as construction work done by other work-scope contractors.

The dual responsibility role of the CM which results from Constructor–XCM deprives the CM system from one of its prime reasons for existing: eliminating as much potential for conflicts of interest as possible. Constructor–XCM and Contractor–XCM have the most potential for generating debilitating conflicts of interest of all the forms and variations.

6.8.6 GMPCM Services

There will be times when owners need to establish the cost of construction prior to the completion of design in order to save time and comply with finance conditions that require a guaranteed cost before funding is approved. The GMPCM form and its variations accommodate these requirements.

GMPCM is really a unique variation of Contractor–XCM, but it stands as a form of CM because it has three variations of its own (Chapter 5, CM System Forms and Variations, explains them in detail). GMPCM provides ACM services but also extracts a guaranteed maximum price for the total project cost from the CM providing services.

Owners should advise the CM that a GMP will be required before entering a contract with the CM. The agreement used can be either a standard GMPCM contract form, or a standard ACM contract form suitable for a GMP amendment should be entered into before design begins.

One approach to establishing the GMP is to have the ACM determine its amount during the latter stages of design, at the time when the ACM is sufficiently confident in providing a GMP for the work to be done. When the GMP becomes part of the contract, the CM converts from an ACM to a Contractor–XCM.

The problem with the veracity of the GMP is that the CM is the one who establishes the amount, and it is very likely that the GMP amount will be set high enough to protect the CM from loss after final costs have been established. This situation presents a potential conflict of interest—will the CM provide a budget that benefits the owner's interests or the CM's interests?

Although the owner only reimburses the CM for the costs expended, the "savings" between the cost expended and the GMP amount represent monies the A/E could not use in the design (because the CM was administering the budget) plus monies the CM included as a self-serving contingency in the GMP. There is no doubt that on a project basis, a GMP figure will be higher than a lump-sum figure.

To protect the owner from underspending the budget or relinquishing desirable features not included in the design, "plus alternates" should be included in the bidding documents. The CM should be asked to provide the GMP on the base design only. When multiple bids are received from contractors on the base bid documents, the difference between the total of the bids and the GMP can be used to purchase selected alternates. Chapter 24, Acquiring CM Services, discusses different CM fee arrangements and explores other facets of GMPCM.

6.9 THE CREDIBILITY OF THE CM SYSTEM

A prime concern during the development of the CM system was the waning performance credibility of the construction industry. Construction users were looking for ways to get better performance, and construction industry practitioners were looking for ways to satisfy construction users. Bringing owner-oriented expertise into the contract structure seemed to be one way to accomplish these goals.

In the early days of CM, many construction industry practitioners considered it a fad. Construction management was a buzzword, a temporary practice that could never compete with traditional practices. In spite of growing pains, CM caught the attention of many users and developed in its own right.

Many construction practitioners felt that CM was a good thing for the industry, that its beneficial practices would be adapted to the general and design–build contracting systems before it lost its popularity. Few industry practitioners believed CM would become a popular alternative to the GC and D–B systems.

As the popularity of CM increased, traditional contracting practitioners looked for ways to compete with the new system. The manufacturing concept of total quality management (TQM) was borrowed by GC and D–B contractors to elevate their image in the eyes of owners. Partnering was introduced as a means of softening the adversarial reputation earned by the GC system. New efforts to promote the design–build system were made. The CM system proved to be a formidable owner oriented competitor in the project-delivery market place.

6.9.1 Credibility Problems

CM is not without problems of its own, most of which are connected with the performance of its practitioners. At its outset, CM was a system looking for acceptance from construction users. CM practitioners diligently searched for ways to improve their practices to better serve owners. Early practitioners especially contributed to its success by growing with the system, experiencing its growing pains, and progressively improving its practice and image.

Later entrants into the CM market place, firms getting into CM to get their share of the business, offered what they perceived to be CM services but which were in fact shallow emulations of services provided by veteran CM practitioners. Unfortunately, latecomers had no other choice.

Effective CM practices were developed by trial and error and were proprietary to the firms that developed them. Although a lot of texts had "CM" in the title, and many articles were written on the subject, none provided a comprehensive how-to scenario that could create overnight CM competency.

Consequently, the average CM performance in the late 1970s and early 1980s was much higher than in the late 1980s/early 1990s. Competition for projects by underqualified CM firms and qualified CM firms who compete by providing minimum services, has lowered the credibility of CM practice.

Efforts by societies and associations, both professional and trade oriented, have focused mainly on the proliferation of the CM system and not the quality of CM services. Statutes and laws have focused on what constitutes CM services and which construction industry group can practice CM, and not the technical proficiency of CM practitioners.

6.9.2 Credibility and Dual Services

The original founding philosophy of CM—the protection of the owner's best interests—has been eroded by CM practitioners who are not satisfied with earning an ACM fee and conspire in many ways to supplement it by providing dual services. Dual services increase the potential for conflict of interest in the CM contract structure and, with the exception of Design–XCM, contractually combine agents with independent contractors, a combination that is difficult to control if the owner's best interest is the objective. Dual services should only be provided when they benefit the owner, not simply to enhance the CM's fee.

CHAPTER 7

ACM Procedures

The previous chapters dealt with the contracting structure, philosophy and origin of the CM system; all of which must be fully understood in order to appreciate CM procedures and the profound level of detail necessary to ensure that each procedure achieves the positive results which the CM system has to offer.

This chapter will cover the management procedures in the ACM form of the CM system. The reason for targeting ACM rather than its subforms and their variations should be evident at this point. ACM performance is the key to performance in every CM form and variation. The added services that convert ACM to XCM, GMPCM, and OCM are traditional construction industry services and require no mention here.

One point that should be mentioned is that many CM practices will seem tedious and perhaps unwarranted to construction industry veterans. This is because traditional project-delivery practices do not normally use the detail and attention on which the CM system depends. Readers unfamiliar with the construction industry (students and most owners) have no perspective from which to form an impression one way or the other. Only readers who are veterans of ACM practice can appreciate the value of the details in each practice described.

The ACM procedures are structured around the CM Body of Knowledge and its twelve CM Areas of Knowledge covered in Chapters 9 through 21. Performance policies and procedures in each of the areas must be developed, recorded, and periodically reviewed to guide team members and their resource persons during the execution of activities that comprise an ACM project. The CM's management level or Second Tier team member should be responsible for organizing and maintaining these procedures.

The last few pages of Chapter 4 contain a broad-scope list of ACM activities; a cursory review of that list would be appropriate before reading on.

7.1 STARTING SERVICES

If CM is to be used, the CM should be hired as soon as possible after the feasibility phase indicates a decision to construct. The ideal start-up situation would be to engage the services of the CM and A/E concomitantly. This facilitates the coordination of contracts and provides the best opportunity to confirm the compatibility of the owner, A/E, and CM.

Team member compatibility and synergistic team effort are essential. Consequently, team members should have respect and high regard for one another as technicians and individuals. They will be working very closely together throughout the entire project. Selecting a CM is a more detailed process than selecting either a contractor or A/E. Information on appropriate procedures for hiring a CM is given in Chapter 24.

7.2 BRAINSTORMING SESSION

As soon as possible after the A/E and CM have been selected, a meeting should be held to introduce resource and support team members to each other and to discuss the project and its ramifications. This is commonly called a brainstorming session. Its purpose is to align project goals, become familiar with owner requirements and mode of operation, and extract as much information on interfacing management activities as possible. The CM should organize, set the agenda and chair the meeting, passing the chair to each team member when presenting and discussing their needs. As many persons as possible from each team member's organization who will interact with one another during the project should be present.

This meeting will set the tone for future team interaction. It is critical that the A/E and CM understand the owner's project-related business policies and procedure preferences before the meeting adjourns.

Sufficient time should be allotted to present each team member's perspectives and learn the perspectives of others. It is not unusual for a brainstorming meeting to consume the better part of a day and involve five to ten persons from each team member organization. This is especially true if it is the first time that team members have worked together on a project.

Information generated from brainstorming allows team members to formulate procedures for communication and interaction during the project. These procedures should be recorded in the CM project manual for future use. Further information pertaining to the brainstorming session and the CM project manual is provided in Chapter 16, Project Management.

7.3 ORGANIZATIONAL MEETING

Assuming that CM and A/E contracts have been signed by the owner and the brainstorming meeting held, the next step should be the organizational meeting. This meeting brings the executive and management level representatives of the owner, A/E, and CM face to face and is critical enough to have the owner's legal representative in attendance as well.

The purpose of this meeting is to establish team member responsibilities in the individual procedures that constitute the CM format. The A/E-owner and CM-owner agreements assign performance responsibilities individually to the A/E and CM but do not provide the details of mutual involvement in team activities. There is no agreement between the A/E and CM regarding interaction, so one must be created by consent of the team members.

This document can best be produced in the matrix form and is appropriately referred to as a responsibility chart. Chapter 16, Project Management, provides more information on this document and has a list of activities and an example to review. The responsibility chart will be replaced some day by a general conditions document when ACM team interaction is universally codified.

The organizational meeting should last as long as it takes to complete the responsibility chart. The owner should be the host and the CM the facilitator. The only item on the agenda is the responsibility chart. The sensitivity of the topic under discussion requires that the facilitator be very well prepared and sensitive to team member contributions.

The facilitator should never use a completed responsibility chart from another project as a base for modification. The matrix may have a suggested list of activities and headings indicating the team members involved, but filling in descriptions of each team member's participation should be developed exclusively at the organizational meeting.

The executive-level people at the meeting should provide input to team and project policies and, along with the owner's attorney (if present), monitor the commitments that are affected by policies already made in the owner-A/E and owner-CM agreements. It is possible that some of the language and the responsibilities in the two agreements will require modification. The executive-level persons should be authorized to do this at the meeting.

The facilitator should bring each listed team activity to the meeting's attention. The management-level people should discuss each activity and come to agreement on the detailed actions each team member is responsible for during the project.

The end result is a matrix-style document that precisely describes the participation of each team member in each activity that is a part of the ACM format throughout the project. The document can be filed as an agreed-to working agreement or become an amendment to each team member's contract with the owner.

7.4 CM PARTNERING

The brainstorming and organizational meetings and the formulation of the responsibility chart are the forerunners of the "partnering" concept promoted in general contracting. The difference between GC and CM partnering is in the documentation. GC partnering produces a voluntary broad commitment to cooperation by the project's major stakeholders that is documented in a noncontractual partnering agreement.

CM partnering produces a voluntary commitment to cooperation, and a detailed contractual commitment to cooperation that clarifies and expands many of the vague standard clauses contained in the owner-A/E and owner-ACM agreements. Cooperation requirements are documented as well as developed and agreed to in CM partnering.

7.5 THE CM PROJECT MANUAL

From the information obtained at the brainstorming and organizational meetings and the responsibility chart, the CM is in a position to document the many procedures to be used by team members during the project. All information pertaining to team interaction and team member performance should be recorded in a CM project manual, a living document that should be reviewed periodically and modified by team action as necessary. Each team member should have a copy of the CM project manual.

The contents of the manual and the procedures described will vary from project to project because they are individualized to reflect specific owner needs and preferences.

The CM's management-level team representative is responsible for maintaining the manual based on the policies set by the executive or First Tier project team representatives and upgraded information gathered at the brainstorming session. Copies of the initial, and all subsequent updates of the manual should be organized in a loose leaf binder.

The CM project manual serves as a reference for executive- and management-level team members, a record of policy and procedure development, and a procedural guide for administrative-level team members and CM operations, resource and support persons who cyclically service the project.

7.5.1 CM Manual Contents

As a minimum, the CM project manual should include:

- Architect's agreement with the owner
- Construction Manager's agreement with the owner
- Other team member agreements with the owner
- Responsibility Chart
- Budget Management Plan
- Contract Management Plan
- Decision Management Plan
- Information Management Plan
- Material/Equipment Management Plan
- Project Management Plan
- Quality Management Plan
- Resource Management Plan
- Risk Management Plan
- Safety Management Plan
- Schedule Management Plan
- Value Management Plan

7.6 MANAGEMENT PLANS

The development of the management plans included in the manual should be a team effort led by the construction manager. Experienced CM firms will have "standard" plans or plans used on other projects which can be used as a beginning point. Unlike the delicate, interactive development of the responsibility chart, the management plans are expected to be the product of the construction manager.

When CM services were first offered, construction managers were indisputably the knowledgeable party. Owners had little knowledge of CM or the scope of CM services. It was common for construction managers to sell their services as a self-determined package of services in competition with other CM firms' self-determined packages. The CM would tell the owner what services would be provided if they were to be selected, and the owner would select the CM that appeared to have the best package of services. By the late 1980s, the process was reversed; owners started telling construction managers what services they wanted and selected the CM on its ability to provide the specified services.

Consequently, most CM firms should now have the capability to provide services that can satisfy the needs of all owners, with the exception of CM firms that exclusively provide ACM or design–XCM services. Putting management plans together should be routine.

Each Management Plan should document its process and procedures in detail and clearly state the interactions of each involved participant during its execution. Actions and interactions must comply with applicable contract requirements and should, in the case of team members, agree with the assignments made in the responsibility chart. Plans must honor procedural requirements and preferences of the owner and include the preferences of the CM and A/E team members as much as possible. As the knowledgeable party, the CM has the responsibility to extract requirements and preferences from both the owner and A/E and should lead the team in formulating and documenting management plans.

To reiterate, the CM should not impose a plan on the team. Plans should be collectively developed by the team, agreed to by the CM and A/E, and approved by the owner. Each plan is a living document that should be adjusted and modified if experience during their application indicates revision would be helpful. To this end, management plans should be on the agenda of all second-level management meetings until satisfactory performance of each plan is achieved.

7.6.1 Developing a Management Plan

Each Management Plan contains procedures for accomplishing the several processes which fall under its title. One of the processes under Material/Equipment Management would be the procedure for acquiring material and equipment for the project. Exactly how this is to be achieved must fit the owner's requirements and the form of CM being used.

Consider the construction of an elementary school for a public school district controlled by an elected school board.

State laws stipulate the manner in which school districts may acquire construction services and purchase materials and equipment, and school board policies modify and add to the state law requirements.

After discussions with the CM consultant they hired, the board decides to use ACM and multiple trade contractors and hires an ACM. Under the procedures of the already written Contract Management Plan, multiple work-scopes are decided upon by the team and written by the ACM. Design is in progress and the contract documents have been customized by the A/E and ACM to facilitate multiple bidding.

From this information, a procedure for Direct Acquisition of Construction Services, Material and Equipment can be developed as part of the Material/Equipment Management Plan. The plan must be reviewed and approved by the team to become part of the CM Project Manual. An example plan is shown in Figures 7.1a, b, and c.

Construction Services: The procedure for acquiring construction services is simply a narrow-scope version of the broad-scope assignments agreed upon in the responsibility chart. Management Plan assignments should be narrow scope. They should

FIGURE 7.1a Example Management Plan (owner requirements).

Material/Equipment Management Plan
for the
Acquisition of Construction Materials/Equipment/Construction Services

1. *Owner Requirements:* State Law, amended or augmented by Board Policy. (Related requirements may be found under other management plans.)

L Material and equipment purchases over $2500 and construction services over $5000 shall be advertised and competitively bid.
L The Advertisement for Bids shall be published in at least one local area newspaper with a daily circulation greater than 50,000.
B Advertisements shall be published in the local newspaper regardless of its circulation.
L Advertisements shall appear twice, at least 10 days apart, with the last at least 10 days before the date bids are received.
L The criterion used to qualify suppliers/contractors for bidding or award shall be obviously objective and incapable of subjective interpretation.
B Either performance or prescription specifications may be used for acquisition purposes.
L All bidders must receive identical bidding information to be deemd competitive.
L A minimum of 3 sealed competitive bids must be received to make an award.
B Bids may be stipulated to be received as either unit price or lump sum proposals.
B Guaranteed maximim price proposals may only be used for construction services less than $5000.
B Expiration time for contract services bids shall be stated as not less than 30 calendar days or more than 90 calendar days.
B Expiration time for material and equipment bids shall be stated as not less than 30 calendar days or more than 60 calendar days.
L Bids shall be opened and read at an official meeting of the Board of Education.
B Bid opening dates and times shall be advertised and open to the public.
B Bids received after the stated time and date shall not be opened and shall be returned to the bidder unopened.
B Contracts for construction services for portions of the total project may be awarded separately providing award requirements are met.
B Contracts for construction services shall not be awarded if the low competitive bid exceeds 5% of the budgeted amount.
B Purchases of materials and equipment shall not be consummated if the low competitive bid exceeds 3% of the budgeted amount.
B All construction service contracts, and those material and equipment purchases over $2500, shall have Board approval prior to award.
B Material and equipment purchases less than $2500 may be consummated by the Superintendent of Schools prior to Board approval.

L = school law; B = board policy.

provide more detail to facilitate the activation and execution of a procedure. In the example (Figure 7.1b) the responsibility chart covered 10 of the 27 listed actions. The 17 actions added in the management plan clarify the procedure.

Material and Equipment: The procedure for acquiring material and equipment can be prescribed from owner requirements plus additional information extracted from the owner. This format is simpler than the construction services format.

Material and equipment purchasing is an ongoing owner function which already complies with school board policy and state laws. The owner is the knowledgeable party, and the procedure adopted by the team must conform to existing owner procedures as closely as possible.

FIGURE 7.1b Example Management Plan (construction services).

2. Acquisition of Construction Services Procedure

Actions designated by:	*Responsibility Chart*	*Management Plan*
A *Process Summary:*		
Write the advertisement	Owner	X
Determine publications/frequency	X	Owner
Arrange to publish advertisement	X	A/E
Pay for the advertisement	Owner	X
Determine contractor qualifications	X	A/E, CM
Write qualification Questionnaire	X	CM
Qualify contractors	X	Owner
Suggest bidders list	X	A/E, CM
Compile bidders list	X	Owner
Develop proposal forms	X	CM
Print proposal forms	X	CM
Distribute proposal forms	X	CM
Print bidding documents	A/E	X
Pay for bidding documents	Owner	X
Distribute Bidding Documents	A/E	X
Arrange for pre-bid meeting site	X	CM
Prepare pre-bid meeting agenda	X	A/E
Notify contractors of pre-bid meeting	X	CM
Conduct pre-bid meeting(s)	CM	X
Write bidding document addenda	A/E, CM	X
Distribute bidding document addenda	A/E	X
Receive proposals	Owner	X
Proposal review for compliance	X	A/E, CM
Proposal review for content	X	A/E, CM
Suggest proposals for acceptance	X	A/E, CM
Select contractors for award	X	Owner
Award contracts	Owner	X

B *Time Performance Guide*

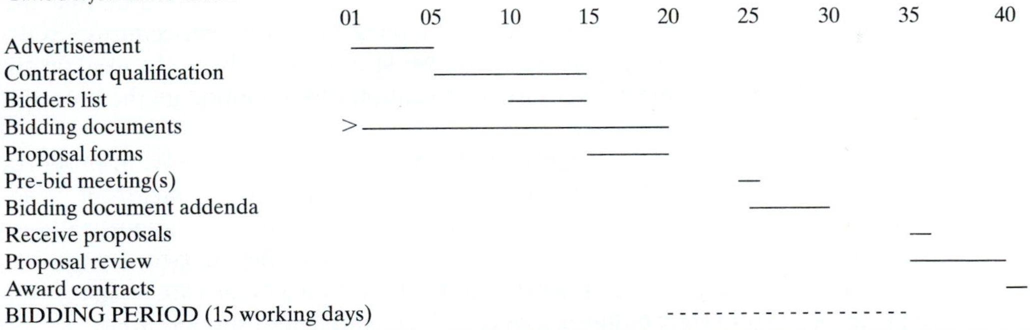

01 05 10 15 20 25 30 35 40

Advertisement
Contractor qualification
Bidders list
Bidding documents
Proposal forms
Pre-bid meeting(s)
Bidding document addenda
Receive proposals
Proposal review
Award contracts
BIDDING PERIOD (15 working days)

7.6.2 The CM's Procedures

In addition to the team procedures recorded in the CM project manual, the CM should determine in-house procedures to be used by CM personnel to accomplish the assigned actions. These procedures must dovetail the team procedures and should be

FIGURE 7.1c Example Management Plan (material/equipment).

3. Acquisition of Materials/Equipment Procedure

Determining purchases:

The CM and A/E shall select the items to be directly purchased by the owner. Selection shall be based on construction schedule requirements (long-lead items) and local contracting practices (items where installation-only contractors are available).

Purchasing procedures:

The A/E and CM shall assist the owner in purchasing long-lead items for the construction of the project.

The A/E shall provide to the owner written technical specification for items to be purchased by the owner.

The CM shall provide to the owner written logistic information covering transport, unloading, and storing owner-purchased items.

The CM and A/E shall assist the owner in determining vendors to be solicited.

The owner will issue the request for quotations including the terms of purchase, delivery, payment and warranties/guarantees, with the assistance of the A/E and CM.

The A/E shall review vendor quotations for compliance to the technical specifications

The CM shall review vendor quotations for compliance to logistic requirements.

The owner will select a supplier and issue the required purchasing agreement (purchase order).

The CM shall arrange for receiving, unloading, storing, and installation of owner purchases.

The CM shall check owner-purchased items for condition and quantity within 24 hours after delivery.

The A/E shall inspect owner-purchased items for technical compliance within 4 calendar days after delivery.

The owner will pay for direct-purchase items according to the terms of the purchase agreement with the vendor.

The owner will maintain the direct purchase files according to the standard operating procedure of the school district.

Original ☐ Update ☐ Update No._____ Date:__________

For Owner: ______________________________, Management Level Representative

For A/E: ______________________________, Management Level Representative

For CM: ______________________________, Management Level Representative

inserted in the CM's copy of the CM project manual as a reference for all CM personnel who will be involved throughout the project.

After involvement in several CM projects, in-house procedures that are frequently used and work best emerge as standard operating procedures (SOP) and become part of the CM's operational menu. Most should be able to be used on all projects without change; some will require modification to accommodate the procedures in the CM Project Manual.

The importance of documented in-house procedures increases in proportion to the work load of the CM, particularly the size and number of concurrent projects in process. When handling several projects simultaneously, the CM's support and resource personnel, and even some operations personnel, switch their day-to-day efforts from project to project depending on what must be done. Few persons are assigned full-time to one project from start to finish, especially in resource and support areas.

Because each project has unique owner requirements, CM personnel must adapt to each project's requirements and execute procedures accordingly. The CM project manual's procedures and the CM's adapted in-house procedures expeditiously direct CM personnel on a project basis, both in the initial phase of the project and later when periodic involvements are required.

Without the manual, resource and support personnel could be misdirected, resulting in wasted time and more importantly, failure in the services provided. The consequences of misdirection could negatively affect the credibility of the CM as a team member and jeopardize opportunities for future CM work.

7.6.3 In-House Procedures

Many in-house procedures take the form of a paper trail which lead CM personnel through the activity they have been assigned and produce a record for the actions of others who follow. Although this process is more obvious in other management areas, such as budget and information management, there are some uses in resource management.

In the acquisition of construction services procedure, the CM is generally designated as the team's liaison with bidding contractors. This can be assumed from the fact that the CM is the knowledgeable party in the area of contractor relations. One of the major efforts of the CM prior to the bid date for construction contracts is generating continued interests among trade contractors to submit bids. The effort is time consuming and intermittently involves several operations and support persons.

In the GC system, it is the responsibility of the general contractor to locate trade contractor and supplier offers for construction, materials, and equipment. In the ACM system, the construction manager does this task in behalf of the owner. The CM's effort is more demanding than the GC's because of the CM's agency obligation.

Under ACM multiple bidding, the CM must generate competitive trade contractor bids in each of as many as 50 or 60 separate work-scopes. If the CM is not successful in locating at least three bidders in each work-scope (according to the owner criteria we assumed in the Figure 7.1a) those work-scopes must be re-bid. The total cost of the project cannot be determined until the re-bidding is complete.

The GC, on the other hand, has no set criteria. The GC would like to have several offers in each trade contractor area, but if it does not happen, the GC can develop estimates for missing trade areas, rationalize single offers, and submit a proposal to the owner anyway. If determined the low bidder, the GC can negotiate missing trade contracts after an award is made by the owner. Although not the best way to commit to a bid price, this often happens in GC bidding.

Not receiving a sufficient number of bids in each work scope area is a risk that must be managed when using ACM with multiple contracts. The value of this risk will differ according to the criteria set by law or policy. The consequences of the risk are not so much monetary as time consuming because of the re-bid requirement. However, the credibility of the CM is sometimes questioned when team bidding goals are not met. A proven in-house procedure that precludes a shortage of trade contractor bids on bid day is essential to successful CM operations.

An example in-house SOP to assure adequate trade contractor competition on multiple bidding projects is shown in Figure 7.2. This procedure would be part of the Resource Management Plan.

7.6.4 Other Procedures

This representative example of the acquisition of construction services, materials, and equipment procedures provides a framework for other procedures to be included in the CM project manual. The point made by the example is the extent of detail necessary to adequately describe a simple procedure if it is to be applied consistently by all persons involved in the project.

Procedures should be firmly based on common sense and good management practices. Consistency is a prime requirement when providing management services for a fee and can only be achieved from a common reference when several persons are involved. The procedures should provide a documented course of action for all operations, resource, and support persons as they switch their input from one project to another.

FIGURE 7.2 Example in-house CM procedure to ensure bidding competition.

Resource Management Plan (ACM In-house)

Bidders and Bidder's Lists

1. As soon as work-scopes (Bid Divisions or BDs) are identified, operations shall assemble a list of potential bidders for each BD.
2. The number of contractors listed for each BD shall be at least 2x the number of bidders required by the owner to make an award.
3. The contractors shall include contractors who have performed well on previous projects in the area of the project.
4. Both the owner and the A/E shall be contacted to determine if there are any contractors they wish to have included on the list.
5. As the BD list is developed, the contractors on the bidder's list shall be contacted and told of the upcoming project.
6. Contractors on the list shall be contacted no less than every 2 weeks after the initial contact to confirm their interest.
7. A list of bidders by BD shall be maintained starting with the first contact. The list shall include phone numbers and contact names.
8. The list of bidders, including weekly updates, shall be included in all copies of the CM Project Manual.
9. Contractors on the list shall be advised when and where the Advertisement for Bids is published (issued).
10. Contacting potential bidders shall become more frequent, according to need, after the Advertisement is issued.
11. If qualification is a requirement, a Qualification Form shall be sent to each bidder, whether requested or not.
12. Assistance shall be offered to the contractors who express problems with Qualification Forms.
13. All contractors on the list shall be contacted 3 days prior to the due date for Qualification Forms to remind them of the due date.
14. Our goal is to assure at least 2x the number of bidders required by the owner to make an award will submit bids.
15. Every effort shall be made to promote the importance of trade contractors to the CM contracting process. They are prime constructors.

Bid List Records

1. Operations personnel shall maintain bidder lists on trade contractors for use on future projects.
2. Bidder lists shall include demographic information and comments on each bidder's bidding performance and construction performance.
3. Bid lists shall be filed by trade(s) and crossindexed by business location and operations area in terms of State, City, and Area Code.
4. Each contractor on the list shall be rated on bidding performace and construction performance on a scale of 1 (low) to 5 (high).
5. Bid lists shall be updated after each bidding experience as well as every year in areas where this CM frequently provides services.

7.7 LIST OF MANAGEMENT PLAN PROCEDURES

It should be understood that procedures interface within a management area and also between management areas. The location of a procedure in the list of management plans is a matter of choice but consistency of location will make them more accessible from project to project.

It is advisable to establish a master index (similar to the Masterformat for CSI specifications) for location of procedure subject matter within management plans that can be used on all projects as well as within the CM firm. Cross-referencing interfacing procedures in the manual make it easier to use.

The following is a list of the procedures that should be included in the CM project manual under the twelve plans. The list is for ACM and is not represented as all-inclusive.

Budget Management Plan: Procedures that prescribe how all project costs are created, managed, tracked, and reported to the owner.

Procedures should begin with the organization of the owner's project's budgeted cost and end with final payments to contractors. In addition to their use in controlling costs, procedures should determine and certify progress payments, track designated owner expenditures, and produce a detailed record of fiscal accountability.

Contract Management Plan: Procedures that lead to a contracting strategy and plan; and the development, review, coordination and documentation of the project's contract and contracting documents.

Procedures should span from the brainstorming session to project completion. They should include contractability reviews, gathering economic/strategic data on the contracting, construction, material/equipment and labor, marketplace, document coordination, and contracting approval procedures during the project.

Decision Management Plan: Procedures that ensure synergistic decision-making, provide checks and balances, and document all decisions.

Procedures should cover all phases of the project. They should establish and document each team member's decision-making hierarchy and prescribe the limits of decision-making authority. They should preassign titular team leadership and explicitly state the responsibilities attached to it during specific project activities and phases.

Information Management Plan: Procedures that communicate information to and between team members, covering all actions, in appropriate formats.

Procedures should cover correspondence, meetings, telecommunications, and the project's day-to-day activities. They should specify the use of agendas, meetings, meeting minutes, reports, letters, memos, and reporting forms, and deal with frequency, format, and content.

Material/Equipment Management Plan: Procedures for bidding, ordering, expediting, receiving, handling, and storing materials and equipment purchased directly by the owner, and, if possible, expediting materials and equipment purchased by contractors hired by the owner.

Procedures should include those items to be installed by contractors and those to be installed by the owner. They should provide positive means for tracking purchases from order to installation (in the case of contractor-purchased items, between purchase and delivery) and provide for security and protection of items.

Project Management Plan: Procedures which implement the selected CM format and prescribe the team activities that guide the project from start to finish.

Procedures should cover all team actions required to move the project forward, between the brainstorming session and the expiration of contractor warranties after owner occupancy. The plan serves as an inventory and guide for actions required to complete the project, some of which are financing, permits, approvals and the actions covered by other Management Plans.

The project management plan procedures provide the activities and durations for the project's overall time schedule.

Quality Management Plan: Procedures which protect the owner's best interest in the selection, specification, and installation of construction materials and equipment and the conformance of contractors to the technical specifications.

Procedures should cover the team's coordinated effort that ensures that the owner's required level of quality is designed into the project and that contractors understand and conform to the quality requirements during construction. Quality procedures should be proactive.

Resource Management Plan: Procedures which are used to acquire construction industry resources for the project. Procedures must precisely follow owner requirements and conform as much as possible to construction industry practices.

Procedures should specify each team member's involvement in the acquisition and management of resources during the project. Resources are contractors, construction equipment, construction support items, and consultants hired by the owner.

Risk Management Plan: Procedures which address the static and dynamic risks inherent to a construction project, with the goal of minimizing the owner's potential exposure to them.

Procedures should be put in place that identify and handle the risks through disposal, assignment, or management. The list of risks is substantial and extends from contractor default and damage to construction to insufficient bids at bid time and the engaging of incompetent trade contractors. Risk-management procedures should be provided for all possibilities.

Safety Management Plan: Procedures to secure the safety of team members when on the project site and to assist contractors in understanding and observing safe site practices.

Procedures should neither accept nor interfere with contractors' legal responsibilities to provide safe working conditions for their employees. They should require team members to observe site safety practices and encourage contractors to comply with safety laws.

Schedule Management Plan: Procedures to effectively use scheduling and scheduling techniques to their maximum practical capacity throughout the project.

Procedures should cover the use of project time from the brainstorming session to owner occupancy. Several scheduling techniques and several levels of scheduling presentation are available. All should be used as appropriate to control and track the project through each of its phases.

Value Management Plan: Procedures to cover the interactions of the team (essentially during the design phase and to some extent during the construction phase) in the areas of designability, contractability, constructability, value engineering, and life-cycle costing.

Procedures during design begin with conceptual design and end with completion of drawings and technical specification. They consist of team reviews of the project's systems being designed, the materials and equipment being specified, and the construction methods required to ensure project goals are met. Procedures during construction consist of team reviews of contract changes.

7.8 THE IMPORTANCE OF CM PROCEDURES

When the CM system is executed to its full potential, it is the result of well-founded, reliable management plans and the timely, competent performance of their procedures. To approach CM from any other perspective rarely produces satisfactory results. It is the timely execution of coordinated strategic procedures that puts the "M" in CM.

The construction manager's controlling area of responsibility and practice is management. Expertise in construction, contracting, and design provides credence to a CM's performance, but proficiency in management makes the system work. This is not to say that technical competency in contracting, construction, and design is not necessary. It is necessary as evidenced by the twelve areas that comprise the CM body of knowledge; however, management ability is a higher priority. While managing, the CM relies on the technical expertise of team members and contractors as a resource; the added technical competency of the CM enhances that resource and raises the CM's standard of performance.

7.8.1 CM Performance Standards

Currently, there is no universally accepted standard of CM performance for ACM services. This can be attested to by the inconsistent results of court decisions where standards of performance influence decisions, and it is highly probable that a universal standard will not be generated in the near future.

There is no consistency between CM firms in the services or the level of the services they provide. Currently, each CM firm establishes its own menu of services and its own standard of performance for those services. It can be said without argument that ACM has found its definition but is still looking for its standard of performance.

Performance standards are established from repetitive experience. What consistently works to owners' satisfaction and acceptance becomes the standard. It is reasonable to assume that improvement will occur with each repetition of an action, providing that consistency is practiced from one action to another. The constantly improving procedures recorded in CM project manuals will eventually set the standards because CM is essentially a practice of procedures.

CHAPTER 8

The CM Organization

Previous chapters discussed the structure of the CM contracting system and introduced the activities in which a construction manager becomes involved. This chapter will identify the expertise required of a CM practitioner and provide an organizational structure from which CM services can be provided.

First and most importantly, a construction manager is not a person. A construction manager is an organization staffed by personnel who collectively possess the management, design, construction, and contracting expertise necessary to credibly execute the CM contracting format.

This does not mean that one person cannot be sufficiently knowledgeable to provide CM services; it simply points out the unalterable fact that one person has neither the time nor the physical capacity to execute the CM format effectively.

The staffing of a CM organization or firm has no model in the traditional construction industry. Neither the staffing of a general contracting nor that of a design firm can fulfill the needs of a construction manager without extensive supplementation. The missing disciplines of these two organizations are obvious, and although the disciplines are available in a design–build firm, the contractor philosophy of the D–B system counters that of the CM system.

The basic departments and the operating (primary functions are bold) and company supporting functions of a typical CM firm are as follows:

1. **Executive—CM Policy**, **CM Philosophy**, **First Tier Project Involvement**, Marketing, Sales and company operations.
2. **Administration—Risk**, **Contract**, **Budget and Information Management**, personnel, payroll, accounting, finances, etc.
3. **Operations—Decision**, **Project**, **Quality**, **Resource**, **Safety**, and **All Other Management Areas**, **Second and Third Tier Project Involvement**.
4. **Resource—Budget**, **Information**, **Material/Equipment**, **Schedule**, and **Value Management**.
5. **Support**—Communications, transportation, clerical, word processing, data processing, records and other support functions.

The executive area provides company leadership, establishes CM philosophy and policy, and is directly involved in CM operations at the start of each project and thereafter if executive (First Tier) intervention is required.

The operations and resource areas are where duties and responsibilities are totally CM oriented. The administration area is typical for any business organization but, through its business related acumen, contributes to CM operations and resources as a resource of its own. The support area has no prime CM responsibility but contributes significantly to the productivity of the other areas.

8.1 CM STARTER ORGANIZATIONS

CM practitioners did not exist until the 1960s. By the 1990s, thousands of firms offered CM services. How did they multiply in such numbers, and so quickly?

There are two answers: one adds credence to construction industry practice, the other adds credence to construction users suspect image of construction industry practice in the 1960s.

The second answer is best disposed of first. Many A/E firms, general contractors, and construction consultants simply added CM capability to their letterheads and business cards without concern for what CM services consisted of or how to function in the CM contracting structure. They felt justified by falsely assuming that construction management was a magnified rendition of the management of construction, something they did in their existing industry involvements. The demand for CM services was increasing and many wanted a share of it. Firms that entered the CM marketplace with this attitude were imprudent to say the least. Their impulsive and unprepared entry into the CM arena created credibility confusion that prevails to some degree today.

The emergence of authentic CM firms was less simple. They took the time to distinguish the differences between construction management and the management of construction, and developed formats and procedures to set CM apart from magnified traditional construction industry management practices.

Contractors, consultants, and A/E firms who understood the requirements of CM realized the expertise of their existing staff had to be augmented by adding disciplines that they didn't have. Contractors needed to gain design expertise; A/E firms had to acquire construction and contracting expertise. Augmentation created a unique organization structure, different from all existing construction practitioner firms.

8.2 PHILOSOPHICAL TRANSFORMATION

The origin or authenticity of CM firms notwithstanding, the change in operating philosophy from contractor to agent was a major hurdle that had to be successfully cleared. The operating philosophy of construction management is so different from that of contracting that contradictions can surface long after a person with a contracting background has become part of a functioning CM organization.

While this may seem a frivolous concern, experience has shown that some very good contracting personnel never become equally good CM personnel. A contractor's position as an independent contractor allows it to make all decisions in its own best interests so long as those decisions comply with its contract with the owner. Agency turns that position around; all decisions must be made in the best interest of someone else (in this case, the owner).

Those with long experience in traditional contracting should be aware of a backsliding possibility. Unwitting reversion can seriously affect decisions made during CM format execution and destroy credibility with team members and possibly the CM firm. CM firms spawned from A/E organizations and consultants are accustomed to agency and are not faced with this problem to the same extent as contractors.

Philosophical transformation from contractor to agent is accomplished only when all personnel in the CM organization think, react, and perform as agents, by holding the best interests of the owner above all other considerations.

8.3 A BASIC CM ORGANIZATION

The staffing of the five basic departmental areas determines the size of an in-house CM organization. In a small CM firm, individuals can perform more than one CM function and operate in more than one departmental area. This reduces the size of the organization but limits the number and size of projects that can be undertaken.

An ideal CM organization would be one with sufficient in-house personnel so that functions and roles would not have to be duplicated. This organization would have to have sufficient work-in-progress to warrant the annual payroll expense. This type of organization should be the goal of every CM firm. It has the capability to provide CM services at an optimum level and a full cadre to expand when necessary.

A less acceptable but often necessary way to operate a CM firm is to use outside sources for some of the resource requirements. The operations, administration, and support areas can be in-house, and the executives can be part of the operations personnel. One problem with this arrangement is the loss of positive control over timeframes. Outside services are not always sensitive to the timely needs of CM operations.

8.4 ORGANIZATION FUNDAMENTALS

The core of the CM organization is the operations and resource departments. There is little opportunity for dual roles in these areas because of time constraints in the operations area, and the realization that the required resource expertise usually does not overlap. However, as we will see later, there is some opportunity for overlap within resource disciplines.

Dual roles are practical in the support and administrative areas where day-to-day activities have reduced time constraints and expertise has overlap. It is also advantageous to have persons in these areas apportion their efforts between the two; it provides a feeling of total involvement in the CM organization's endeavors.

The only department with no dual role constraints is the executive area. In fact, most in an executive role will function in one or more of the other four areas in an ideal CM organization. There is a responsibility to operations and to running the company, but neither is sufficiently time consuming to prevent involvement in other areas.

A CM organization by department and function is shown in Figure 8.1.

As mentioned, the staffing of the ideal CM organization is controlled by three factors: the ability to promptly provide required services, and the number and size of projects handled concurrently. The organization chart in Figure 8.2 shows the expertise of personnel assigned to the five departments.

8.5 AN OPTIMUM CM ORGANIZATION

Each individual CM firm should optimize its organization from the perspective of services provided and the market area served. This is necessary to earn profit as a CM practitioner. Optimizing a CM organization is not an easy task, especially when CM is a new business venture, or when operating in a market area that constantly fluctuates in construction volume.

Unlike contractors, who do not require a high overhead organization in proportion to the dollar volume of work they do, CM organizations have high overhead and

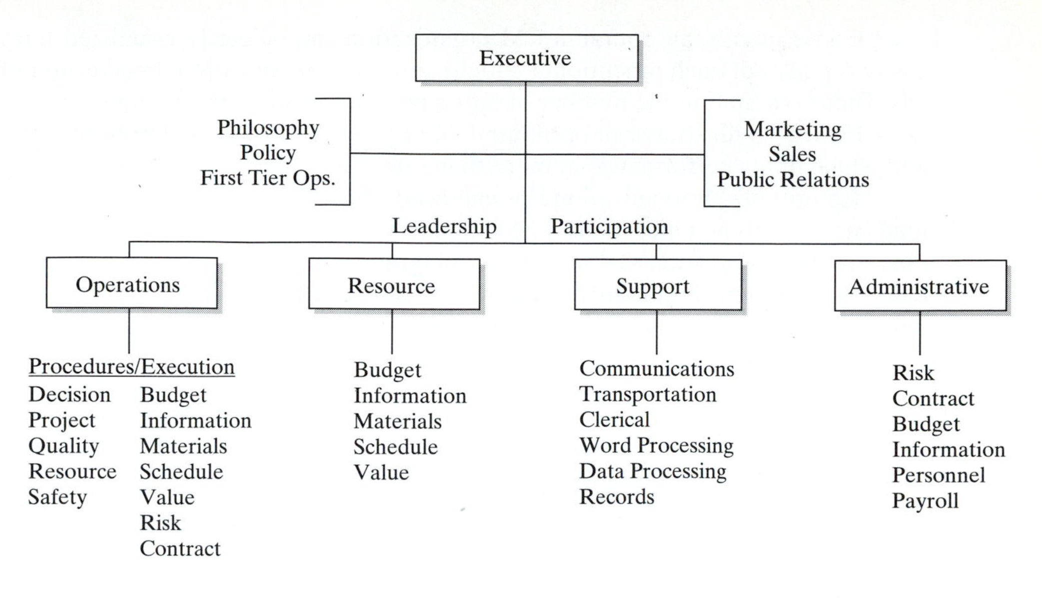

FIGURE 8.1 CM organization by area and function.

cannot conveniently increase and decrease the size of their organization in response to a fluctuating market place.

CM organizations also have to keep more personnel on staff because of the wide variety of expertise required and the pace at which services must be provided. As mentioned previously, although one person may have the required expertise to perform all required CM services, the pace of projects makes it physically impossible to satisfactorily accomplish tasks in a reasonable time.

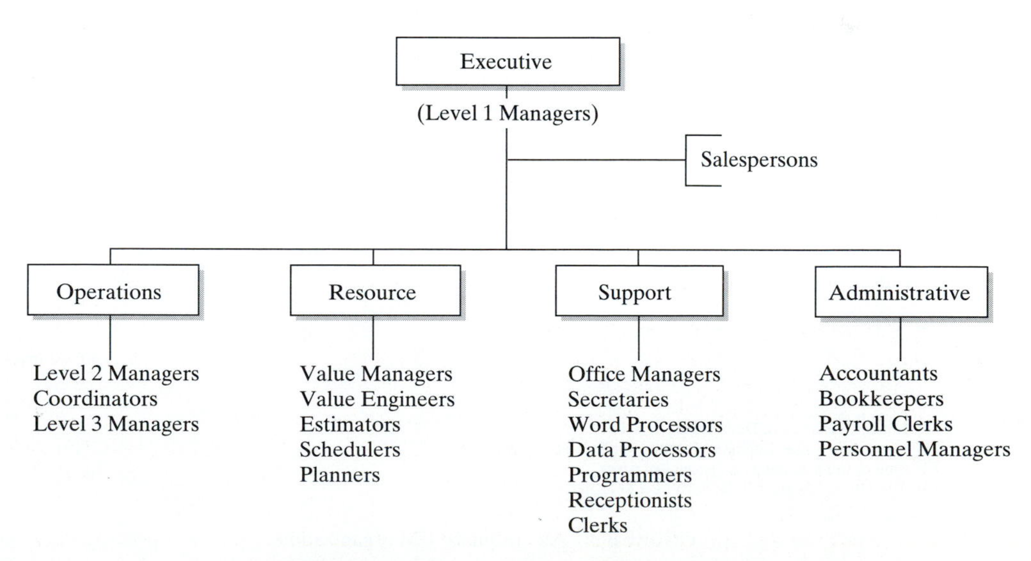

FIGURE 8.2 CM organization by area and personnel.

Consequently, an optimum CM organization can be easily visualized but not so easily replicated. Each practitioner should approach organizational make-up individually. There is a limit to the number of hats a person can wear at one time.

Figure 8.3 illustrates an optimum CM organization with total in-house capability and which limits performance in more than one area to executives only.

An optimized organization is a balanced organization, one from which the work load capacity of the Operations department is determined by the size of resource, support and administration staffing. In the diagram, staffing is set at a minimum for an ideal organization (minimum, because each discipline must be represented by at least one expert; ideal, because all disciplines are in-house).

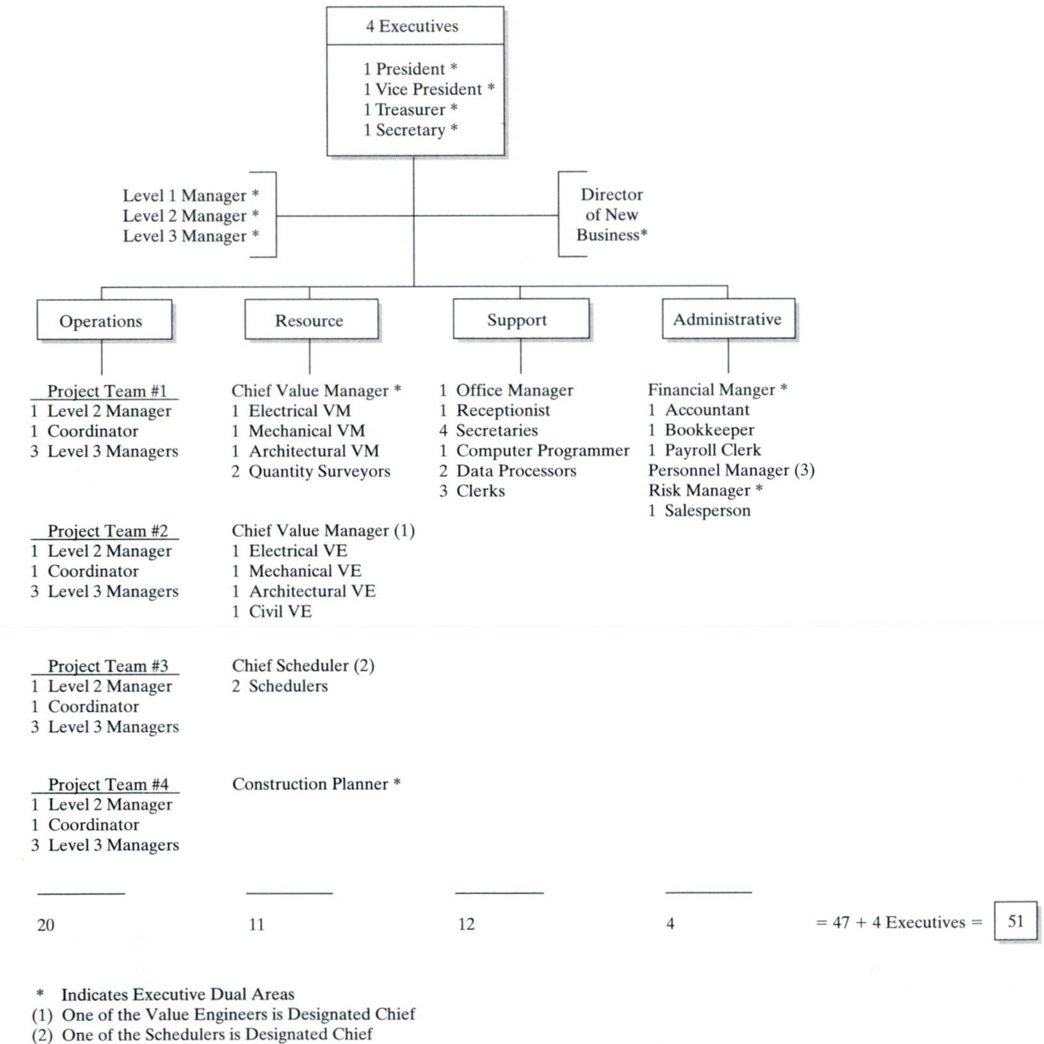

FIGURE 8.3 An optimized CM organization.

8.5.1 Further Optimization

The staffing of the organization as shown can be reduced in size from 51 to 48 if the high priority practical expertise of the electrical, mechanical, and architectural value managers includes lesser priority design expertise as well. Combined Value Manager (VM) and Value Engineer (VE) expertise is common in these three areas. The civil VE position could be covered by one of the executives, reducing total staffing to 47, the probable minimum in this organization.

The operations department consists of four teams, each capable of handling three projects at one time. The work load of the organization is therefore twelve concurrent projects, all in different stages of completion. These projects are considered optimum projects—projects of a size and complexity that can be accomplished by one full-time third level manager, or Field CM, per project within a twelve month period.

Projects larger or smaller than the optimum project will decrease/increase the number of projects that can be handled concurrently. The total dollar work load of concurrent projects is not affected.

8.5.2 The Optimum Project

The dollar value of an optimum project can be theoretically determined by summing the annual salaries of the total organization; multiplying by a salary-burden/profit factor to determine required gross revenues, dividing gross revenues by the average project CM fee, and dividing by twelve projects. Both the salaries and burden factor will vary geographically; fees will vary according to project size and complexity.

8.5.3 Example

Assuming annual salaries will be $2 million, a burden/profit factor of 1.7 and an average CM fee of 7%, the required annual volume of construction would be $48.6 million and the optimum project value $4.05 million over the course of one year.

This type of example will assist in determining the potential of a CM organization in the marketplace. A variety of possible project size and complexity combinations can be substituted, and experience-based burden/profit factors and salaries can be used to inject reality into the calculation.

Each organization must determine its own salary burden factor or multiplier that converts employee wages to company direct personnel expense, and a company's profit requirement is a matter of choice modified by competition. The average CM fee percentage does not depend on competition because the example is based on providing full services as represented by the staffing. However, the fee percentage includes all money paid by the owner to the CM for services, including those incurred at the site that are sometimes paid as reimbursables.

8.5.4 Optimizing Small CM Operations

The optimum ideal CM organization (Figure 8.3) utilizes its in-house resources, support, and administrative personnel at full capacity when utilizing the operations personnel shown in the diagram at full capacity. It is highly unlikely that a newly formed

CM organization, or an existing contracting, consulting, or A/E firm just entering the market, will have the opportunity to begin CM services with a project work load large enough to warrant the assembly of a complete in-house organization.

It is probable that such organizations will start up gradually, acquiring new personnel, borrowing personnel from their primary business endeavor, and using outside sources on a fee basis as experience is gained and additional projects acquired.

Gradual CM organization start-up is a critical time for new practitioners. It combines a commitment to provide untried services for a fee with a learning experience directly involving those services (a combination that experience has shown to be problematic regardless of the type of services involved).

The only caution that can be offered is to accept the situation as it is and proceed with the owner's best interests in mind. Again, the management of construction is not construction management, and the difference should be clearly understood and foremost in mind when accepting a fee from an owner for CM services.

Unfortunately, too many owner-CM agreements have been hastily entered into when the owner was not sufficiently knowledgeable about CM and the CM was not properly staffed to provide full-value services. In most of these instances, the CM is responsible for any disappointing project conclusions.

It is worth noting that CM organization start-ups, past and present, account for a significant share of the dissatisfaction that owners, A/E firms, and trade contractors have expressed about the performance of the contracting system.

8.5.5 Optimizing Large CM Operations

To increase operations, either by number of projects or total volume, additional resource, support, and administrative personnel must become involved. Sometimes organization expansion can be as difficult as organization start-up; the need to balance resources to operations is still a problem.

However, functioning CM firms have the advantage of learning-curve experience; they know the capacities of their resource, support, and administrative people and the demands of the projects that constitute their work-load. These knowns provide positive direction for expansion, something new organizations do not have.

Every CM firm should have an entry-level training program to help new employees ease into the firm's CM philosophy and procedures. Procedure manuals are a big help, but mentor relationships complete the job. If at all possible, staffing structures within departments should provide for ongoing training and development for advancement. If the firm is large enough to absorb the cost of an interdepartment training program, that will undoubtedly produce the best results.

8.6 MARKETING AND SALES

Optimization should be the goal of every competitive CM organization; it produces consistent high performance, good reputation, and profitability. However, for continuing success, at whatever work-load level the CM firm decides to settle into, marketing and sales are important factors.

Even the smallest CM firm must sustain a proactive sales effort to get work. CM firms, like A/E firms, get most of their work by solicitation, locating and contacting owners who are anticipating projects. Even though many CM projects today are advertised and openly invite proposals, CM selection is usually (and properly) based on performance ability, not fee.

Consequently, a CM's name and reputation should be made known to owners prior to involvement in a selection process. The support department of the CM organization in Figure 8.3 lists one salesperson and assigns responsibility to an executive level manager. As an organization expands, additional salespersons are needed to proportionately increase workload.

Marketing and sales are covered in Chapter 23, but because general contractors are not accustomed to having salespersons on staff, the importance of this function is noted here for added emphasis.

8.7 QUALITY OF CM PERSONNEL

Every business organization wants to hire the best person for an open position. A proven way to reduce the chances of hiring the wrong person is to anticipate and document the requirements for each position.

A job description and a list of responsibilities not only helps in the personnel search process, it serves as a reminder to those already on the payroll and as a measure for periodic performance reviews. Job descriptions and lists of responsibilities should be living documents, reviewed at least yearly and updated as the requirements of a position change or new positions are added.

Firms that are planning to enter the CM field should consider planning the organization on paper before making in-house personnel reassignments or hiring new personnel. The potential of a CM organization can be summarily evaluated by reviewing its organization chart, job descriptions, and lists of responsibilities. However, to measure the potential effectiveness of a CM organization, the competence of the persons filling the positions must be evaluated as well.

The following are examples of selected job descriptions: the generic titles First, Second, and Third Level Persons should be replaced with the preferences of the individual CM practitioner.

Level 1 Person (Executive)—Job Description

General: The purpose of this position is to supervise the CM operation, to initiate project direction, to formulate each project, and to guide each project through the Level 2 Person (Manager). Complete responsibility for use of the CM format on each project is implied.

Background: Proven experience as a Level 2 Person (Manager). The position calls for extensive knowledge of the construction, design, and contracting aspects of the industry and their relationship with the CM format and its techniques and capabilities. Experience as a manager and administrator at a high business level is required.

Skills: All of the skills required for successful performance as a Level 2 Person, plus the ability to work effectively with owners, architects, and engineers during all stages of a project at the highest level of responsibility.

Supervision: Performance in this position is under the direction of the President.

Function: The position is one of individual authority and action performed within the parameters of corporate direction. The position is responsible for programming and supervising each project from the initial interview with a prospective client through occupancy. The position demands ingenuity and innovation plus administrative/organizational abilities in working with clients and in the operation of the CM program. For all practical purposes, this position is self-motivating and autonomous.

Work Day: As required to adequately perform the requirements of the position.

Compensation: This is a classification 0 position and the corresponding pay range is applicable.

Benefits: All of the standard employee benefits.

Level 2 Person (Manager)—Job Description

General: The purpose of this position is to provide the management function on one or more CM projects through the use of organizational resources, a Coordinator, and the on-site representation of a Level 3 Person (Administrator). Complete responsibility for the conduct of the project from start to finish is implied.

Background: The position calls for extensive knowledge of construction means, methods, and techniques and experience in construction/contracting procedures. Knowledge through experience with contracting, equipment, estimating, specifications, bidding procedures, labor relations, contracts, insurance, bonding, scheduling, billing, customer relations, and construction materials is mandatory. The degree of knowledge and experience in the above categories is a major key to performance in this position.

Skills: Plan reading, value engineering, surveying, cost estimating, scheduling, cost accounting, expediting, computer literacy, and efficient, accurate communications are some of the skills required to adequately perform the position.

Supervision: Performance in this position is under the direction of a Level 1 Person.

Function: The position is one of individual authority and action, performed within the parameters of the CM program. The projects assigned to this position are to be carried out expeditiously and accurately in a manner acceptable to the client. The position demands anticipation in the organization of all aspects of the project, plus the ability to call upon and use the required resources, both within the organization and beyond. For all practical purposes, this position is self-motivating.

Work Day: As required to adequately perform the requirements of the position.

Compensation: This is a classification 1 position and the corresponding pay range is applicable.

Benefits: All of the standard employee benefits.

Level 2 Person's Responsibilities

- Prepare for and attend organizational meetings
- Develop the responsibilities of the team members' responsibility chart and the CM project manual
- Develop activities and timing of the program (program schedule)
- Determine the available basic parameters of the project
- Coordinate program activity dates with budget managers
- Assist in developing the program schedule
- Review the responsibility chart and CM project manual with the Level 1 Person (executive)
- Obtain all required budget information from the owner
- Establish the preliminary budget
- Provide the construction budget to the value management (VM) people
- Provide adequate design concepts and information to chief VM
- Provide value management information to value managers
- Decide whether or not each phase of the project is ready to bid
- Review front end specification with Level 1 Person prior to issuance to team
- Comply with and enforce the program schedule
- Establish bid division (work-scope) index
- Write bid division (work-scope) descriptions
- Ensure adequate bidder participation
- Conduct pre-bid meetings
- Conduct post-bid interviews
- Provide adequate coverage of bid openings
- Advise the team of bidding procedures
- Make award recommendations to A/E
- Formulate and prepare proposal forms
- Assist in development of the detailed construction schedule
- Review and approve the milestone schedule
- Modify detailed schedule for use as the initial Short Term CAP.
- Conduct preconstruction meetings
- Establish on-site liaison with team and contractors

- Assist the A/E and owner in contract preparation
- Advise the owner of project insurance requirements
- Monitor expiration dates on insurance policies
- Conduct monthly project meetings
- Develop project completion schedule
- Develop and update owner cash flow requirements
- Generate and approve contractor's contract cost breakdowns
- Review the billing summary percentages at project meetings
- Review payment book for completeness prior to release to A/E
- Keep project information board current
- Assist in interviewing Level 3 Persons (administrators) and coordinator candidates
- Recognize problems and bring them to the Level 1 Person's (executive) attention
- Visit the site as deemed necessary
- Manage and monitor CM performance throughout construction
- Attend owner meetings and team meetings
- Administer the bulletin and change order procedure
- Promote performance that adheres to the conditions of the contract documents
- Be involved in a continuing public relations effort
- See that a satisfactory quality control program is in effect
- Ensure that Short Term CAP is properly utilized and updated
- Provide final budget and schedule statistics for compilation and the record
- Review the construction budget with the team prior to bidding
- Provide continuing review with the team of the project budget
- Advise owner to obtain legal review/advice on contact documents
- Maintain owner contact during warranty period
- Coordinate contractor call backs during warranty period
- Generate finalization of owner acceptance of the project
- Assist Level 1 Person (executive) according to his directions

---

CM Coordinator—Job Description

General: The purpose of this position is to provide support and back-up for the Level 2 Person. It is intended that this position will eventually lead to a position of Level 2 Person. There is no set responsibility other than to accomplish the routine work assigned by Level 2 Person's, and there is no time requirement in the position prior to promotion to a higher position.

Background: The position calls for sufficient knowledge of construction methods and contracting procedures to relieve the Level 2 Persons of routine duties and to take on additional tasks that cause an overload on their time.

Knowledge of subcontracting, equipment, estimating, specifications, bidding procedures, labor relations, contracts, insurance, bonding, scheduling, billing, and other construction and contracting related aspects is mandatory. The degree of knowledge in the above categories is the key to progress in this position.

Skills: Plan reading, quantity take-off, estimating, scheduling, technical knowledge of construction materials, cost accounting, drafting, expediting, computer literacy, and efficient communication are some of the skills required to adequately perform the work involved in the position.

Supervision: Performance in this position is under the direction of a senior Level 2 Persons. Work assignments shall be accepted from any Level 2 Person designated by the senior Level 2 Person.

Function: The position is one of support and assistance. The various duties assigned by the Level 2 Persons shall be carried out according to their instruction in both a timely and efficient manner. The position requires neatness, organization and perception in order to accomplish the daily demands of the work load. Handwriting must be legible. Day-to-day reporting shall be automatic, complete and timely. Special assignments on other than routine work will be the rule and not the exception.

Work Day: As required to adequately perform the responsibilities of the position.

Compensation: This is a classification 4 position and the corresponding pay range is applicable.

Benefits: All of the standard employee benefits.

Coordinator Responsibilities

Basic *(Entry Level Position)*

- Learning the CM format
- Complete familiarization with the project assigned
- Receive and record project phone calls
- Establish and maintain job files
- Handle mailings
- Assemble bidders lists
- Confirm bidders
- Plan distribution coordination
- Assemble contract documents
- Encourage contractor attendance at pre-bid meetings
- Take minute notes at pre-bid meetings
- Procure bidder prequalifications
- Clerical handling of bonds, insurance, and contracts
- Sort, record, and file bids, bonds, and insurance
- Assemble final bid tabulation

- Arrange time and place for post-bid meetings
- Assemble project information list
- Take minute notes at preconstruction meetings
- Initiate the field files
- Expedite and collect Financial Management Control System (FMCS) information
- Expedite response to bulletins
- Check and assemble monthly pay book
- Be aware of the proceedings at every monthly meeting
- Visit project sites at least once each month
- Read and react to daily job reports
- Cooperate with CM support persons
- Expedite submission of final documents for final payment
- Assist the Level 2 Persons according to their directions

Advanced *(Experienced Level)* All of the above, plus

- Quantity take-offs for change orders
- Become knowledgeable with the program schedule
- Become knowledgeable with the responsibility chart
- Become knowledgeable with the project budget
- Become knowledgeable with the CM-Owner Agreement
- Follow development and become knowledgeable with the contract documents
- Prepare computer data entry sheets for budget, billing, and expediting report
- Carry out bidder solicitation procedures
- Preliminary checking of bonds and insurance certificates
- Assemble contract packages for forwarding to A/E
- Become familiar with and remain knowledgeable of all project-related meetings
- Assist the Level 3 Person (Administrator) in developing the expediting list
- Maintain warranty tickler file
- Look after the welfare of a project in the absence of a Level 2 Person.
- Function as a Level 3 Person on a temporary basis when required
- Become knowledgeable with FMCS and VM procedures
- Read and become knowledgeable with the CM presentation booklet information
- Attend in-service seminars and self advancement meetings as required

---

Level 3 Person (Field CM, Field Administrator)—Job Description

General: The purpose of this position is to provide the coordinating function in the field on a specific CM project under the direction of a Level 2 Person.

Essentially, this position serves as the eyes, ears, and voice of the team members on the project site.

Background: The position calls for thorough knowledge of construction means, methods, and techniques, and experiences in contracting procedures.

Knowledge from experience of contracting equipment, construction materials, project scheduling, progress payments, billing procedures, quality control, expediting techniques, contractor performance, and client relations is mandatory.

The degree of knowledge and experience in the above categories is a major key to successful performance in this position.

Skills: Plan reading, surveying, scheduling, expediting, cost estimating, organizing, inspecting, coordinating, communicating, computer literacy, and planning are some of the skills required to adequately perform the position.

Supervision: Performance in this position is under the assignment and direction of a Level 2 Person.

Function: The position is one of coordination and control as prescribed by the project schedule, drawings and specifications. This position represents the team in the field and is called upon for information and action pertaining to the project. The position can only be successful through planning, coordination and surveillance on a persistent basis. Total communication with the team members, especially the Level 2 Person, is mandatory. Although the position lacks the burden of major decision making, the position assumes responsibility for the proper and timely construction of the project within the established budget.

Work Day: As required to adequately perform the requirements of the project. Most projects operate on an eight-hour day/five-day week and all projects require on-site Field CM attendance according to the established work schedule.

Compensation: This is a classification 2 position and the corresponding pay range is applicable.

Benefits: All standard employee benefits.

Special Requirements: Persons filling this position must be currently qualification in "First Aid to the Injured" as certified by the American National Red Cross.

Level 3 Person (Field CM) Responsibilities

Field CM *(Experienced Level)*

- Learn the CM format
- Become completely familiar with the project assigned
- Adequately staff the project in conformance with the contractors' work schedule
- Inform the Level 2 Person prior to any planned absences
- Inform the Level 2 Person immediately when unable to staff the project
- Attend meetings when required
- Present at all times as a member of the CM team
- Accept direction from the Level 2 Person in charge of the project

- Recommend improvements of whatever nature to improve the general welfare of the project
- Perform duties in strict conformance with CM procedures
- Do not permit a contractor to work on-site without valid proof of insurance
- Call any dimensional discrepancies to the attention of the A/E as soon as possible
- Keep a record of all project discrepancies of whatever nature as they are discovered
- Keep an accurate and legible project diary
- Adequately record all significant events and phone
- Complete and submit logs on a daily basis
- Update the billing summary percentages just prior to the monthly meetings
- Obtain periodic compliance reports when required by the project
- See that all required legal notices are properly posted at the job site
- Be completely familiar with the site survey and layout control
- Coordinate surveys and control points for the project with all contractors
- Check layout, line, grade, and quality of workmanship whenever possible
- Establish and control site use by contractors
- Plan, conduct and record Weekly Progress meetings
- Participate in Monthly Project meetings
- See that all project site security measures are properly implemented
- Coordinate general condition and construction support items
- Coordinate contractor performance
- Observe and assess contractor workmanship
- Obtain quality control interpretations from the A/E
- Accumulate as-built information
- Issue on-site safety "citations" to violating contractors
- Wear a hard hat and avoid exposure to hazardous conditions
- Observe all safety regulations as a construction participant
- Become thoroughly familiar with the initial Short Term CAP
- Become thoroughly familiar with the milestone schedule
- Review and confirm or revise the Short Term CAP weekly
- Immediately notify Level 2 executive of Short Term CAP failures
- Participate in Completion Schedule meetings
- Establish, maintain and update the Expediting Report
- See that all activities can and do commence according to the Short Term CAP
- Generate sufficient contractor effort to maintain scheduled progress
- Complete/submit extra work logs daily when extra work is being performed
- Do not assume the responsibility to authorize extra work expenditures under any circumstances
- Verify the extent of labor and materials expended on claimed extra work

Value Manager—Job Description

General: The purpose of this position is to provide authentic, timely cost information and alternative approaches to proposed design concepts, with the intention of controlling the economic factors of CM projects. This is a resource position.

Background: The position calls for extensive knowledge of construction materials and equipment and the applications thereof, including their values and availability in specific geographical areas. Experience as a construction estimator is essential.

Skills: Plan reading, quantity analysis, conceptual estimating, computer literacy, and economic forecasting, as well as a thorough understanding of construction means, methods, and techniques is mandatory. Effective communication skills are especially important.

Supervision: Performance in this position is under the assignment and direction of the Chief Value Manager.

Function: To provide cost-oriented information to all CM projects in the format required by the Level 1 and 2 Persons in charge. To review drawings and specifications, evaluate material, labor and equipment, suggest alternative products and procedures, and advise on economic factors affecting each specific project. To provide self-motivated monitoring of all projects with respect to cost-related factors.

Work Day: As required to perform the requirements of the position.

Compensation: This is a classification 3 position and the corresponding pay range is applicable.

Benefits: All the standard employee benefits.

Value Manager Responsibilities

- Have a continuing interest in the cost factors affecting every project
- Establish all construction costs
- Extract all pertinent information pertaining to each project
- Attend brainstorming meeting
- Develop a cost for the project's Construction Concept
- Define and record the elements of the project's Construction Concept
- Establish construction costs based on schematic information
- Provide intense VM between schematic and design development drawings
- Communicate value management possibilities directly to A/E
- Establish construction costs based on design development documents
- Attend design development team workshop
- Encourage timely A/E involvement/cooperation during design development stage

- Assist in establishing long-lead items throughout design development
- Continue value management during working drawing stage
- Establish construction costs based on working drawings and specs
- Distribute costs into Bid Division format
- Review Bid Division descriptions
- Attend pre-bid meetings when held
- Review/comment on contract documents as to clarity and completeness
- Attend post-bid interviews upon request of Level 2 Person
- Review bulletins when requested to do so by the Level 2 Person
- Review all low bidders' proposals
- Review completed project costs
- Update value management capabilities on a continuing basis
- Establish cost criteria at specific project locations
- Establish cost criteria for various project types
- Communicate pertinent information to the rest of the organization
- Make periodic visits to projects during construction
- Maintain an ongoing interest in every project
- Provide input to the expediting report

The preceding job descriptions and lists of responsibilities take time to prepare and maintain, but their value is significant and time spent on them is worthwhile. As stated, an organization chart plus job descriptions and responsibility lists provides a blueprint for the firm to follow.

The variable content of CM services makes a blueprint necessary. An entry CM firm should determine the services that are within its start-up capability, design an organization that accomplishes those services, and write the job descriptions and responsibilities that will make the organization function as planned. An organized CM firm expansion can be accomplished in the same manner.

Pioneer CM firms could not use this approach; they had no idea of what the blueprint should look like, and their services fell far short of what CM services consist of today. However, they had no trouble adding services as they became obvious through experience.

Scheduling and value engineering were the forerunner CM services. The original strength of the CM system was vested in these two areas. It was only after repeated CM experiences that many other service opportunities were recognized and became equally, if not more valuable, to the owner. What is currently referred to as multiple contracting, contractability, financial accountability, checks and balances, value management, partnering, quality management and risk sharing are areas not planned for in CM's original scope of services.

8.8 THE FUNCTIONING CM ORGANIZATION

During the progressive phases of a project, short-term assignments (responsibilities given to CM personnel, within their areas of expertise, for the duration of one phase or more) are required of CM personnel. Operations, administration, or support persons are usually the ones designated.

Figures 8.4 through 8.7 designate assignments during various phases of the project. Short-term assignments are identified with a "T" in the upper left corner of the chart boxes.

Not all phases have been diagrammed. The predesign phase is not complex enough to warrant an assignment chart. Predesign consists of the brainstorming and organizational meetings, each appropriately attended to accomplish their purposes.

On projects where the team is in place during a feasibility phase, the assignments would be customized to reflect the owner's needs. A typical chart is not appropriate.

The postconstruction phase is also not complex enough to warrant charting. The postconstruction or warranty phase consists of follow-up on owner satisfaction and latent problems which must be corrected by contractors.

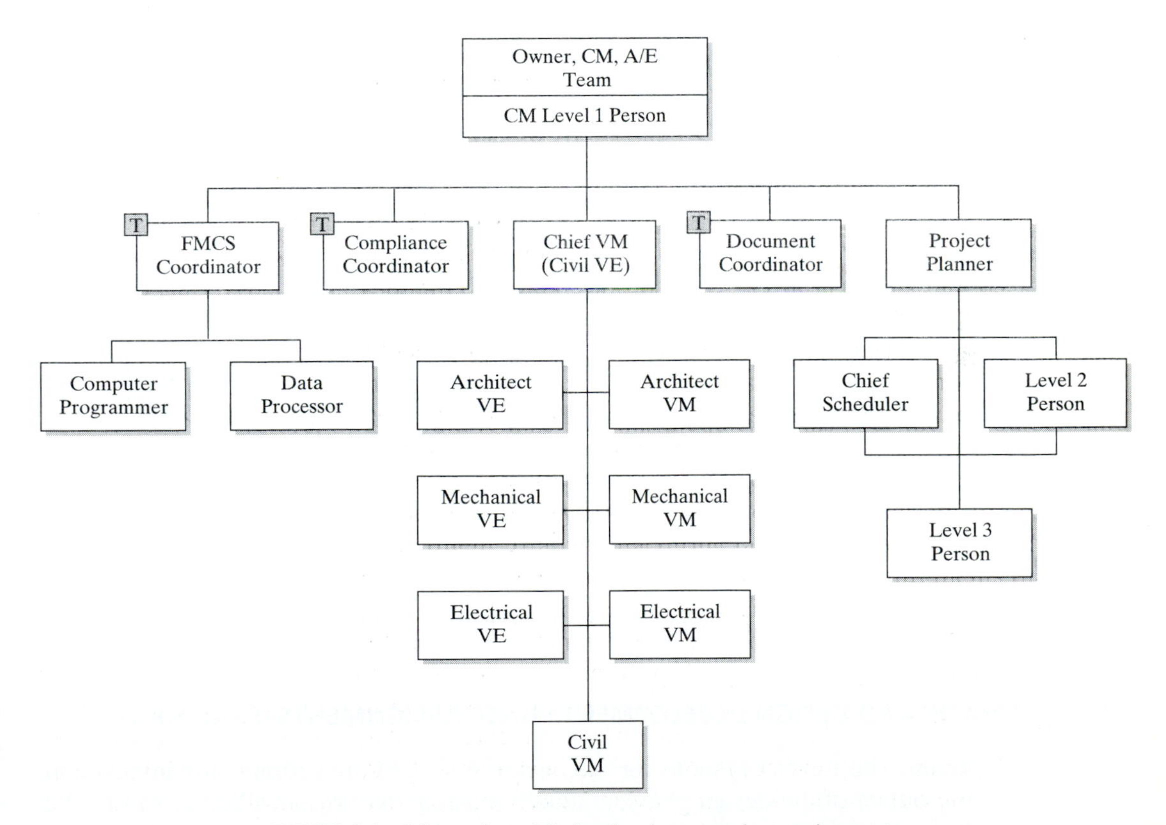

FIGURE 8.4 Schematic and design development phase assignments.

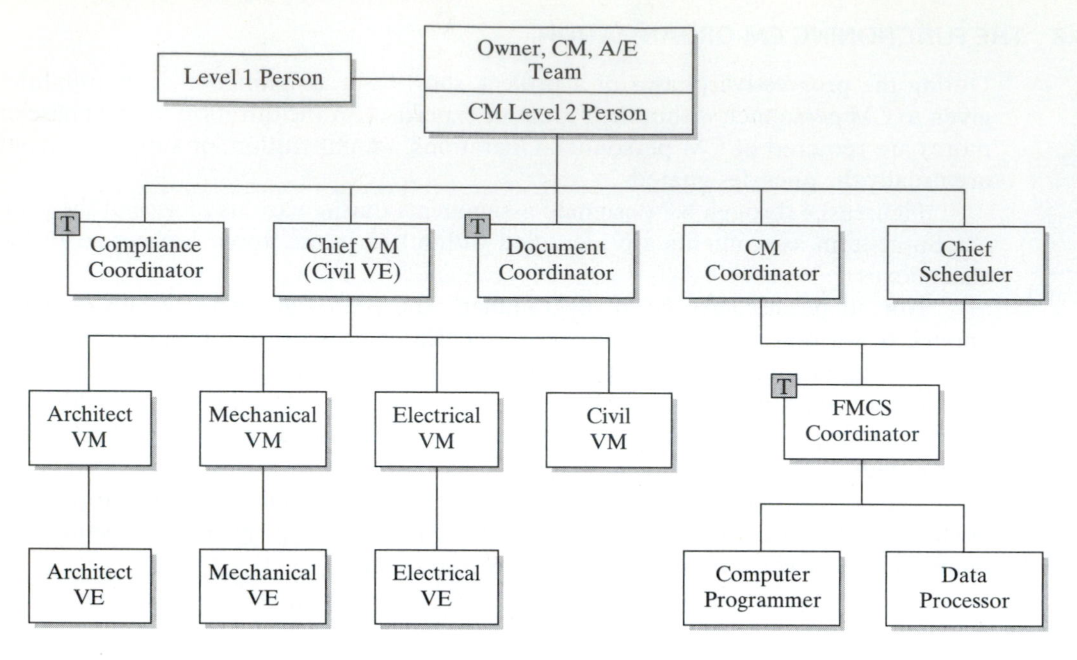

FIGURE 8.5 Construction documents phase assignments.

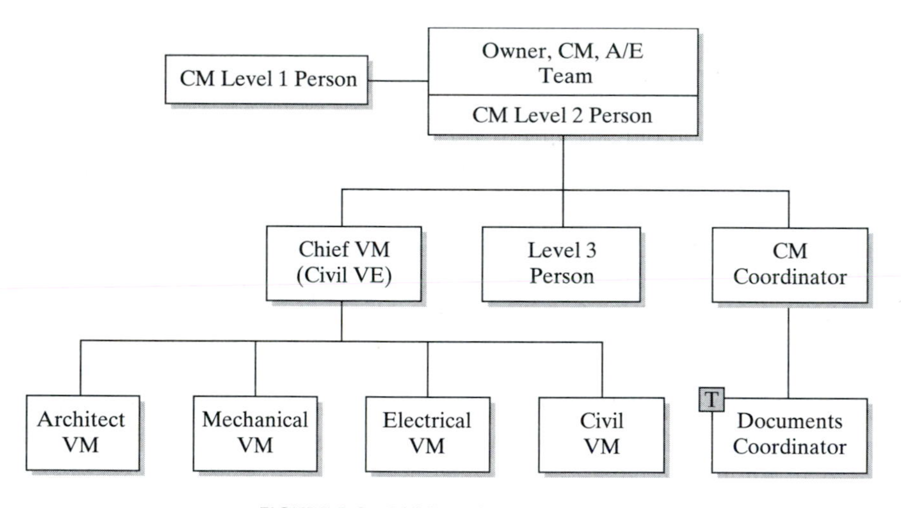

FIGURE 8.6 Bidding phase assignments.

8.9 SCHEMATIC AND DESIGN DEVELOPMENT PHASE ASSIGNMENTS (FIGURE 8.4)

Team: The Level 1 Persons for the owner, A/E, CM are prominently involved at the outset of the design phase, gradually turning over responsibilities to Level 2 Persons as design moves toward the contract document phase. The CM's Level 3 Persons should be involved as soon as available.

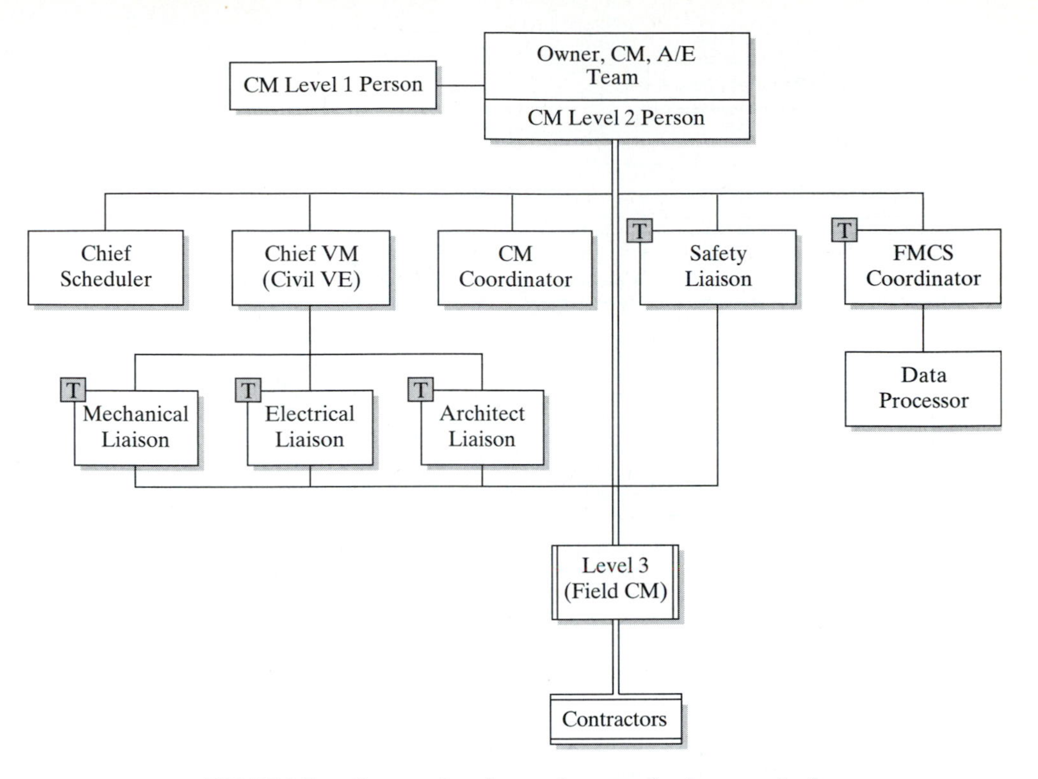

FIGURE 8.7a Construction phase assignments (optimum project)

Team activities are directed to design and quality decisions to ensure timely design progress and contractability decisions which will facilitate the bidding process. CM activities are the responsibility of the Level 1 Person.

Level 1 Person: The Level 1 Person is responsible for organizing the project and arranging the proper and timely involvement of CM resources as required. Much of the responsibility can be delegated to the Level 2 Person. Close contact with the owner and A/E should be maintained during the early part of the design phase.

FMCS Coordinator: The assigned is responsible for the design/adaptation and utilization of the financial management control system. Responsibility is achieved from team direction and with the assistance of the computer programmer, data processor, chief scheduler, chief value manager, and project planner.

Compliance Coordinator: A design professional familiar with the building codes, federal, state, and local design and procedure requirements (such as the EPA, Fire Marshal, Health Department and permit issuers) in the area of the project. The assigned assists the A/E and owner in compliance matters to minimize redesign and delays.

Chief Value Manager: A design professional who coordinates value management input during this formative phase of design. Assistance is provided by appropriate value engineers and managers. The function of the in-house VE/VM team is to convert design into cost, make suggestions and provide alternatives from a

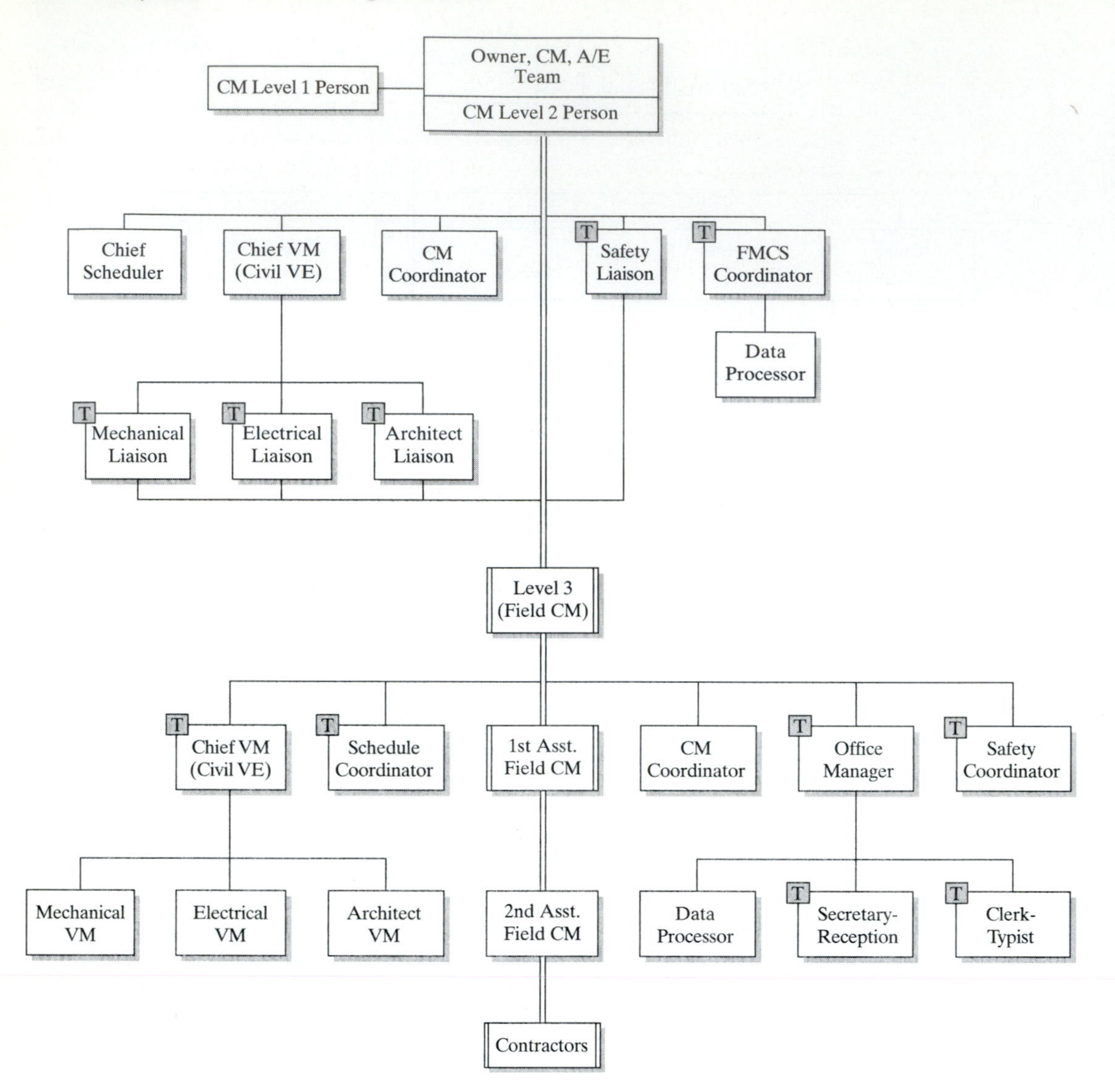

FIGURE 8.7b Construction phase assignments (larger/more complex than optimum project)

value perspective. The VE/VM team does not perform design but rather critiques the work of the A/E as design progresses.

Documents Coordinator: The responsibility of this assignment is to assure that all drawings, specifications, contract forms, and other necessary documents are in process; that they are credible, accurate, properly formulated, and appropriately reviewed. A major responsibility is to see that the CM format is properly integrated into the contract documents.

Project Planner: A person experienced in contracting, the construction industry, construction equipment, and particularly in construction means, methods, and techniques. The responsibility of this position is to plan the physical construction of the project in a feasible way. The project planner is assisted by the Level 2 Person and the Level 3 Person, if available. The project plan is subsequently used to produce construction schedules and allocate work areas to contractors on-site.

Chief Scheduler: Prepares/maintains the program schedule, the first of five major schedules used, with input from the team, the project planner, and Level 1 Person.

8.10 CONSTRUCTION DOCUMENTS PHASE ASSIGNMENTS (FIGURE 8.5)

Team: The Level 1 Person turns the project over to the Level 2 Person. Activities during this phase focus on value management and budget management as the contract drawings and technical specifications are developed and finalized. Preparations for bidding include bidders lists, bidder qualification, work-scope descriptions, proposal forms, and document review.

Level 2 Person: Assumes a stronger management role; responsible for guiding the project with the help of resource, support, and administration personnel. A major assignment is to keep the project budget on track and preparing bidding procedures.

CM Coordinator: Provides assistance to the Level 2 Person and assumes responsibilities and assignments accordingly. Principal responsibilities include bidders lists, proposal forms, bidder contact and motivation, bidding procedures, and keeping track of progress as a project moves toward the bidding phase.

Chief Scheduler: Maintains the program schedule and prepares the detailed construction and milestone schedules.

Compliance and Documents Coordinators: These two coordinators continue the assignments that originated in the schematic and design stage.

Chief Value Engineer: Concentrates on value management with the assistance of the value managers and the support of the value engineers. The construction budget must be assembled prior to issuing bidding documents to contractors.

8.11 BIDDING PHASE ASSIGNMENTS (FIGURE 8.6)

Team: Final bidding document review, receive proposals, evaluate proposals, make award recommendations.

Level 2 Person: Conducts the pre-bid, post-bid and preconstruction meetings and contractor interviews, assists with bid tabulations, leads the proposal evaluation team, and makes award recommendations to the A/E.

Level 3 Person: Must become actively engaged in the project at this time; previously has been a contributor to the CM team as available. Assists in pre-bid, post-bid, and preconstruction meetings.

Documents Coordinator: Assigned to final review of bidding documents before distribution to contractors. Assist the A/E in formulating pre-bid and post-bid addenda.

Chief Value Manager: Coordinates the involvement of value managers at pre-bid and post-bid meetings, contractor interviews, and proposal evaluation meetings.

8.12 CONSTRUCTION PHASE ORGANIZATIONS (FIGURES 8.7a AND 8.7b)

Projects vary significantly in size and complexity, making it impossible to diagram a model organization that will fit them all.

When staffing an optimum organization, it was pointed out that an optimum size project had to be assumed. Figure 8.7a is a typical construction phase organization for the optimum CM organization diagrammed in Figure 8.3.

Figure 8.7b is a typical construction organization for a larger, more complex, project than the optimum; so large and complex that resource personnel had to be assigned to it full-time. When this occurs, the construction phase organization is said to be field based, as opposed to office based, as in the case of an optimum project. Resource personnel are physically located in the field office rather than the main office, at least for the time their expertise is constantly required.

8.13 CONSTRUCTION PHASE ASSIGNMENTS (OPTIMAL)

Team: Level 2 and 3 Persons are the principal team participants during construction. Level 1 Persons are available if needed. Prime concerns include progress, reporting, contractor payments, changes, and budget control. Close contact is maintained between team members by telecommunications, letters, reports, and meetings.

Level 2 Person: Essentially runs the project. Designates site responsibilities to the Level 3 Person (field CM); is assisted by the CM coordinator. Uses services of resource, support, and administrative personnel as required. Visits site regularly. Stays in touch with A/E and owner Level 2 team counterparts. Attends weekly, monthly, and special meetings. Initiates information dissemination. Anticipates project requirements. Provides liaison with contractors' home offices.

Level 3 Person: (field CM): Essentially runs all activities related to on-site construction. Holds preconstruction, progress, and exit meetings. Organizes contractor safety meetings. Coordinates day-to-day contractor activities. Resolves document conflicts. Keeps daily records. Reports to the Level 2 Person.

CM Coordinator: Assists the Level 2 Person as directed. Serves as the information and communications center for project management and administration activities. Attends meetings and visits the site according to responsibilities assigned.

Safety Liaison: Visits the site periodically. Distributes published safety information to contractors. Attends contractor on-site safety meetings when invited. Maintains a file of contractors' project safety plans. Does not make safety inspections or monitor contractor safety conformance.

Chief Scheduler: Monitors all scheduling activities. Confers with and advises managers on project progress. Does special scheduling studies if requested. Attends and contributes to monthly meetings. Monitors on-site scheduling. Maintains planned and executed schedule comparisons.

FMCS Coordinator: Produces the project's financial management report for each progress payment. Gathers the required information from the Level 2 Person, CM coordinator, and team Level 2 Persons for each progress payment.

Chief VM: The Value person, in this case the civil VE, assigned as the chief VM for the project. Utilizes the services of other value manager and value engineer disciplines as required. Mainly used to review changes and estimate costs of changes.

NOTE: The assigned positions (those with the T in the upper left corner of the position box) can be filled by personnel who are already part of the project organization and have capacity/expertise to shoulder assignment responsibilities or by CM organization personnel with the required expertise, who currently have additional capacity available. When making assignments, expertise should be the priority closely followed by capacity.

8.14 CONSTRUCTION PHASE ASSIGNMENTS (MORE COMPLEX)

The assignments and responsibilities of a larger/more complex project are similar to those on an optimum project. The exceptions accommodate the location of operations, resource, support, and administrative persons in the field office.

The supervision hierarchy changes due to the location of personnel, but the responsibility hierarchy remains the same; i.e., the scheduling coordinator is responsible to the chief scheduler but functions under the administrative control of the Field CM.

8.15 OFFICE-BASED VS. FIELD-BASED PROJECTS

One type of CM fee reimburses the CM for all field costs; another includes all field costs in the CM fee. It is easy to see how a reimbursed arrangement could influence where CM personnel are located.

The following comments do not take into consideration the type of fee the CM will be receiving for services. As an agent, looking after the owner's best interests, the CM should not let it become a consideration. The CM should provide the most efficient organizational structure possible.

The inherent strength of a CM organization is located in the CM's main office. The expertise and ability of the Field CM notwithstanding, it is the operations, resources, support, and administrative personnel that make the CM system work effectively on a project.

A CM organization has a matrix management structure, and the best available talent, except for Field CMs, is located in the main office; their expertise should be directly available on as many projects as possible. Their assistants should be sent into the field as an extension of their mentor's expertise and to strengthen their own expertise. Consequently, an office-based organization should be a priority option on every project.

When planning a field-based organization for a project, the minimum number of services should be transferred to the field office, and only for the duration necessary. This requires good judgment in selecting services and the personnel to provide them.

Even though a matrix management structure and modern communications permit field persons to have quick, direct contact with main office expertise, many field decisions will be made without contact. Consequently, assignments to the field should be made prudently and not (as many in the business world are) on the basis of convenience.

8.15.1 CM Organization Control

In further support of office-based operations, a look at concurrent CM activities might help. A CM firm has several projects in process at one time. Each progresses on its own schedule and (it can be safely assumed) in varying stages of feasibility, design, bidding, award, construction and commissioning, or move-in. They are projects of varying size and complexity, some requiring field-based organizations and others, office-based organizations.

Although it is possible to control a CM firm's total work load, it is not possible to time-control each project's acquisition or regulate the schedule of each project's phases. Each project has a time and phase agenda which cannot be altered and smoothly integrated into an optimum CM organization work load. Effectively handling the operational peaks and valleys is one of the major challenges of a CM firm. (More will be said about this in later chapters.)

However, the resulting demands and pressures from the uneven project work load, which is mainly placed on resource persons, substantiates the fact that resource persons must not only be specifically talented, but those most talented must also be constantly available.

Figure 8.8 is an example of how a moderately large and complex project can be office-based rather than field-based, although a hurried inclination might be to do otherwise.

The project organization in Figure 8.8 utilizes two Level 2 Persons, two CM Coordinators, one Level 3 Person, and three assistant Field CMs, who are Level 3 Persons on assignment. A project planner is assigned to assist in this area because of the project's size.

Size and complexity mandate an expanded on-site organization but only in the area of operations. Although there will be more construction going on each day over a larger area, the number of contractors will probably be the same as for an optimum project. Only the number of workers employed by the contractors will increase. Consequently, contract administration activities should not intensify.

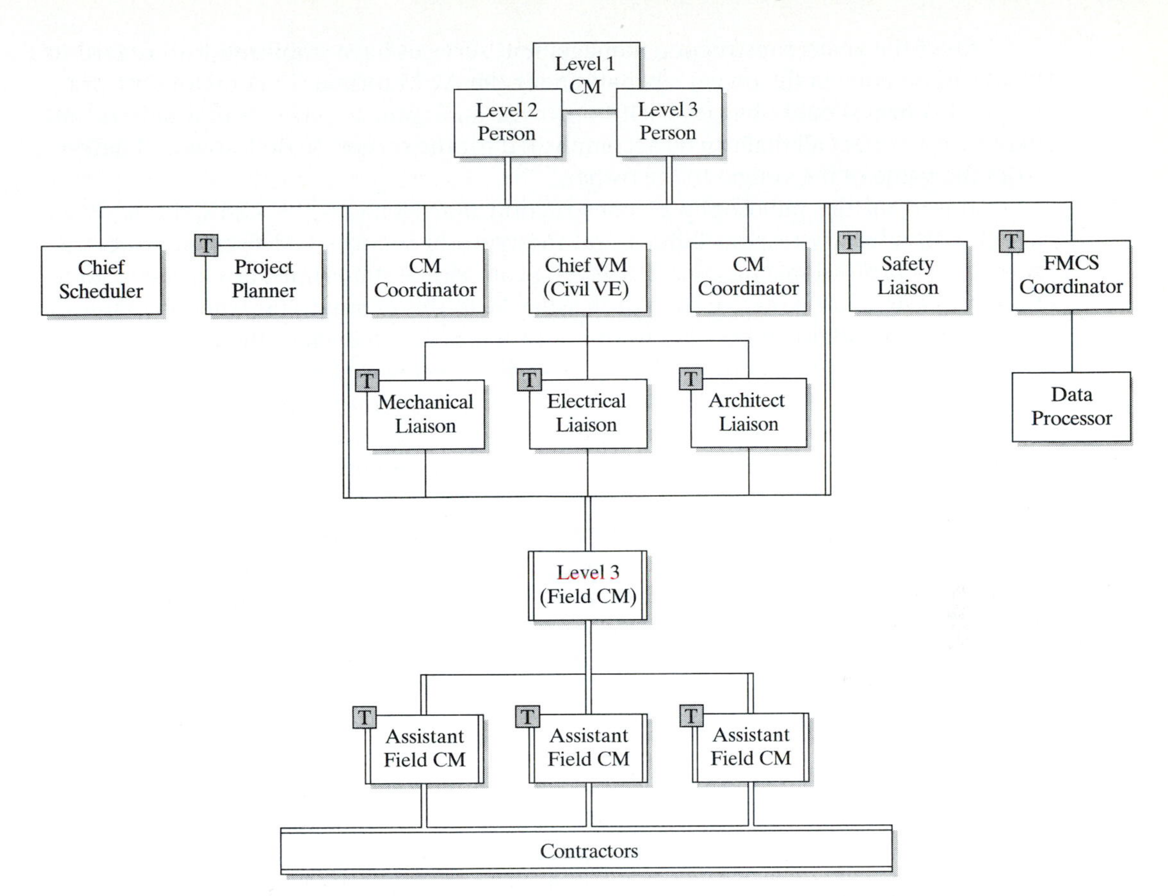

FIGURE 8.8 Moderately large/complex construction phase (office-based).

8.16 CONSTRUCTION ORGANIZATIONS IN RETROSPECT

The organizational make-up of a construction management firm is unique in the construction industry. Neither a contractor nor an A/E firm can match its functions or personnel. However, a design–build contracting firm's organization comes closest in many ways. The required contracting and construction expertise is present, but the design expertise is more than required.

Because of the way they must be constituted for their principal endeavors, none of the three organizations can function in the other's area without considerable modification. As previously stated, an A/E firm has the proper owner-serving philosophy but is short in construction and contracting expertise. A contractor is short in design expertise and is hindered by an independent contractor business philosophy. A design–build contractor has more than enough expertise to do what has to be done but, like the contractor, is hindered by an independent contractor philosophy.

Over the years, construction management services have stabilized with regard to what can be done in the owner's behalf under the ACM format. To conscionably practice CM, a firm should obligate itself to provide an organization, in both numbers and expertise, to effect all that can be accomplished for the owner. To do less would detract from the value of the system to the owner.

Based on the philosophy of construction management, the CM's status as an agent with fiduciary responsibilities, and the types of services CM firms provide, CM should be a professional practice. However, professional status must be earned in addition to having criteria that marks a profession. With this in mind, construction management firms should do all possible to present themselves to owners, the public, and the construction industry as professionals. This will be achieved by creating an organization of highly qualified people who know how to cooperate within the organization and how to work with others. Technical competency is extremely important but not enough; CM personnel must be people-persons with communication skills of the first order.

Normal CM activities place persons at all levels of the organization in direct contact with owners, architects, engineers, accountants, bankers, attorneys, consultants, and other professionals, as well as with contractors and suppliers. This exposure is unlike the exposure of any other construction industry practitioner. When hiring personnel, personality, communication skills, appearance, and the general demeanor of prospects must all be considered.

8.17 SPECIAL CONSIDERATIONS

Those in the CM organization whose job description includes management require additional talents in this area. Management methods can be taught; management ability cannot be as easily acquired. A competent manager is not produced by seniority, decree, or title. In addition to the qualities listed for all CM personnel, persons selected as managers must have the special ability to manage. These persons must be sought after on a continuing basis, both within the organization if and when they mature and outside the organization if the need arises. Promotion to manager is not based on technical ability.

A CM organization must manage itself as well as the projects comprising its work load. As alluded to earlier, internal management ability is a prerequisite for competent external management ability. If the CM organization is not well managed, it is probable that the projects will not be well managed either, and there are significant management differences between the two.

The management of the CM organization has the benefits of continuity and time on which to reflect and build an effective management structure; projects do not. The management structure of a project has to be right the first time; there may be time for minor adjustments but not for major ones. Therefore, the astuteness that shapes a CM organization over a period of time could be totally different from the insight that must quickly formulate a unique management structure for a project. This is where CM experience rather than expertise is more important.

8.18 DESIGNING A PROJECT MANAGEMENT STRUCTURE

The management structure must accommodate the project from design to occupancy; it should be based on the contractability study made after the initial meetings of the team. The most critical management design decisions relate to the construction phase, when the CM has major individual involvements both on and off the site.

Preconstruction phases are also critical, but the management structure during those phases is built around the program team and has fewer interacting parties. A review of Figures 8.4 through 8.8 will help to establish management design priorities.

The management structure design process is straightforward. The Level 1 Person assigned to the project and who led the team through the contractability study should meet with at least two Level 2 Persons and review the project phase by phase. Experience will provide the answers for the preconstruction phases; experience and innovation will eventually design a management structure for the construction phase.

The organization is charted by area and personnel (Figure 8.2) to assure that all of the functions are covered and responsibilities assigned. Personnel are then designated to fill positions and other assignments made.

When options are available, designating and assigning personnel to the management team is critical. Compatibility between CM personnel and owner and A/E personnel is extremely important and should play an important role in management assignments.

Although great care was taken when employing CM personnel and all persons should be compatible, the fact remains that owner and A/E personnel were not part of any compatibility criteria when CM personnel were hired. The Level 1 person should make the assignments based on his/her observations of owner and A/E personnel during the early meetings.

The final step is to draw and distribute the organization chart for review and approval or modification by the team, and include its details in the various management plans which can now be developed.

CHAPTER 9

The CM Body of Knowledge

Some current opinions and decisions notwithstanding, construction management practice differs significantly from the licensed practices of engineering, architecture, or general contracting. To assume that demonstrated proficiency in one of these traditional practices automatically indicates proficiency as a construction manager is erroneous.

In many instances, this assumption is made simply because comprehensive defining criteria for CM expertise is not available. In other instances, the assumption is self-serving, especially for A/E firms and contractors who summarily decide to become construction managers. Perhaps the the most serious use of the assumption has been by government units who have, for one reason or another, enacted legislation to control CM practice in their jurisdictions.

Construction management is and will remain to be for the foreseeable future a multi-faceted service provided by a variety of CM firms with different opinions of what constitutes full CM services. However, during the past 20 years, CM practice has substantiated the fact that construction management firms must be multi-discipline organizations compared to traditional construction industry practitioners such as architects, engineers and general contractors.

During the past two decades, CM has surfaced a body of knowledge based on the scope and types of services provided by accomplished CM practitioners. The broad range of expertise required to provide those services confirmed the earlier multi-discipline assumption and surfaced component areas of knowledge that explicitly describe the various disciplines and establish specific expertise.

9.1 WHO IS THE CONSTRUCTION MANAGER?

This chapter asserts a premise that individual CM practitioners perform CM services for a CM firm, and CM firms provide CM services to owners. The fundamental support for this premise is that the range of expertise required to perform CM services and the demand for timely CM performance on projects precludes performance of complete CM services by an individual, except on the simplest of projects. A CM practitioner is a firm, not an individual.

It follows that a CM firm only has access to the CM body of knowledge through its personnel, each of whom must be expert in one or more of the CM areas of knowledge. If a firm has personnel who collectively possess expertise in all CM areas of knowledge, it can be assumed that the firm at least has the potential to provide owners with complete CM services.

9.2 THE CM AREAS OF KNOWLEDGE

Twelve areas of knowledge have been established in this chapter which provide guidelines to the expertise required of CM firms. It is very likely that additional areas of knowledge will surface as CM practice gains experience with the needs of owners. CM's flexible contracting structure invites productive innovation, and its project team concept guarantees performance credibility. The first of possible new areas of knowledge could be capital financing and real estate.

In this book, the CM body of knowledge has been divided into twelve areas to better facilitate an understanding of the scope of CM knowledge and explain how each area fits into the execution of complete services. Each area has a definition and description, and each has a representative bibliography which loosely spans the area of knowledge. The twelve areas are clearly identifiable, represent the critical thrusts of CM services, and are units of expertise that can be taught and learned in academia, in continuing education and by self-study and experience.

Some of the areas are highly specific and technical, such as scheduling management and value management; others tend to be obscure and more general, such as decision management. However, none is more important than another—each contributes in its own way to complete performance, and optimum expertise in all twelve areas is essential to successful CM practice.

The twelve Areas of Knowledge which collectively comprise the CM Body of Knowledge are:

Budget Management
Contract Management
Decision Management
Information Management
Material/Equipment Management
Project Management
Quality Management
Resource Management
Risk Management
Safety Management
Schedule Management
Value Management

Chapters 10 through 21 are devoted, singularly, to each area of knowledge. They expand on the requirements and provide examples of representative procedures and practices. Before viewing each area in detail, however, it is best that the reader acquire a feeling for the expanse of the complete body of knowledge by absorbing the crux of its component parts. Although the definitions of the areas do not overlap, their descriptions and representative bibliographies may. When reading the material in this chapter, remember that future chapters will amplify what is concisely stated here and was alluded to in the management plans in Chapter 7.

An appropriate comment here is that there are no areas of knowledge which specifically cover communication skills, problem-solving skills, personal demeanor, and ethical behavior, all of which are essential to CM performance. (The author assumes that the important qualities represented in these four areas would be automatically dealt with during the hiring process of any service organization.) Although these characteristics are teachable, they represent abilities which can be gleaned from proficiency in one or more of the twelve areas and, in the best cases, are inherent to an individual's character rather than acquired from a text.

9.2.1 Budget Management

Definition

The budget management area of knowledge encompasses all project-related cost aspects of CM practice. The CM has the responsibility to confirm, generate, track, report, and substantiate all budgeted costs from the first estimate to the final accounting. The conceptual budget for the project, prepared by the construction manager before design begins, becomes the team's line-item financial guide as the design process moves toward the bidding phase. After bids are received, the amounts of accepted contractor proposals replace estimated line-item amounts and become the construction phase budget. As construction proceeds, payments to contractors, contract changes, and budgeted expenses are accounted for in detail. Every aspect of project cost is estimated as early as possible and substantiated as it occurs.

Description

The CM must have the expertise to forecast project costs from preliminary information without the aid of detailed drawings and must accurately estimate construction costs from completed contract documents. The CM must have the ability to progressively and accurately transform a combined line item conceptual budget into a detailed line-item construction budget and breakout estimates for phased construction and fast-tracking if required. The CM must be able to identify and predict project-related expense in addition to construction costs and deliver accurate budget information to the client in a timely manner and an acceptable format.

The budget management area of knowledge includes but is not limited to: conceptual and construction estimating, feasibility studies, comparative cost studies, communicative estimating techniques; construction labor, material and equipment costs; equipment and labor production rates, material technology, industry standards, labor practices; construction means, methods, and techniques, cost accounting, general accounting, construction industry economics, and communication skills.

Representative Bibliography

Beeston, Derek T. 1983. *Statistical Methods for Building Data*. London and NY: E. & F. N. Spon.

Collier, C. A. and D. A. Halperin. 1984. *Construction Funding: Where the Money Comes From*. 2d ed. NY: John Wiley & Sons.

Collier, Keith. 1985. *Fundamentals of Construction Estimating and Cost Accounting*. NJ: Prentice-Hall.

Cost data sources such as published by R. S. Means Company, Inc. Updated periodically. Kingston, MA: Construction Consultants and Publishers.

Halpin, Daniel W. 1985. *Financial and Cost Concepts for Construction Management*. NY: John Wiley & Sons.

Neil, James M. 1982. *Construction Cost Estimating for Project Control*. Englewood Cliffs, NJ: Prentice-Hall.

Stone, P. A. 1980. *Building Design Evaluation: Cost-in-Use*. London and NY: E. & F. N. Spon.

Tumblin, C. R. 1980. *Construction Cost Estimates*. NY: John Wiley & Sons.

9.2.2 Contract Management

Definition

The contract management area of knowledge encompasses the involvement of the CM in the operational and administrative provisions of the contracts used on the project. CM expertise includes the recommendation of standard contract forms and the performance responsibilities to be included in contracts, but does not extend to the writing of contracts or in any way infringe upon the legal profession. This area is signally important because the CM system is a unique contracting system, the success of which depends on a workable realignment of traditional contracting roles and participant responsibilities. It is the CM's responsibility to establish a contracting format for the project and see that each contractor's operational and administrative requirements are definitively included.

Description

The construction manager must be able to evaluate each project from a contracting perspective based on the unique conditions and requirements of the project and local construction industry practices. The CM must assess the available possibilities, recommend a contracting structure to the client, assist in developing contracts, review contract documents for suitability, and coordinate their provisions on the project.

The contract management area of knowledge includes an understanding of contracts, contract language, standard contract documents, contract law, and construction contracting in the area of the project. A thorough knowledge of traditional contracting procedures, CM contracting procedures, and the reasonable possibilities for contracting innovation is necessary. Excellent communication skills are required.

Representative Bibliography

Collier, Keith. 1987. *Construction Contracts*. NJ: Prentice-Hall.

Haltenhoff, C. E. 1986. "The Forms and Variations of the CM System," in the Proceedings, *Construction Management, A State-of-the-Art Update*, ed. C. E. Haltenhoff, 1–15. Boston, MA: ASCE.

Lambert, Jeremiah D. and Lawrence White. 1982. *Handbook of Modern Construction Law*. NJ: Prentice-Hall.

Loulakis, Michael C., et al. 1995. *Construction Management: Law and Practices*. Colorado Springs, CO: John Wiley & Sons.

Report A-7, "Contractual Arrangements," *The Construction Industry Cost Effectiveness Project*. 1982. NY: The Business Roundtable.

Standard Contract Documents. Updated periodically. McLean, VA: Construction Management Association of America.

Standard Contract Documents. Updated periodically. Washington, DC: American Institute of Architects.

Standard Contract Documents. Updated periodically. Washington, DC: Associated General Contractors of American.

Standard Contract Documents. Updated periodically. Washington, DC: The Engineer's Joint Contract Documents Committee, NSPE, ACEC.

9.2.3 Decision Management

Definition

The decision management area of knowledge encompasses the development and handling of the interrelationship of the project and construction teams and the relationship of their respective members. This area of knowledge is the least technical, but one of the most important when providing CM services. It is the CM's responsibility to consistently extract decisions from the team which are in the best interest of the owner without alienating any team members in the process. Team members must approach decisions and make decisions cooperatively, respecting each other's project function, expertise, and operational capacities. Decisions which become contentious must have a prescribed path for resolution.

Description

The construction manager must effectively establish each team member's involvement in the team as early as possible and guide the decision-making process throughout the project. The CM must have a clear understanding of the services being provided by team members and their business operation and style, induce a productive, low-profile system of decision-making checks and balances, generate synergistic team action on decisions, create a productive team management hierarchy, and see that a fair process is available when decisions are contended.

The decision management area of knowledge includes a generous understanding of business structures, organizations, practices, procedures, motivations and philosophies, plus detailed insight into the design profession and contracting business operations. High-level communication skills and ethical standards are required, and an understanding of human resource management and alternative dispute resolution practices is important.

Representative Bibliography

American Arbitration Association. Updated periodically. *Construction Industry Arbitration Rules*. NY: American Arbitration Association.

Bell, David E. 1988. *Decision Making*. NY: Cambridge University Press.

Drucker, Peter F. 1967. *The Effective Executive*. NY: Harper & Row.

Foxall, W. B. 1972. *Professional Construction Management and Project Administration*. NY: Architectural Record and the American Institute of Architects.

Hirschhorn, Larry. 1991. *Managing in the New Team Environment*. MA: Addison-Wesley.

Janis, Irving L. 1977. *Decision Making: A Psychological Analysis of Conflict, Choice, and Commitment*. NY: Free Press.

Lifsen, Melvin W., and Edward F. Shaifer, Jr. 1982. *Decision and Risk Analysis for Construction Management*. NY: John Wiley & Sons.

Richards, Max De Voe. 1972. *Management: Decisions and Behavior*. IL: R. D. Irwin.

Sweet, Justin. 1977. *Legal Aspects of Architecture, Engineering and the Construction Process*. St. Paul, MN: West Publishing Co.

9.2.4 Information Management

Definition

The information management area of knowledge encompasses the collection, documentation, dissemination, safe keeping, and disposal of verbal and graphic project-related information. The team structure and the use of multiple contracts significantly increases the information available to the owner. The volume of infromation, generated for project accountability purposes and by team member participation in decision-making checks and balances, requires a multilevel, need-to-know reporting structure and an efficient information storage and retrieval system.

Description

The construction manager must be able to properly and effectively communicate and manage the information generated by the project. The CM must be familiar with all forms and means of communications, especially computer-based information systems; recommend and install those which will be most appropriate in all areas where project information will be communicated, for each level of management. The CM must be proficient in setting agendas, chairing meetings, recording meeting minutes, presenting oral and written reports, and perceiving the need-to-know requirements of team members at each management level.

The information management area of knowledge requires high-level communication skills, including personal conversation, correspondence, technical writing, meeting leadership, note taking, meeting recording and reporting, report form design, information management systems, business protocol, computer systems, and a strong sense of ethics.

Representative Bibliography:

Becker, Hal B. *Functional Analysis of Information Networks: A Structured Approach to the Data Communications Environment*. NY: John Wiley & Sons.

Bernold, Leonard E. "Bar Code-Driven Equipment and Materials Tracking for Construction." ASCE, *Journal of Computing in Civil Engineering* 4 (Oct. 1990): 381–395.

Business Roundtable. 1982. Report A-6, "Modern Management Systems." *The Construction Industry Cost Effectiveness Project*. NY: The Business Roundtable.

Giannotti, Alejandro C. "Project Information Management System—Another Approach." ASCE, *Journal of Management in Engineering* 9 (Jan. 1993): 2–63.

Horton, Forest W. 1985. *Information Resources Management: Harnessing Information Assets for Productivity Gains in the Office, Factory and Laboratory*. NJ: Prentice-Hall.

Houp, Kenneth W. 1996. *Reporting Technical Information*. Scarborough, Ontario: Allyn & Bacon Canada.

McCullouch, Bob G. "Automated Construction Field-Data Management System." ASCE, *Journal of Transportation Engineering* 118 (July/Aug. 1992): 517–526.

Shuman, Chester A. "Is Your Automated Information System Obsolete?" ASCE, *Journal of Management in Engineering* 10 (March/April 1994): 24–26.

Woodall, Jack. 1997. *Total Quality in Information Systems and Technology*. FL: St. Lucie Press.

9.2.5 Material/Equipment Management

Definition

The material/equipment area of knowledge encompasses all activities relating to the acquisition of material and equipment from specification to installation and warranty. The CM format facilitates direct owner purchase of material and equipment for the project. The advantages (and disadvantages) of direct owner purchases must be evaluated and decisions on direct purchase items extracted from the team in a timely manner. The planning, specifying, bidding, acquisition, expediting, receiving, handling and storing of direct purchases, must conform to the owner's purchasing policies and accurately reflect the requirements of the project schedule.

Description

The CM must be able to identify materials and equipment which are in short supply, have long-lead delivery times, or would provide an economic advantage to the owner for direct purchase items early in the design phase of the project. The CM arranges for their proper procurement, timely delivery, and physical disposition at the site. The CM must expedite owner purchases from their determination of need to delivery at the site, and must be familiar with procurement procedures and strategies and understand the material and equipment marketplace.

The material/equipment area of knowledge includes but is not limited to technical specifications, purchasing techniques, bidding and negotiations, transportation, expediting, inspection, materials handling, storage and warehousing, the Commercial Code, lien statutes, and material and equipment costs and values. Excellent communication skills are required.

Representative Bibliography

Ammer, Dean S. 1968. *Materials Management*. Homewood, IL: R. D. Irwin.

Bell, Lansford C., and George Stukhart. "Attributes of Materials Management Systems." ASCE, *Journal of Construction Engineering and Management* 112 (March 1986): 14–21.

Charles, J. A. 1989. *Selection and Use of Engineering Materials*. MA: Butterworth.

Dobler, D. W., L. Lee, Jr., and D. N. Burt. 1984. *Purchasing and Materials Management*. 4th ed. NY: McGraw-Hill.

Edgar, Carroll. 1956. *Fundamentals of Manufacturing Processes and Materials*. MA: Addison-Wesley.

Fabrycky, W. J., and J. Banks. *Procurement and Inventory Systems*. NY: Van Nostrand Reinhold.

Heinritz, Stuart F. 1965. *Purchasing Principles and Applications*. NJ: Prentice-Hall.

Olin, Harold B. 1980. *Construction Principles, Materials and Methods*. Chicago: The Institute of Financial Education/Interstate Printers and Publishers.

Pollock, Daniel D. 1993. *Physical Properties of Materials for Engineers*. FL: CRC Press.

Tersin, Richard J. 1976. *Materials Management*. NY: North Holland.

White, Philip D. 1978. *Decision Making in the Purchasing Process: A Report*. NY: AMACOM.

9.2.6 Project Management

Definition

The project management area of knowledge encompasses all of the operations aspects of project delivery, including determining, formulating, developing, installing, coordinating and administering the necessary elements from the beginning of design to the termination of warranty and guarantee periods. The CM has the responsibility to make the selected CM process work, to coordinate the efforts of the team and the performing contractors in achieving their common goal.

Description

The CM must provide discreet leadership and expertise in carrying out responsibilities, and effectively organize the other five required elements—design, contracting, construction, construction coordination and contract administration—into a functioning management format that fully develops the team's resource potential and prudently orchestrates participation throughout the project.

Project management requires a thorough understanding of the design process, the contracting process, the construction industry, and all forms and variations of the construction management system; how the CM process works, what the required activities are and what is required of each activity and each team member. A good grasp of communications skills and ethics is necessary.

Representative Bibliography

Ahuja, H. N. 1984. *Project Management: Techniques in Planning and Controlling Construction Projects*. NY: John Wiley & Sons.

"Construction Management Responsibilities During Design." Committee on Construction Management. ASCE, *Journal of Construction Engineering and Management* 113 (March 1987): 90–98.

Fisk, Edward R. 1992. *Construction Project Administration*. 4th ed. Englewood Cliffs, NJ: Prentice-Hall.

Foxall, W. B. 1972. *Professional Construction Management and Project Administration*. NY: Architectural Record and the American Institute of Architects.

Halpin, Daniel W. 1980. *Construction Management*. NY: John Wiley & Sons.

Haltenhoff, C. E. 1986. "The Forms and Variations of the CM System," in the Proceedings, *Construction Management, A State-of-the-Art Update*, ed. C. E. Haltenhoff, 1–15. Boston, MA: ASCE.

Hendrickson, Chris, and Au Tung. 1972. *Project Management for Construction*. NJ: Prentice-Hall.

Melvin, Tom. 1979. *Practical Psychology in Construction Management*. NY: Van Nostrand Reinhold.

Reiner, Lawrence E. 1972. *Handbook of Construction Management*. NJ: Prentice-Hall.

Spinner, Pete M. 1992. *Elements of Project Management*. Englewood Cliffs, NJ: Prentice-Hall.

9.2.7 Quality Management

Definition

The quality management area of knowledge encompasses all elements of CM project delivery that contribute to the quality of the end product. Quality is stipulated by the client, designed into the project by the A/E, reviewed by the team, and constructed into the project by contractors. During design, quality has varying levels from high to low. Once specified, quality must conform to the levels specified. Quality management is a continuing process originating with client decisions and ending with contractor conformance.

Description

A CM must be capable of designing, installing and directing a quality management system that fits the needs of the project. The CM must have knowledge of construction materials and products, understand their use and capabilities, and the available means, methods, and techniques applicable to their installation.

The CM must know how to interpret contract drawings, technical specifications and shop drawings, understand field and laboratory testing procedures,and stay current on construction means, methods, and techniques and the latest procedures for assessing material and installation quality.

The quality management area of knowledge includes technical specification writing, materials testing and measurement procedures, product and material characteristics and capabilities, manufacturing tolerances, contractor installation capabilities, building codes, and design standards. This area also requires excellent communucation skills and performance ethics.

Representative Bibliography

ASCE. *Quality in the Constructed Project*, Proceedings. Ed. A. J. Fox and H. A. Cornell. Chicago, IL, Nov. 13–15, 1984.

Clifton, James R. 1982. *In-Place Nondestructive Methods for Quality Assurance of Buildings*. IL: U.S. Corps of Engineeers; VA: National Technical Information Service.

Crosby, Philip B. 1984. *Quality Without Tears: The Art of Hassle-Free Management*. NY: McGraw-Hill.

Crosby, Phillip B. 1979. *Quality If Free*. NY: New American Library.

Gatton, Thomas M. 1990 *Methodology for Development of Expert Systems Quality Assurance in Construction*. IL: U.S. Corps of Engineers; VA: National Technical Information Service.

Juran, J. M., and Frank M. Gryna. 1980. *Quality Planning and Analysis*. 2d ed. NY: McGraw-Hill.

Lamprecht, James L. 1992. *ISO 9000: Preparing for Tegistration*. WI: ASQC Quality Press; NY: Marcel Dekker.

Olin, Harold B., et al. 1980. *Construction Principles, Materials and Methods*. Chicago: The Institute of Financial Education/Interstate Printers and Publishers.

Rosen, Harold J., and Tom Heineman. 1990. *Construction Specification Writing*. 3d ed. NY: John Wiley & Sons.

Shearer, Clive. 1984. *Practical Continuous Improvement for Professionals*. WI: ASQC Quality Press.

9.2.8 Resource Management

Definition

The resource management area of knowledge encompasses the selection, organization, direction and use of all project resources, both human and physical. The CM contracting structure places all consulting, design, management, contracting, construction and construction services in a cooperative or team environment, focusing team coordination activities on the construction manager. Additionally, the CM's own multifaceted resources must be maintained in the flow of the project. These ubiquitous obligations make resource knowledge and resource management essential parts of successful CM performance.

Description

The CM must have the capability to proficiently understand, organize, and motivate the project's resources in order to extract the best results. The CM must possess good judgment and excellent communication skills, and exhibit leadership qualities in one-on-one and team situations.

The resource management area of knowledge includes but is not limited to an understanding of human resource disciplines, physical resource capabilities, organizational structures, human nature, conflict management, motivational factors, productivity factors, and overall human relationships.

Representative Bibliography

Adams, J. R., and B. W. Campbell. 1982. *Roles and Responsibilities of the Construction Manager*. Drexel Hill, PA: Project Management Institute.

Business Roundtable. 1982. Report A-2, "Construction Labor Motivation." *The Construction Industry Cost Effectiveness Project*. NY: The Business Roundtable.

Day, David A. 1973. *Construction Equipment Guide*. NY: John Wiley & Sons.

Dinsmore, P. C. 1984. *Human Factors in Project Management*. NY: American Management Association.

Habercom, Guy E. 1976. *Construction Equipment: A Bibliography With Abstracts*. VA: National Technical Information Service.

Helander, Martin, ed. 1981. *Human Factors/Ergonomics for Building and Construction*. NY: John Wiley & Sons.

Melvin, Tom. 1979. *Practical Psychology in Construction Management*. NY: Van Nostrand Reinhold.

Neely, E. 1975. *Construction Equipment Cost Guide*. VA: National Technical Information Service.

Nunnally, S. W. 1987. *Construction Methods and Management*. 2d ed. Englewood Cliffs, NJ: Prentice-Hall.

Shapiro, Howard I., et al. 1991. *Cranes and Derricks*. NY: McGraw-Hill.

Truskie, S. D. "The Driving Force of Successful Organizations." *Business Horizons* (May–June 1984): 43–48.

United States General Accounting Office. Health Education and Human Services Division. 1996. *Information on the Davis-Bacon Act*. MD: The Office, Gaithersburg.

9.2.9 Risk Management

Definition

The risk management area of knowledge encompasses the dynamic and static risks that are part of every capital expansion program. Dynamic risks (risks directly tied to team decisions) and static risks (risks simply inherent to a construction environment) must be identified, evaluated and disposed of in a manner which will minimize economic loss to the owner in the event a risk with attached liability occurs.

Description

The construction manager must be able to anticipate and analyze static and dynamic risks as well as identify, evaluate, and recommend manners of their disposal in the best interest of the owner. Disposal can be accomplished by elimination, assignment, or by accepting and managing them to minimize the consequences if they accrue.

The risk management area of knowledge includes surety bonding and insurance in the static risk area, and contracting and construction processes and procedures in the dynamic risk area. The CM must understand construction related insurance coverages, performance bonds, labor and material bonds, bid security, and other forms of available surety protection.

The CM must thoroughly understand the dynamic risks inherent to construction contracting, contracting procedures, construction planning, construction means, methods, and techniques, and be able to evaluate potential consequences and offer advice for minimizing. This area also requires excellent communication skills and high ethical standards.

Representative Bibliography

Ahuja, Hira N., and Arunachalam, Valliappa. "Risk Evaluation in Resource Allocation." ASCE, *Journal of Construction Engineering* 110 (Sept. 1984): 324–336.

Ashley, David B., et al. "Critical Decision Making During Construction." ASCE, *Journal of Construction Engineering and Management* 109 (June 1983): 146–162.

Ashley, David B. *Construction Project Risk Sharing*. Technical Report 220. CA: The Construction Institute, Department of Civil Engineering, Stanford University. June 1977.

Associated General Contractors of America, and National Association of Surety Bond Producers. 1980. *The Basic Bond Book*. Washington, DC.

Bramble, Barry E. 1995. *Resolution of Disputes to Avoid Construction Claims*. Washington, DC: National Academy Press.

Cornes, David L. 1983. *Design Liability in the Construction Industry*. NY: Granada.

Derk, Walter T. 1987. *Insurance for Contractors*. 5th ed. Chicago, IL: Fred S. James & Co.

Lambert, Jeremiah D., and Lawrence White. 1982. *Handbook of Modern Construction Law*. Englewood Cliffs, NJ: Prentice-Hall.

Lifsin, Melvin W., and Edward F. Shaifar. 1982. *Design and Risk Analysis for Construction Management*. NY: John Wiley & Sons.

Ronco, William C. 1996. *Partnering Manual for Design and Construction*. NY: McGraw-Hill.

Schwartzkopf, William. 1992. *Calculating Construction Damages*. NY: Wiley Law Publications.

9.2.10 Safety Management

Definition

The safety management area of knowledge encompasses safe practices at the construction site in accordance with the prevailing regulations in the area of the project. The CM has the responsibility to promote safe site conditions by example and urge contractors to have organized safety procedures in force. Although each contractor bears the responsibility for the safe practices of its own employees, the CM has the responsibility to coordinate safety requirements common to all contractors and to see that safety provisions are included in construction contracts.

Description

The CM must be familiar with construction practices, safety programs, practices, procedures, and administration and safe environmental conditions at the construction site, as well as an understanding of the 1970 Occupational Safety and Health Act (OSHA) and the safety/health regulations in the area of the project. Excellent communications skills and high ethical standards are required.

Representative Bibliography

Associated General Contractors of America. 1977. *Manual of Accident Prevention in Construction*. Washington, DC.

Business Roundtable. 1982. Report A-3, "Improving Construction Safety Performance." *The Construction Industry Cost Effectiveness Project*. NY: The Business Roundtable.

Commerce Clearing House, Inc. 1989. *OSHA Safety and Health Standards for the Construction Industry* (29 CFR Part 1926) with Amendments as of 6/1/1989. Chicago, IL.

Gans, George M. Jr. "The Construction Manager and Safety." ASCE, *Journal of the Construction Division* 107 (June 1981): 219–226.

Koehn, Enno, and Kurt Musser. "OSHA Regulations Effects on Construction." ASCE, *Journal of Construction Engineering and Managment* 109 (June 1983): 233-244.

Laufer, Alexander, and William B. Ledbetter. "Assessment of Safety Performance Measures at Construction Sites." ASCE, *Journal of Construction Engineering and Management* 112 (Dec. 1986): 530–542.

Lee, D. H. K. 1964. *Heat and Cold Effects and Their Control* (Public Health Monograph No. 72). Washington, DC: U.S. Department of Health Education and Welfare.

Stanton, William A., and Jack H. Willenbrock. "Conceptual Framework for Computer-Based Safety Control." ASCE, *Journal of Construction Engineering and Management* 116: 393–398.

U.S. Department of Labor. 1984. *Code of Federal Regulations*, Title 29, Part 1920–end. Washington, DC.

U.S. Department of Labor. 1983. *OSHA Safety and Health Standards: Construction Industry* (OSHA 2207). Washington, DC.

U.S. Department of Labor. 1983. *OSHA Safety and Health Standards Digest: Construction Industry* (OSHA 2202). Washington, DC.

United States OSHA. 1980. *OSHA Requirements for Construction Equipment and Operations Standards*. CA: Equipment Guide-Book Co.

9.2.11 Schedule Management

Definition

The schedule management area of knowledge encompasses all aspects of scheduling throughout the project. Scheduling is the management tool that best represents the controlled operations philosophy of the CM contracting system. It combines the element of time with the project's resources from the start of design to owner occupancy. Scheduling eliminates or mitigates potential time-resource crises by predicting start and finish dates for intermediate project milestones. The use of scheduling is a means to an end, not an end in itself. It is a form of communication that should be presented in the simplest form with just enough detail to convey its message.

Description

The CM must be able to apply scheduling as a major management tool, proficient in its use and applications, and sensitive to its ability to plan and predict. The CM must have the capability to design the scheduling management plan for the project, select appropriate scheduling formats and techniques for specific project applications, and extract information from team members and contractors for scheduling purposes.

The schedule management area of knowledge includes a thorough understanding of scheduling techniques, from bar charts to precedence diagramming, including their use in planning, predicting, analyzing and tracking project activities and events. The CM must be computer literate in the area of scheduling, understand the fundamentals of modeling, be proficient in matching scheduling applications to requirements, and possess excellent communications skills.

Representative Bibliography

Archibald, Russel D. 1966, 1967. *Network Based Management Systems (PERT/CPM)*. NY: John Wiley & Sons.

Bennett, F. Lawrence. 1977. *Critical Path Precedence Net Works; A Handbook on Activity-on-Node Networking for the Construction Industry*. NY: Van Nostrand Reinhold.

Bertsekas, Dimitri T. 1991. *Linear Network Optimization: Algorithms and Codes*. MA: MIT Press.

Callahan, Michael T., et al. 1992. *Construction Project Scheduling*. NY: McGraw-Hill.

Fondahl, John W. 1962. *A Non-Computer Approach to Critical Path Method for the Construction Industry*. Stanford, CA: Stanford University Department of Civil Engineering.

Kelly, F. P., S. Zachary, and I. Ziedins. 1996. *Stochastic Networks: Theory and Application*. Oxford, U.K.: Clarendon.

Moder, J. J., C. R. Phillips, and Edward W. Davis. 1985. *Project Management with CPM, PERT and Precedence Diagrams*. 3d ed. NY: Van Nostrand Reinhold.

O'Brien, J. J. 1984. *CPM in Construction Management*. 3d ed. NY: McGraw-Hill.

Weist, Jerome D., and Ferdinand K. Levy. 1969 *A Management Guide to Pert/CPM*. NJ: Prentice-Hall.

Whitehouse, Gary E. "Critical Path Program for a Microcomputer." ASCE, *Civil Engineering* (May 1981): 54–56.

9.2.12 Value Management

Definition

The value management area of knowledge encompasses a project's cost versus value issue. It has three value components: designability, constructability, and contractability. Designability relates value to overall project design. Constructability relates value to construction materials, details, means, methods, and techniques. Contractability relates value to contracting options, contractual assignments, and contracting procedures. The CM is expected to extract maximum value for the owner from the constructability and contractability options which are available.

Description

In the area of designability, the CM must be capable of extracting the optimum overall design. In the area of constructability, the CM must be capable of securing the owner's prescribed level of value from the design and construction of the project. In the area of contractability, the CM must be capable of translating the owner's goals and the project's characteristics into a contracting structure that will extract the most value from the construction industry. The CM must consider the four major areas of owner concern (time, cost, quality, and business interruption) when recommending courses of action to the team.

The value management area of knowledge includes an understanding of contracting structures, formats, and procedures and the construction industry as a supplier of services. A thorough knowledge of construction materials and equipment, value engineering techniques, and life-cycle cost analysis is essential. Support knowledge in the areas of design, materials technology, estimating, scheduling, and procurement is important. Of major importance are communication skills and ethics.

Representative Bibliography

Ahuja, Hira N. 1980. *Successful Construction Cost Control*. NY: John Wiley & Sons.

Construction Industry Institute, Constructability Task Force. 1987. *Guidelines for Implementing a Constructability Program, Publication 3-2*. Austin, TX.

Dell'Isola, Alophonse J. 1973. *Value Engineering in the Construction Industry*. NY: Construction Publishing Co.

Jelen, Frederic C., and James H. Black, eds. 1983. *Cost and Optimization Engineering*. 2d ed. NY: McGraw-Hill.

Kleinfeld, Ira H. 1986. *Engineering and Managerial Economics*. NY: Holt, Reinhart and Winston.

Miles, L. D. 1961. *Techniques of Value Analysis and Engineering*. NY: McGraw-Hill.

O'Brien, James J. 1976. *Value Analysis in Design and Construction*. NY: McGraw-Hill.

O'Connor, James T., et al. "Constructability Concepts for Engineering and Procurement." ASCE, *Journal of Construction Engineering and Management* 113 (June 1987): 235–348.

Olin, Harold B., et al. 1980. *Construction Principles, Materials and Methods*. Chicago: The Institute of Financial Education/Interstate Printers and Publishers.

Stein, Benjamin, and John S. Reynolds. 1992. *Mechanical and Electrical Equipment for Buildings*. 8th ed. NY: John Wiley & Sons.

Zimmerman, Larry W., and Glen D. Hart. 1982. *Value Engineering*. NY: Van Nostrand Reinhold.

9.3 ACQUIRING CM KNOWLEDGE

Although the CM system has been widely used for more than 25 years, a unique source for acquiring its body of knowledge has not yet emerged. This text documents the CM system, explaining its fundamentals and practices, but only outlines the knowledge required to perform services as part of a CM firm.

The representative bibliographies at the end of each area of knowledge list sources where CM knowledge can be found. Few sources relate exclusively to the CM system; most explain the methods and tools as they could be used in any system. To acquire all of the knowledge in any one area, a person must seek it out from several sources and culminate the learning process with appropriate experience.

It will take several more years before the CM body of knowledge will be taught at the university level in a similar fashion as general contracting. One reason for this is that the CM system was a product of informal and unorganized industry research rather than university research. Consequently CM has no comprehensive formal documentation on which faculty can base a credible CM curriculum.

Without documentation, it will require considerable time for existing construction faculty to gain sufficient confidence and acquire enough knowledge of the CM system to be competent in the complete body of knowledge.

Another reason why the CM body of knowledge is not yet available at universities is that many existing university construction programs are supported in one way or another by general contractors. CM is not the contracting system to which GCs subscribe; they consider CM a dubious alternative, one to resort to only when an owner prescribes it.

As long as general contractors exclusively support construction education at universities, CM education will be minimized in the curriculum.

9.3.1 Some Common Ground

Many university construction programs are titled "Construction Management" but should more appropriately be titled "The Management of Construction." Contractually and philosophically, they are simply not the same. However, there is equity between the two in the technical aspects of the management tools used and the structure and characteristics of the construction industry. The technical knowledge of scheduling, estimating, value engineering, life-cycle costing, cost accounting, general accounting, safety, quality, contracts, specifications, materials, equipment, productivity, bonding and insurance are interchangeable between the systems. The manner in which this knowledge is applied in the two systems differs considerably from one to the other.

Examples of the differences can be found in estimating and cost accounting. General contractors rely on construction estimating (estimating based on construction documents where quantities and qualities are accurately conveyed). Construction programs teach construction estimating to the level required by construction managers but do not teach conceptual estimating (estimating based on preliminary information before design begins) to an acceptable level. In the CM system, both types of estimating are equally important.

In the case of cost accounting, an ACM has no need to track unit costs during construction. Unit costs are developed by and for contractors to determine how they are performing in relationship to their estimates for labor, material, and equipment. Construction managers require expertise in budget tracking, where most construction costs are lump sums, and predicting the end result is based on estimated costs to complete including changes and extras generated during the course of construction.

University construction programs dwell on unit price cost accounting rather than budget accounting, and although construction managers must understand unit price cost accounting, budget accounting expertise is the important cost accounting requirement for CMs.

9.3.2 Problem-Solving Ability

A major function of a CM is problem solving *before* a problem occurs. The CM's management expertise can be measured by the number of crises which arise: the fewer the problems, the better the management.

Construction managers must have problem-solving ability to properly serve an owner, and these skills cannot be acquired from a textbook. They must be developed. From this perspective, civil engineering programs that emphasize construction provide a better problem-solving environment than schools of construction. Civil engineering students spend a major portion of their training solving technical problems, so that the process of problem solving becomes second nature.

Construction schools, on the other hand, dwell more on contracting procedures and construction means, methods and techniques. Problem solving is part of the curriculum but not to the extent it is in an engineering program. This is not to say that students from construction programs cannot solve problems as well as civil engineers—it simply indicates that civil engineers have a better potential in problem solving as a result of their basic learning environment.

9.3.3 Gleaning CM Knowledge

Without an organized source available, expertise in the multidiscipline CM body of knowledge must be gathered wherever it is found. The various societies and associations connected with the construction industry are excellent sources. Several of them have certification programs complete with short courses and competency examinations.

The Project Management Institute has a program leading to certification as a project manager. The Society of American Value Engineers has a 40-hour course of instruction which certifies attenders as value managers. The Construction Specifications Institute has a certification program available in the area of specification writing. While most of these groups are not specifically CM oriented, the unique expertise and management tools they foster are directly transferable to the CM system.

The Construction Management Association of America (CMAA) has a certification program which is based on its version of a CM standards of practice. This program does not examine proficiency in either the tools used by construction managers in providing services or the technical aspects of design and construction. It does provide insight to procedural performance within the standards CMAA has established.

9.3.4 Assessing A CM Firm's Proficiency

Chapter 24, Acquiring CM Services, provides guidelines for an owner to use when selecting a CM firm. In addition to requesting the usual demographic and experience information, it is suggested that an owner closely examine the technical qualifications of the firm's personnel. It is assumed that the services of a CM firm can rise no higher than the performance of the individuals performing the services.

Chapter 8, The CM Organization, points out that a construction manager is not a person. A construction manager is an organization staffed by personnel who collectively possess the management, design, construction, and contracting expertise necessary to credibly perform the services the firm is engaged to provide.

A model program for the certification of construction management firms and individuals was developed by a CMAA committee in 1987. The structure of the model established CM as a service provided by an organization of qualified individuals. Although the model was rejected by CMAA for certification purposes, it remains a provocative reminder of how CM is actually practiced in the industry today. The essence of the 1987 model certification program has been included as Appendix B.

The required technical abilities of individuals providing CM services are not covered in this text nor are they pointedly listed in the bibliographies listed after each area of knowledge. The acquisition of these abilities is through education and experience, preferably a combination of both. It is left to schools of engineering, architecture, construction, and business to provide the basics for eventual technical expertise.

To acquire a perspective of the required technical abilities that comprise a CM organization, refer to Question 2.0 in the suggested Initial Requests for Proposals (Chapter 24) and the list of questions in Appendix A.

Currently, no single source for acquiring the complete body of CM knowledge exists. It must be acquired from several sources and galvanized with experience. Schools of engineering, architecture, construction, and business only provide the base on which to build. Acquiring the CM body of knowledge is a gleaning process.

CHAPTER 10

Budget Management

During the course of a project, there are two points when the CM's efforts will be sternly evaluated by the owner. The first point is the opening of proposals for the work. If the cost of the project is within the conceptual budget, the owner will note that the CM firm performed adequately with regard to its cost commitment. The second is the date of completion. If the project is completed on schedule, the owner will note that the CM firm performed adequately with regard to its time commitment. If either goal is not met, extenuating circumstances or changed conditions notwithstanding, the owner will not be easily placated.

This reality must be understood by every CM that undertakes a project for an owner. Chapter 20, Schedule Management, provides insight to improve the CM's chance of completing a project on time. This chapter provides information to enhance the CM's ability to complete the project within budget.

10.1 GENERAL

Budget Management, Information Management, and Value Management are inseparable CM activities. They are presented separately simply because each has a unique area of knowledge, but all factor into successful project economics. Budget management contributes estimating and cost control; information management, distribution of cost information; and value management, technical alternatives to cost problem areas.

A major duty owed to the owner by the CM is expert estimating and control of project costs. Of the priority items of project delivery (cost, time, quality and minimum business interruption), cost is the owner's major consideration. On most projects, cost is the deciding factor whether to build or not, and cost expansion is usually the prime owner concern throughout design and construction.

The CM's cost control goals can only be achieved by superior estimating and competent value decisions. Budget management and value management are inseparable priority activities. The quality of the CM's performance of these services determines the financial success of a project. Every CM firm must have exceptionally qualified estimating personnel to carry out these responsibilities to their expected level.

Although CMs are expected to be skilled estimators, they are asked to do no more than their peers who work for contractors or design professionals. What sets them apart from their peers is their required versatility—the ability to do both conceptual and construction estimating and to do both equally well. This is a combination of estimating expertise that contractors and design professionals do not require and usually do not have.

Budget management requires a CM to closely monitor project costs during the design phase. If a reasonable budget is adopted at the start of a project, competent estimating and value management decisions during the design phase can keep the budget in line. When a budget figure is confirmed as correct, after construction bids have been taken, the CM is often assumed to be a super-estimator. This is not a correct assumption; the CM is actually a capable budget manager and an expert but not super-estimator.

10.2 THE BUDGET MANAGEMENT AREA OF KNOWLEDGE

The budget management area of knowledge encompasses all project-related cost aspects of CM practice. The CM has the responsibility to confirm, generate, track, report, and substantiate all budgeted costs from the first estimate to the final accounting. The conceptual budget for the project, prepared by the construction manager before design begins, becomes the team's line-item financial guide as the design process moves toward the bidding phase. After bids are received, the amounts of accepted contractor proposals replace estimated line-item amounts and become the construction phase budget. As construction proceeds, payments to contractors, contract changes, and budgeted expenses are accounted for in detail. Every aspect of project cost is estimated as early as possible and substantiated as it occurs.

The CM must have the expertise to forecast project costs from preliminary information without the aid of detailed drawings and must accurately estimate construction costs from completed contract documents. The CM must also have the ability to progressively and accurately transform a combined line-item conceptual budget into a detailed line-item construction budget and breakout estimates for phased construction and fast tracking if required. The CM is also called upon to identify and predict project-related expenses in addition to construction costs and deliver accurate budget information to the client in a timely manner and in an acceptable format.

The budget management area of knowledge includes but is not limited to conceptual and construction estimating, feasibility studies, comparative cost studies, communicative estimating techniques; construction labor, material and equipment costs; equipment and labor production rates, material technology, industry standards, labor practices; construction means, methods and techniques, cost accounting, general accounting, construction industry economics, and communication skills.

10.3 PROJECT BUDGETS

Generally speaking, every proposed project has a proposed budget. We live in a cost-oriented society where a capital expenditure must have justification in the form of a return. The cost expended on a manufacturing plant must be returned through the sale of the products produced. The cost of an office building or apartment complex must be returned through rents and leases. The cost of court houses, public schools, roads and bridges must be returned through a public benefit.

To keep the cost of manufactured products competitive, maintain rental rates within the market, or return a public benefit from a new courts building, the capital investment must be a known quantity. An assured return on investment or a public benefit is dependent upon accurately predicting the costs involved, and construction cost is a significant expense, third in line to financing and operating expenses during the life of a facility.

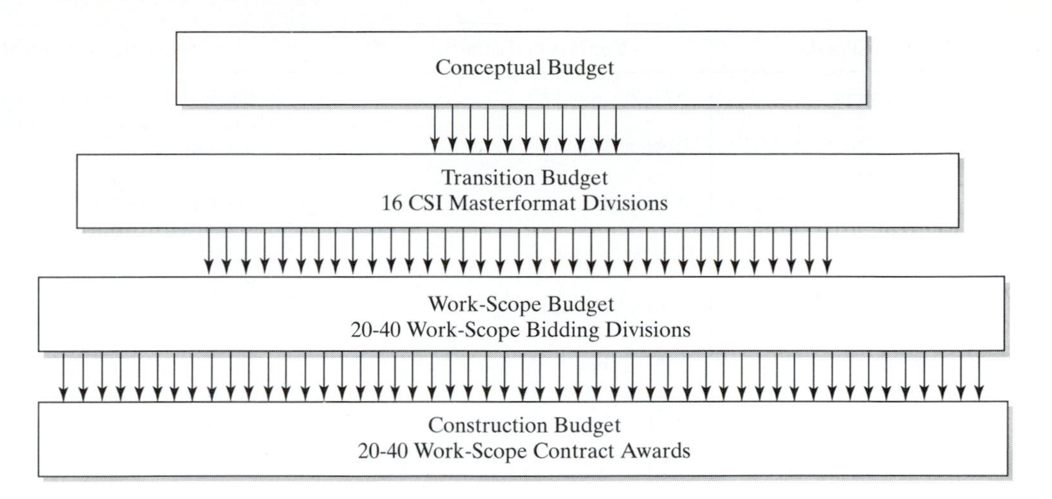

FIGURE 10.1 Conceptual to construction estimating flow chart.

10.3.1 The CM Budgeting Process

The estimating and budgeting expertise of the construction manager is expected to guide the project from its conceptual phase to its completion. The four components that contain the process are: the Conceptual Budget, the Transition Budget, the Work-Scope Budget, and the Construction Budget. (See Figures 10.1 and 10.2.) Both estimating and budgeting are continuous CM activities prior to construction; budgeting and some estimating continues to project completion.

The first budget to be developed is the Conceptual Budget. The broad-scope costs, the products of this budget, are refined to narrow-scope costs as design provides definite drawings and specifications. Estimating techniques switch from experienced-based unit pricing of areas and volumes to quantity take-offs and derived labor, material, and equipment costs.

The Transition Budget is a controlled, step-by-step estimating maneuver that gradually reallocates broad-scope conceptional costs to narrow-scope construction costs. One way to make this transition is to use the CSI Masterformat's Divisions 1 through 16* as the depository framework for transition estimating. This step also provides an opportunity to check the conceptual budget using modeling techniques that are based on the 16 divisions.

After the work-scope list is determined and the work-scope definition is available, the CM can allocate costs from the CSI Division depository to the various work-scopes that will be bid by contractors. The sum of the work-scope estimates, plus an estimate for general condition or construction support items, comprises the work-scope budget—the bottom-line target cost for constructing the project.

The final budget, the Construction Budget, is the summation of all accepted work-scope proposals after bidding or negotiations are completed. It is important for owners to understand that the summation, not the cost of individual work-scopes, is the number to compare with the work-scope budget. (The estimates of each work-scope should

*Published by the Construction Specifications Institute and Construction Specifications Canada.

Conceptual Estimate	Conversion Estimate	Work-Scope Estimate
Site Development Site Utilities Facility Architectural Facility Structural Facility Electrical Facility Mechanical	CSI 2, Sitework CSI 3, Concrete CSI 4, Masonry CSI 5, Metals CSI 6, Wood/Plastics CSI 7, Thermal/Moisture Protections CSI 8, Doors/Windows CSI 9, Finishes CSI 10, Specialties CSI 11, Equipment CSI 12, Furnishings CSI 13, Special Construction CSI 14, Conveying Systems CSI 15, Mechanical CSI 16, Electrical	Earth Work to Rough Grade Gravel Work to Final Grade Asphalt Paving Site Concrete Site Utility Systems Water Well Building Concrete Pre-Cast Concrete Masonry Hollow Metal Door Frames Door Hardware Plastic Covered Doors Carpentry Glass and Glazing Structural Steel Fabrication Structural Steel Erection Roofing and Sheet Metal Plumbing Heating/Ventilating/A.C. Fire Protection Systems Balancing Electrical Communication System Lath/Plaster/Drywall Painting Resilient Tile Flooring Glazed Wall Tile Terrazzo Wood Flooring Carpeting
Construction Contingency	Construction Contingency	Design Contingency
		Interface Contingency
		Scope Change Contigency
		Escalation Contigency
Construction Support Items	CSI 1, General Requirements	Construction Support Items

Figure 10.2 Converting a Conceptual Construction Budget to a Work-Scope Budget.

not be considered as individually competitive.) However, work-scope estimates are very helpful when reviewing work-scope proposals received from bidders prior to awarding contracts.

10.3.2 Conceptual Estimates

Budget management begins when the owner establishes a maximum project cost in the form of a conceptual estimate or feasibility budget. How one is established varies by project type; its credibility depends on the available information and the combined experience of an owner, A/E, and CM. If the owner has complete definitive information, the A/E and CM should be able to estimate the probable cost of the project within 20 to 30% (10 to 15% above or below the total of work-scope proposals received from bidding contractors).

If the A/E and CM have been hired for the purpose of conceptual estimating, information about the project is best collected through interviews between the A/E, CM and the owner. The owner should be represented by a person who understands the project and has a reasonable knowledge of its eventual use and physical requirements.

The A/E and CM should be represented by Level 2 Persons; the CM should also involve value management personnel from the various construction disciplines. The information extracted should be recorded in writing and in sketches as it is gathered and agreed to by the three parties as it is recorded.

It is best if the CM and A/E not bandy numbers about until they have extracted sufficient information from the owner to support a 15% plus or minus range conceptual budget. Owners have a tendency to remember the lowest figure mentioned, regardless of context. If adequate information is not forthcoming from the owner, the owner should be completely briefed on the problem and advised that an increased plus or minus range will be applicable to the budget.

It is important to the success of the venture that the budget be well done and well documented. Although theoretical, from everyone's perspective it is the number that must be sustained throughout the project-delivery process. Unlike the "probable cost of construction" on a general contracting project, the Conceptual Budget on a construction management project is fully expected to survive the design process and the competitive bidding process.

The secret of an accurate conceptual estimate is communication between the owner, A/E, and CM. The CM's value manager or estimator must know the information required and the questions to be asked and correctly interpret the answers provided by the owner and A/E. It is not mandatory that the CM be experienced in the facility type being estimated; the A/E probably is, so design information is readily available to the CM from the A/E. Familiarity makes conceptual estimating easier, but no more accurate.

Each project has its own inherent characteristics that influence construction costs. If the project is a school, the A/E and owner must inform the CM of school operation, administration, and maintenance, as well as the design criteria and construction requirements that influence cost. If the project is a condominium, the A/E and CM must be informed of the cost-quality trade-offs the owner will accept to ensure the expected return on investment. If a manufacturing facility, the owner must convey to the A/E and CM production methods and the environmental requirements that affect production costs. Although actual cost compilation is the responsibility of the CM, credible conceptual estimating is definitely a team effort.

Typically, a Conceptual Construction Budget has sub-budgets covering at least six categories: site development, site utilities, and facility architectural, structural, electrical, and mechanical. The estimate for the facility categories are developed using unit costs applied to areas and volumes of the various parts of the building. Site development and site utilities estimates are developed as lump sums, based on site-specific needs, the typography of the site, and the project's location on the site. If there are other major cost areas that can be split-out for estimating purposes (such as process piping, vertical transportation, a swimming pool or stage equipment), they should be included in the budget as separate categories.

10.4 CONTINGENCIES

A conceptual estimate consists of a detailed record of decisions made, a few figures, and considerable thought. Conceptual estimates should never be adopted with unjustified confidence; contingency dollars should be included in every conceptual budget.

Contingencies are budgeted dollars exclusively dedicated to compensate for unforeseeable costs, and there are two kinds of contingencies: one covers indeterminate construction market costs and estimating infirmities; the other, unpredictable project conditions and circumstances. Both are capable of analysis and close estimation.

The contingency for indeterminate construction market costs and estimating infirmities should be included as part of each work-scope estimate line item. These are costs that cannot be definitively estimated because of the minimum information available when conceptual budgeting takes place.

Contingency amounts for unpredictable project conditions and circumstances are estimates of probable cost increases that can not be logically allocated to construction cost line items. They should be included in the budget as individually accessible line items.

The areas of unpredictable project conditions and circumstances that should be assigned individual line item contingencies are: design, work-scope interfaces, cost escalation, and scope changes. The design contingency belongs to the A/E, the interface contingency belongs to the CM, the escalation contingency belongs to the team, and the scope change contingency belongs to the owner. The amount of each should be set and used exclusively by the designated party without exception.

The design contingency should cover the cost of design oversight. It is practically impossible to document without error a three-dimensional facility using two dimensions. The A/E is in the best position to determine what the extent of oversight could be on a specific project. Although the CM document review process minimizes oversight in contract documents, some chance still remains.

The CM contingency should cover work-scope interface gaps that could exist between work-scope descriptions. From a project cost perspective, gaps are certainly more desirable than overlaps, so the CM approaches the writing of work-scope descriptions from that perspective. Consequently, when used, this contingency does not indicate added cost.

The owner contingency should cover scope changes. The owner may want a contingency to cover vague or uncertain decisions pointed out during conceptual estimating. Sometimes this is to cover manufacturing or process equipment that can not be fully specified during the feasibility phase. There will be scope changes, and a contingency appropriate to the possibilities should be included in the conceptual budget.

The team contingency should cover escalation on projects with a long construction duration, especially projects that are phased over a period of time. Inflation, increased material and equipment costs, and changing labor rates could adversely affect the budget when the later phases are bid. The team should manage these risks by studying the possibilities and prudently earmarking funds to cover the probabilities.

When a custodian deems that a contingency can be eliminated, the remaining amount should be transferred to the owner's scope change contingency line item. The owner can use the balance remaining for upgrading finishes, expanding site development, improving the quality of furnishings and equipment, or simply retain it as an unexpended amount.

10.5 THE CONCEPTUAL CONSTRUCTION BUDGET

To this point, only construction costs have been budgeted. The following chart of accounts consists of 10 line items. The alpha designations used here are for demonstration purposes. As the budget transforms into a more definitive format, an expanded coding system will be required.

A Site Development
B Site Utilities
C Facility Architectural
D Facility Structural
E Facility Electrical
F Facility Mechanical
G Design Contingency
H Work-Scope Interface Contingency
I Scope Change Contingency
J Escalation Contingency

As design progresses, the few broad-scope line items of the Conceptual Budget will be divided, and in some cases combined, into a large number of narrow-scope line items with more definition. The responsibility of the CM is to effect this transformation, using a logical process that can be understood by all team members. It is safe to say that the only line items that will survive intact from the conceptual budget format to the eventual construction budget format are the contingency line items.

10.6 BUDGETING OTHER COSTS

Before the Conceptual Budget is prepared, the owner should determine which project costs are to be included and tracked in the budget. The conceptual estimate customarily covers construction costs only; there are many other costs connected with a project that could be maintained in a single budget. Items such as real estate, utility services, furnishings, equipment, finance costs, legal fees, and design costs may be included in an expanded budget at the owner's option. However, before changing the construction budget to a project budget, the format should be reviewed for compatibility with the owner's accounting system.

10.7 BUDGET ACCURACY

The accuracy of the budget, knowns and unknowns notwithstanding, should never be compromised. All items in the Conceptual Budget must be estimated as thoroughly and accurately as possible. The amount of the conceptual estimate often determines the fate of the project.

Optimism should be a constant area of concern. It is not uncommon for owners (as well as A/Es and CMs) to view the project in the most favorable way during conceptional planning. Realism becomes obscured when the thought of a new facility to occupy, a new project to design, and a new project to manage looms ahead.

The checks and balances of the project team will be severely tested, especially if the cost of the project is just a bit beyond the reach of available funds. Conceptual estimating must be an open and fair exchange of cost information, and the team must commit itself to an accurate and realistic conceptual budget, regardless of the consequences of a go or no-go decision.

It is common for a project to "grow" during design as a result of decision concessions made to the budget during conceptional planning. If the budget is to be maintained throughout the project, growth must be effectively controlled during design. Many fringe items that were not deemed necessary when establishing economic feasibility will suddenly became desirable after the project is funded. A detailed list of early design, material, and equipment decisions will help contain the urge to grow.

10.8 CONTINUOUS UPDATING

As design develops and is graphically defined, the team must maintain a close rein on scope and quality. This requires ongoing value management by the CM. Space equates to dollars, and it is the responsibility of the team not to expand the building area if at all possible. When design passes the schematic phase, an assessment of original space accuracy can be made. However, area can still grow until the design development is complete.

Refinement of the budget continues as the various elements of the project proceed through design development. Construction line items become more detailed as drawings and technical specifications are finalized. Quantity surveys are not accurate enough during the schematic phase, and the estimator must rely on area and volume calculations. As design development proceeds, budget refinement is kept current using drawings and outline specifications.

With more definition comes more refinement, until all costs originally based on areas and volumes are converted to costs based on quantities and derived labor, material, and equipment. At that time, the Work-Scope Budget is completed.

10.9 SCHEMATIC PHASE

In addition to the constant updating of the estimate during the flow of design information, specific team meetings should be scheduled for budget update decisions. The first meeting should be held at the completion of the schematic phase. At this point, the area of the building or structure is essentially set, and the relationships of space have been established. Approximate grades have been determined and overall dimensions generally fixed. Most site decisions have been made, and the project is delineated at least in plan.

This is usually a contractual progress payment point in both the A/E's and CM's agreement with the owner. Schematic design documents and an updated budget should be presented to the owner for review and ultimate approval. To reach approval, the owner should be led through the documents and budget by the A/E and CM and note their comments on both.

The continuous contact of the second level team members and resource persons during the schematic design phase makes surprises unlikely. However, the meeting provides the opportunity for first level owner persons who have not been involved in day-to-day decisions to review what has been done. With approval of the schematics, the owner authorizes the A/E, in writing, to proceed with the design development phase.

10.9.1 Schematic Budget

The budget provided at the schematic review meeting is somewhat problematic because it is a combination of area/volume numbers and quantity/cost numbers; budget review occurs early in the transitional estimating period. Most of the Schematic Budget is derived from updated previous conceptual estimating—the rest is based on construction estimating. However, the credibility of the budget has improved somewhat, as a result of the CM's mature insight into the project and the A/E's interpretation of design parameters.

10.10 DESIGN DEVELOPMENT

The design development phase is the most productive step in transition estimating. Horizontal and vertical dimensions are fixed; the structural system has been selected; wall sections have been decided upon; a preliminary finish schedule permits quantitative estimation of finish trades; and with the exception of mechanical and electrical, most schematic budget lump sums can be detailed.

Electrical and mechanical design is so dependent on architectural and structural design development that little detail is shown in these systems until the contract document phase is underway. Additionally, they usually experience the largest cost growth of all systems during design and represent an increasingly larger portion of the construction budget than in the past. These two conditions should be allowed for when estimating during the conceptual and schematic phases.

The design development phase provides construction definition that allows many line-items to emerge. Credible transition from area/volume numbers to quantity/cost numbers makes this possible.

10.10.1 Design Development Review

The completion of the design development phase is another milestone in the A/E progress payment requirement. Design continuity is temporarily interrupted to provide a document and budget review by the owner. The review of this phase replicates the schematic review, the difference being that more detailed documentation is available for the owner to review and ultimately approve before authorizing the A/E and CM to proceed with the contract documents phase.

On fast-track projects, the initial work-scopes to be bid are developed in the design development phase and prioritized during the early part of contract document phase. This allows the CM to develop their work-scope estimates as soon as they have adequate definition, which could be during design development. Therefore the design development budget could be a combination of refined conceptual and schematic lump sums based on area/volume estimates and a few work-scope lump sums based on quantity/cost estimates.

10.11 RATED ESTIMATING

During the contract document phase, the transition from area/volume estimating to quantity/cost estimating of work-scopes must be finalized. Budget reporting to team members is important. To facilitate this, a form of rated estimating should be introduced when the contract document phase begins (if it isn't already in place) as this phase provides the last opportunity to economically employ value management techniques for budget adjustments.

Several people in the A/E and CM firms and the owner's organization will be interested in the status of the estimate as it is developed, especially the CM's value managers. It is helpful for them to know the estimator's perceived credibility of each line item, to be able to quickly identify the existence and extent of line items still subject to cost fluctuation. Rated estimating is a simple application that provides that

Line Item	Budget	Estimate	Rating	Rated Estimate
Mass Excavation	$ 80,000	$ 86,000	1.3	$111,800[a]
Building Excavation	26,000	21,000	1.0	21,000[b]
Roads and Parking	120,000	122,000	1.0	122,000[c]
Drainage Structures	22,000	25,000	1.2	30,000[d]
Site Mechanical	48,000	45,000	1.1	49,500[e]
Site Electrical	23,000	22,000	1.1	24,200[f]
	$319,000	$321,000		$358,500
Estimate Rating:		(358,500 ÷ 321,000) = 1.117[g]		

[a]The Mass Excavation estimate has a high potential for change and must be reduced $6,000. A value management review may be required before the rating changes to 1.0.

[b]The Building Excavation estimate rating of 1.0 indicates the budget has a current $5,000 underrun and can surrender estimate dollars to line items estimates with overruns.

[c]The Roads and parking estimate rating of 1.0 indicates that the budget has a $2,000 overrun that could be eliminated by a value management review.

[d]The Drainage Structures estimate has a medium potential for change and must be reduced by $3,000. A value management review may be necessary before the rating changes to 1.0.

[e]The Site Mechanical estimate has low potential for change, and when the rating is 1.0 could have a $3,000 underrun; and at that time can surrender estimate dollars to line items with estimate overruns.

[f]The Site Electrical estimate has a low potential for change, and when the rating is 1.0, could have a $1,000 underrun; and at that time can surrender estimate dollars to line items with estimate overruns.

[g]The estimate has a low potential for change.

FIGURE 10.3 An example of rated estimating.

information. An example of a simple rated estimating format is shown in Figure 10.3. Other more sophisticated formats are available or can be devised.

At some point, all line-items will consist of quantity and cost units. A rating should be assigned to the quality or accuracy of these estimates. The rating is simply the estimator's opinion of the accuracy of quantity and value units. How accurate is the quantity survey for masonry? How accurate is the material cost used as the multiplier? Is the labor rate firm? How real is the productivity factor?

A computerized system, using nominal spreadsheet software, could rate estimate line items on an accuracy base of unity. The extent of each estimate element (quantity, material cost, labor rate, and productivity) deemed uncertain by the estimator is reduced to a scale of 0.1 to 0.3 and added to the base of 1.0.

A line item considered by the estimator to have a high potential for change would be rated 1.4 by the estimator; one with a medium potential would be rated 1.2, and one with a low potential would be rated 1.1.

When the estimate line-item cost is multiplied by its rating, the result is a rated value of the estimate line item. The sum of line items in the rated estimate column divided by the sum of line items in the unrated estimate column produces the rated value of the estimate. The goal is a rating of 1.0.

Unity rated estimate line-item costs that are less than their budget counterparts can surrender dollars to unity rated estimate line items that exceed their budget counterparts. Unity rated estimate line-item costs that exceed their budget counterparts should be subjected to value management reviews before receiving dollars from other estimate line items. At no time during the project should the original budget line-item costs be changed.

The key to success of any rating system is consistency by the parties doing the rating. Without consistency, the system would be counterproductive and could be seriously misleading. When the CM decides to use rated estimating, a commitment of dedication to the system on the part of the CM's estimating resource person is absolutely mandatory.

10.12 CONTRACT DOCUMENTS

The completion of the contract document phase is yet another milestone in the A/E progress payment requirement and, more importantly, in the forward movement of the project. This review of documents and budget by the owner is the final requirement before issuing bidding documents to contractors. The review replicates the two previous ones except the delineation and documentation is in its final form for bidding purposes and includes the general and special conditions, proposal forms, and instructions to bidders. The owner's legal counsel should be involved in this review.

10.13 WORK-SCOPE ESTIMATES

The ultimate goal of CM estimating is to produce a construction budget that can withstand the test of competitive bidding in the construction marketplace. Work-scope estimates will provide a comparison budget for the proposals received in each bid division of the project. It should be understood by the owner and the A/E that the construction budget is only competitive in its aggregate amount and is subject to the plus/minus accuracy established by the conceptual estimate.

Individual Work-Scope Budgets were not established by the CM as competition for bidding contractors. However, they should be within a reasonable range of the work-scope proposals received from bidding contractors. The proposals received from the lowest qualified bidders in all work-scope divisions will establish the minimum competitive price for the project.

In essence, the function of the CM estimator is to establish costs as accurately as possible during every phase of design, so that the financial direction of the project is clear to all team members, and the value management process can be applied in a timely and effective manner. Although the CM's work-scope estimates are assembled by the estimator using the same techniques that contractors use, it should be understood that the CM is not a bidder.

During the bid opening and after bids have been received, work-scope estimates can be extremely useful. For example, if only one bid is received in a particular work-scope division and it is permissible for the owner to award on the basis of a single bid, the estimator's figure for the work-scope can determine whether the single bid should be opened or if a re-bid of that work-scope is preferred. It should be handled discreetly.

Before opening the single proposal, it is suggested that the owner announce a maximum amount that would be acceptable for award (perhaps the work-scope estimate plus 5%). If the proposal is within the range, the bidder could consent to having the bid opened or ask that it be returned unopened. This is a fair and equitable way to handle a single proposal situation; if an owner cannot make an award, bids should not be exposed to anyone for any reason.

Additionally, the estimator's quantities, prices, and person-hour allocations are valuable data during post-bid interviews with apparent low bidders. Comparing the bidder's calculations with the estimator's calculations provides credibility to the bidding process. Comparisons confirm the interpretation of the work-scope as well as the quantity survey and the labor costs. Even though estimating (especially rated estimating) has a high potential for accuracy, errors sometimes occur. Comparisons can set the record straight before awarding contracts.

10.14 CONSTRUCTION BUDGET MAINTENANCE

Prior to the award of contracts, the budgeting process was essentially an estimating process. The Conceptual Budget was preserved through the combined use of estimating and value management. Once contract awards are made and the Construction Budget established, comprehensive construction estimating is only occasionally needed.

The sum of all contracts awarded establishes the construction budget. The budget line items are the individual lump sum contract amounts of the work-scope contractors. The only line items that remain estimates are the contingencies and the construction support items. Construction estimating is only needed to price or verify changes when and if they occur.

Each Construction Budget line item is subdivided by a listing called a Schedule of Values (a breakdown of each work-scope into identifiable units of work that are measurable in terms of quantity and cost). Schedules of value are not new; they have been used on lump sum contracts for progress payment determination for many years. See Chapter 11, Contract Management for more on Schedules of Value.

Converting the quantity of work done to date by a contractor on a subline item in the schedule of values to a percentage of the total quantity of work in the subline item, and multiplying it by the budgeted amount in that subline item, produces the dollar value of work done as of the end date of the accounting period, by the contractor on that line item.

This amount is called the "cost to date." Subtracting the cost to date of the previous accounting period from the current cost to date provides the "cost for the period"—the dollar value of the work done by the contractor during the current accounting period.

Subtracting the cost to date from the total value of work in the subline item provides the "cost to complete," providing no changes have been made or are planned for the line item quantity. Because changes are always a possibility, the cost to complete is usually listed as the "estimated cost to complete."

The total for the subline-item is the "budgeted cost." However, due to the fact that there could be changes, an additional heading is included for the "estimated final cost"—the cost to date plus the estimated cost to complete.

A basic spreadsheet program can be used to produce these figures, one for each Schedule of Values and one for the total project. Although both spreadsheets work the same, the headings for each should be as descriptive as possible for better communication. The information extracted from the Schedules of Value makes it unnecessary to use all the headings needed in the total project spreadsheet. Typical budget maintenance spreadsheet headings are shown in Figure 10.4. Schedules of Value are also covered in Chapter 11, Contract Management.

All of a contractor's Schedule of Value subline items must be manually updated for each progress payment period, and the line items produced by the sum of the subline items for each contractor should be automatically transferred to the Total Project spreadsheet under the matching headings.

The only other manual input to the budget maintenance program should be the estimated cost to complete. This data can be entered to either contractor schedule of values subline items or total line item, whichever is preferred.

10.15 SCHEDULES OF VALUES

Each contractor's Schedule of Values should be meticulously assembled by the CM and ultimately approved by the team. The breakdown of the lump sum cost should represent the true value of each subline item as close as possible. The CM's work-scope estimate should be the source on which the subline items are established. There is no need for contractor input as there is when using the GC system.

Because the Schedule of Values is primarily a source document and secondarily a report document, some of the column headings required for reporting purposes in the

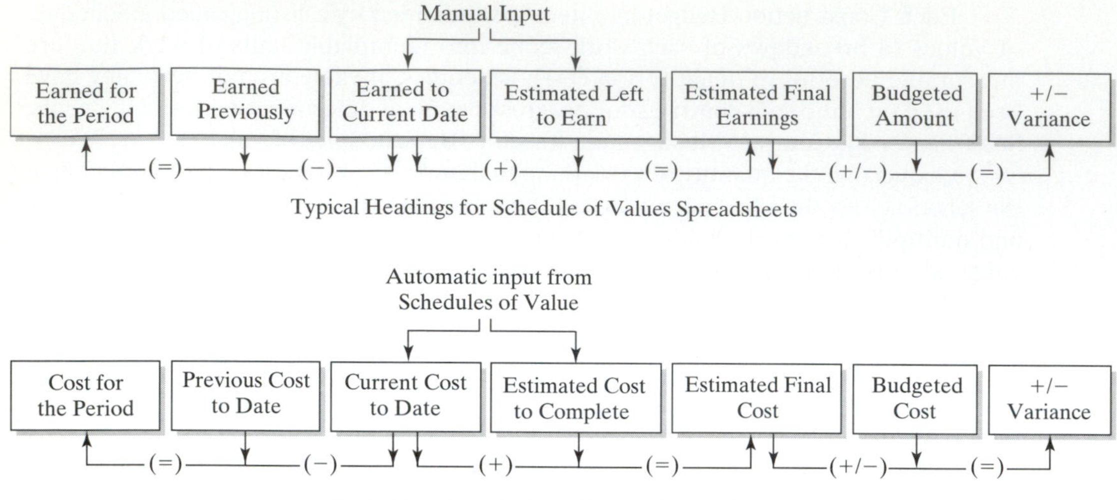

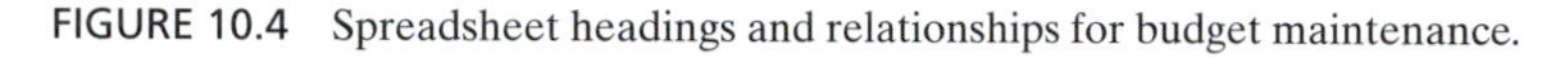
FIGURE 10.4 Spreadsheet headings and relationships for budget maintenance.

total line item spreadsheet can be omitted in the schedule of values. The information in the Schedule of Values is the option of the team. Its main purpose is to determine contractor payments.

A typical comprehensive Schedule of Values is shown in Figure 10.5. A more common form of schedule is shown in Figure 10.6.

Column (1) lists the subline item's mark numbers. Considerable thought should be put into the numbering system, to permit different printout sorts for transferring information within the system and for reporting/record purposes. The subline item mark numbering system should be tailored to meet an individual CM's cost accounting and reporting requirements (and many times, to meet the owner's requirements).

In the illustration, the first pair of digits is the work-scope division. The second pair is the type of line item (1: contract, 3: change order, 5: owner purchase, 7: owner supply, 9: extra work). The next pair is the subline item number. The last pair is reserved for internal CM cost-tracking purposes.

Column (2) is the description of the subline item. Column (3) is the Change Order column where Change Order amounts are listed as they are included. Column (4) is the percentage complete used on the previous assessment of work done.

Column (5) is the manually entered percent complete, as determined by the CM and A/E, by visually evaluating the work done by the contractor as of the period ending date of 0-06-96.

Column (6) is the team's breakdown of the contractor's lump sum contract amount. The allocations to the subline items include housekeeping, final cleanup, and punchlist (items that must be performed to receive payment). The total of the column is the contractor's contract amount, not including changes.

Column (7) is the amount earned to date by the contractor. The subline item amounts are the product of the percentages in column (5) and the budget amounts in column (6). The credit for the vinyl tile deduct is taken when the C.O. is entered.

Schedule of Values Report

Page 1 of 1
Period Ending: 06-28-96
Date This Report: 07-01-96

Project No. 9606
Payment No. 6

Work-Scope No. 15
Work-Scope: Floors, Walls, Ceilings
Contractor: Boden & Jones, Inc.
Original Contract Amount: $185, 617

(1) Item	(2) Description	(3) C.O. Amount	(4) % Complete Prev.	(5) % Complete Cur.	(6) Budget Amount	(7) Earned To Date	(8) Previous Earnings	(9) Est. to Be Earned	(10) Est. Fnl. Earnings	(11) +/− Variance
15010102	Vinyl Tile		0	0	10,000	0	0	10,000	10,000	
15010206	Ceramic Tile		0	0	12,000	0	0	12,000	12,000	
15010303	Partitions		50	70	115,000	80,500	57,500	34,500	115,000	
15010408	Acoustical Ceilings		15	15	16,8000	2,520	2,520	14,280	16,800	
15010508	Gyp Board Ceilings		0	25	7,800	1,950	0	5,850	7,800	
15010603	Acoustical Walls		0	10	2,400	240	0	2,160	2,400	
15010712	Housekeeping		0	0	9,217	4,500	3,000	4,717	9,217	
15010823	Final Cleanup		0	0	3,000	0	0	3,000	3,000	
15010930	Punchlist		0	0	6,400	0	0	6,400	6,400	
15011005	Decorative Channel		0	0	3,000	0	0	3,000	3,000	
15031105	Decorative Channel	900	0	0	0	0	0	900	900	+ 900[a]
15031201	Skylight Liners	1,400	40	100	0	1,400	560	0	1,400	+1,400[b]
15031303	Relocate Bulkhead	138	50	100	0	138	66	0	138	+ 138[c]
15031403	Add Stud Line	119	100	100	0	119	119	0	119	+ 119[c]
15031502	Deduct Vinyl Tile	(300)	0	100	0	(300)	0	0	(300)	− 300[c]
15031603	Add Bulkhead	400	100	100	0	400	400	0	400	+ 400[c]
		X			X	X	X	X	X	X
Work-Scope Div. Totals:		$2,657			$185,617	$91,467	$64,165	$96,807	$188,274	$2,657
C.O. Percent Variance:		+1.43%								
Work-Scope % Complete:						49%	34%			
Earned this Period:		(91,467 − 64,165) = $27,302								

[a]Interface gap.
[b]Design omission.
[c]Scope change.

FIGURE 10.5 A typical comprehensive schedule of values.

Schedule of Values Report

	Project No. 9606	Work-Scope No. 15
Page 1 of 1	Payment No. 6	Work-Scope: Floors, Walls, Ceilings
Period Ending: 06-28-96		Contractor: Boden & Jones, Inc.
Date This Report: 07-01-96		Original Contract Amount: $185, 617

(1) Item	(2) Description	(3) C.O. Amount	(4) % Complete Prev.	(5) Cur.	(6) Budget Amount	(7) Earned To Date	(8) Previous Earnings	(9) Earned for Period
15010102	Vinyl Tile		0	0	10,000	0	0	0
15010206	Ceramic Tile		0	0	12,000	0	0	0
15010303	Partitions		50	70	115,000	80,500	57,500	23,000
15010408	Acoustical Ceilings		15	15	16,8000	2,520	2,520	0
15010508	Gyp Board Ceilings		0	25	7,800	1,950	0	1,950
15010603	Acoustical Walls		0	10	2,400	240	0	240
15010712	Housekeeping		0	0	9,217	4,500	3,000	1,500
15010823	Final Cleanup		0	0	3,000	0	0	0
15010930	Punchlist		0	0	6,400	0	0	0
15011005	Decorative Channel		0	0	3,000	0	0	0
15031105	Decorative Channel	900	0	0	0	0	0	0
15031201	Skylight Liners	1,400	40	100	0	1,400	560	840
15031303	Relocate Bulkhead	138	50	100	0	138	66	72
15031403	Add Stud Line	119	100	100	0	119	119	0
15031502	Deduct Vinyl Tile	(300)	0	100	0	(300)	0	0
15031603	Add Bulkhead	400	100	100	0	400	400	0
		X			X	X	X	X
Work-Scope Div. Totals:		$2,657			$185,617	$91,467	$64,165	$27,302
C.O. Percent Variance:		+1.43%						
Work-Scope % Complete:						49%	34%	
Estimated Final Earnings:		(+2,657 + 185,617) = $188,274						

FIGURE 10.6 A typical schedule of values.

Column (5) shows 100% complete and Column (10) shows the deduct in the estimated final earnings.

Column (8) is the subline item amount paid to the contractor previously. Column (9) is the estimated amount the contractor will earn to complete the work-scope and changes.

Column (10) contains the subline item amounts the contractor will earn when the work of the work-scope and changes is completed. The sum of the column is the current value of the contractor's contract.

Column (11) shows the variances from the budget for each subline item. The change order amounts are identified by notations. The sum of the column equals the change order amount.

The more common form of Schedule of Values in Figure 10.6 can be compared to the comprehensive form in Figure 10.5. It contains less detailed information but is sufficient for payment purposes. Columns (9), (10) and (11) have been eliminated and a new column (9) heading added that lists the current period earnings by subline item.

10.16 THE BUDGET REPORT

The Schedules of Value line item report for all work-scopes, including Change Order status and estimated completion costs, provides the current cost status of the Construction Budget. The budget report is the prime budget maintenance document. An example is shown in Figure 10.7 on page 162.

If a comprehensive Schedule of Values (Figure 10.5) is used, the estimate for the cost to complete ($96,807) can be transferred from the bottom line, Column (9), to the budget report, Column (6), work-scope 15. If the Schedule of Values (Figure 10.6) is used, the estimate of costs to complete must be calculated and entered on the budget report form.

The budget report form also provides tracking for the three contingencies as well as other noncontracted costs, such as construction support, CM fees, or any other project-related costs the owner chooses to include in the budgeting process. Information on the budget report can be expanded to the extent required by the owner.

However, care should be taken to keep the tracking system as simple as possible. As with scheduling, budget management systems can be overdone. The computer is a powerful tool, and there is a tendency to use its power to full capacity whether the task at hand requires it or not. The schedule of values and the budget reports shown here are simple spreadsheet applications; keep them that way.

10.17 BUDGET CREDIBILITY

The credibility of the budget during formation and maintenance depends on estimating expertise and timely communication. The most difficult part of budget management occurs prior to the start of construction, when construction knowledge, estimating experience, and communication skills are the only tools to accomplish what must be done. Shelf-cost models and proprietary cost records are helpful but only if used and interpreted correctly. When a budget fails, it is almost always due to human miscalculation or delinquent communications.

However, multiple contracting within the CM system generates construction cost data to an extent and degree of accuracy never before available. The CM is privy to authentic construction quantities and costs that have the potential for making conceptual estimating almost as accurate as construction estimating. If the CM accumulates cost records in a productive uniform format from every project, it will not take long for a dependable conceptual cost modeling system to emerge. The potential in this area is almost unlimited.

Construction Budget Report

Period Ending: 06-28-96
Date This Report: 07-01-96

Page 2 of 2
Project No. 9606
Report No. 6

(1) Work Scope Nos.	(2) Work-Scope Description	(3) Previous Cost to Date	(4) Current Cost to Date	(5) Cost for The Period	(6) Estimated Cost to Complete	(7) Estimated Final Cost	(8) Budgeted Amount	(9) Variance (+/−)	(10) Adjusted Budgeted Amount
13	Concrete Flatwork	121,800	121,800	0	0	121,800	121,800		121,800
14	Masonry	135,400	140,200	4,800	23,643	163,843	158,000	+5,843	163,843
15	Floors, Walls, Ceilings	64,165	91,467	27,302	96,807	188,274	185,617	+2,657	188,274
16	Painting	0	1,800	1,800	30,900	32,700	30,000	+2,700	32,700
17	Landscaping	0	0	0	29,000	29,000	26,000	+3,000	29,000
SC	Work-Scope Contingency	(4,050)	(4,450)	(400)	31,550	31,550	36,000	−4,450	31,550
DC	Design Contingency	(6,500)	(8,250)	(1,750)	17,750	17,750	26,000	−8,250	17,750
IC	Interface Contingency	(1,200)	(1,500)	(300)	6,500	6,500	8,000	−1,500	6,500
CS	Construction Support	7,500	8,600	1,100	34,400	43,000	43,000	0	43,000
		X	X	X	X	X	X	X	X
		328,865	363,867	35,002	270,550	634,417	634,417	0	634,417

FIGURE 10.7 One page of a typical budget report spreadsheet.

C H A P T E R 1 1

Contract Management

Contract management in the CM system extends far beyond the traditional responsibilities of contract administration in the GC system. Contract management regards contracts as management tools that should be used resourcefully to improve the construction process as well as protect the owner's best interests.

The standard construction documents, used in the GC system of contracting, attempt to favor the parties who issue the documents. Their provisions (sometimes at the insistence of owner's attorneys) are under continuous revision as loopholes detrimental to the owner are uncovered.

These documents are currently to the point where exculpatory provisions, words, and phrases make it almost impossible for the parties to perform their responsibilities in a productive way. The construction process has been neglected and protectionism has become paramount.

The CM contracting system brought about the need for new contract documents that would be appropriate for the forms and variations of the new system. Figure 11.1 lists nine forms or variations of forms that require contract documents. To fill this need, three families of CM documents had to be developed, one for each of the forms, plus supplementary documents to implement the variations.

With this number of basic contracting options, and the opportunity to use different contracting formats for different contractors under the multiple-bidding format,

	Project Delivery Elements					
Contracting System/ Form/Variation	Design	Project Management	TC Contracting	Construction	Construction Coordination	Construction Administration
General Contracting	AE	AE	GC	GC/TC	GC	AE
D–B Contracting	DB	DB	DB	DB/TC	DB	O/DB
Agency–CM (ACM)	AE	AE/CM	O	TC	CM	AE/CM
Design–XCM	AE	AE	O	TC	AE	AE
Constructor–XCM	AE	AE/CM	O	CM/TC	CM	AE/CM
Contractor–XCM	AE	AE/CM	O/CM	TC	CM	AE/CM
Constr./Contr.–XCM	AE	AE/CM	O/CM	CM/TC	CM	AE/CM
GMPCM	AE	AE/CM	O	TC	CM	AE/CM
Constructor–GMPCM	AE	AE/CM	O	CM/TC	CM	AE/CM
Contractor–GMPCM	AE	AE/CM	O/CM	TC	CM	AE/CM
Constr./Contr.–GMPCM	AE	AE/CM	O/CM	CM/TC	CM	AE/CM

CM: Construction Manager GMP: Guaranteed Maximum Price AE: Architect/Engineer TC: Trade Contractor
X: Extended Services GC: General Contractor O: Owner D–B: Design–Build Contractor

FIGURE 11.1 Responsibility Distribution: CM, GC and D–B.

CM contract management becomes a more demanding task than with either the GC or D–B systems. The complex contracting requirements and various contracting options positions the construction manager as the knowledgeable party on the project team.

11.1 THE CONTRACT MANAGEMENT AREA OF KNOWLEDGE

The contract management area of knowledge encompasses the involvement of the CM in the operational and administrative provisions of the contracts used on the project. CM expertise includes the recommendation of standard contract forms and the performance responsibilities to be included in contracts but does not extend to the writing of contracts or in any way infringe upon the legal profession. This area is important because the CM system is a unique contracting system, the success of which depends on a workable realignment of traditional contracting roles and participant responsibilities. It is the CM's responsibility to establish a contracting format for the project and to see that each contractor's operational and administrative requirements are definitively included.

The construction manager must be able to evaluate each project from a contracting perspective based on the unique conditions of the project and local construction industry practices. The CM must assess the available possibilities, recommend a contracting structure to the client, assist in developing contracts, review contract documents for suitability, and coordinate their provisions on the project.

The contract management area of knowledge includes an understanding of contracts, contract language, standard contract documents, contract law, and construction contracting in the area of the project. A thorough knowledge of traditional contracting procedures, CM contracting procedures, and the reasonable possibilities for contracting innovation is necessary. Excellent communication skills are required.

11.2 CONTRACT PROVISIONS

The CM system has several forms and variations as described in Chapter 5, CM System Forms and Variations. Contracting formats are available for all variations, one or more of which will closely (if not exactly) meet an owner's contracting preferences and requirements. The following are a few examples of owner options.

Guaranteed Maximum Price CM (GMPCM) provides an upfront construction cost commitment that can be used to arrange project financing before design is 100% complete.

Agency CM (ACM), coupled with comprehensive multiple contracting, produces maximum project economy by extracting competition at the lowest possible contracting levels. Multiple contracting permits pre-qualification of trade contractors and improves the potential for obtaining construction quality.

Contractor Extended Services CM (Contractor–XCM) relieves the owner from holding multiple contracts and sheds almost all responsibility for overall construction cost.

Selective bonding of contractors is a value management option that can save bonding costs when dealing with reputable contractors. This option is usually only available on private sector projects.

11.3 CONTRACT DOCUMENTS

When using the general contracting system, contract options are minimal. The owner selects either a standard series of contract documents issued by the AIA, NSPE, or AGC or uses a proprietary series developed and published by the owner. These documents have usually stood the test of time and can be used over again with minimal modifications. They consist of: (1) the contract for construction, (2) the general conditions, (3) the supplementary and special conditions, and (4) accessory documents, such as the instructions to bidders, subcontracts, proposal forms, and surety, insurance, and request for payment forms.

When using contract documents, it is imperative that all documents used are from one series; they are coordinated within the series and refer to provisions one to another. Using documents from different series should not be considered and adding new documents and forms should be done with great caution.

Aside from methods of payment (lump sum, unit price, guaranteed maximum price, and cost plus options) and added features such as liquidated damages or bonus-penalty payment clauses, the contracting process is fairly constant from one project to another and should be well understood by all project participants.

Accountability for using the right contract documents, and using them correctly, is the owner's. However, as the knowledgeable party regarding contract documents, the A/E is expected to guide the owner through the selection and modification process and, with the counsel of the owner's attorney, assemble and issue the documents for the project.

11.4 CONTRACTABILITY

When using the CM system, there are many contracting options that produce very different contracting structures, all of which should be considered by the team in the context of the owner's preferences. The selective use of contracts and the opportunity to establish an ideal contracting plan and install the best contracting structure has added a new term to the construction contracting glossary: *contractability*.

The basic concepts of contractability are not new. Most have been used for as long as construction services existed. However, the emergence of the CM system, with its broad variety of contracting opportunities, demanded closer attention to contracts and contracting strategies. The development of a productive contracting plan is recognized as a prime management strategy for improving project delivery performance.

The basic premise of contractability is that each project has a unique set of circumstances that can be beneficially exploited through optimum contracting arrangements. When properly formulated and installed, the appropriate contracting plan will accommodate most of the owner's criteria and greatly improve the opportunity for project success.

The owner criteria that lead to the decision to use the CM system must be prioritized, because the inherent features of the construction industry may not allow all of them to be included in one CM contracting plan. Trade-offs similar to those made during constructability reviews may also have to be made during contractability reviews.

The success of the team's contractability effort depends equally on a complete and accurate understanding of the construction industry and construction contracting and a complete and accurate understanding of the owner's requirements and limitations. Owners should not attempt to make contractability decisions without the expertise of a qualified CM and A/E, and CMs should not make contractability recommendations without complete, detailed input from the owner and A/E. The time to initiate contractability discussions is the brainstorming session (described in Chapter 7, ACM Procedures).

11.4.1 Contractability Limitations

Contracting strategies and contracting plans have limitations; every option is not available under all project circumstances. Industry protocols, local area practices, and common sense, enter into the contractability decisions made. For example, arbitrary bid packaging on multiple contract CM projects might nullify the cost advantages of competitive bidding. Fast-tracking when not really needed could increase contracted costs. Prequalification of trade contractors is effective on private sector work but may not be permitted in the public sector, and negotiated contracts have public sector restrictions or bans.

The dollar difference between the positive and negative consequences of a contract strategy's success or failure positions contractability as a high-level responsibility. Not all construction industry users, A/E's, or CM practitioners have the necessary expertise to effectively optimize a CM project. This should be a consideration when an owner selects a CM.

11.4.2 Project Delivery Elements

When performing a contractability review or formulating a contractability plan, all service elements or components necessary to deliver a project must be considered individually and each contractually assigned to members of the project team. The required service elements are project management, design, contracting, construction, contract administration, and construction coordination, covered in Chapter 1.

When using the CM system, the distribution of responsibility for these elements depends on the form of CM used. The agency form of CM assigns the A/E sole responsibility for design but shares responsibility for project management and contract administration with the CM. Trade contractors are responsible for construction. The CM has sole responsibility for construction coordination and shares contracting responsibilities with the owner.

Figure 11.1 (on page 163) shows the distribution of responsibility for the forms of the CM system. The GC and D–B systems have been included to demonstrate the numerous constructability options that are available in the CM system but not in the other two.

11.5 CONTRACTS AND CONTRACT DOCUMENTS

Contractability studies produce innovative contracting strategies that require significantly modified standard contract documents. Consequently, major contractability decisions should be made prior to the owner signing service agreements with the A/E and CM. Otherwise, provisions to accommodate the contractual effects of constructability decisions should be included in the modified standard service agreements.

Contractability stresses the review of all proposed contract documents prior to committing to use. The most acceptable of all definitions for CM is innovative contracting. There are three major forms and several variations of two of the forms to choose from. Acceptable standard documents, capable of convenient modification are published for some but not all variations.

Unlike general contracting and design–build contracting, where longstanding standards of performance have been established by precedent, CM contracting has no universally accepted standards from which owners can determine performance requirements or results, let alone formulate specific contract provisions to produce a desired result. The absence of standards requires considerable analysis to determine the specific provisions that should be added to CM contract documents. Resultantly, a much higher level of contract and contracting expertise is required on CM projects than on general contracting or D–B projects. The construction manager is expected to provide this expertise with input and guidance from the A/E.

11.6 MULTIPLE CONTRACTING IMPLICATIONS

A major feature of CM contracting is the use of trade contractors as prime contractors. Chapter 22, Multiple Bidding and Contracting, covers this feature in depth. However, multiple prime contracts provide contract management opportunities that should be expanded upon under this heading.

To those unfamiliar with multiple prime contracting, the mere implications of the concept often provokes negativism and serious doubt. Yet when properly employed, the practice provides significant financial returns to the owner. In addition to economy, multiple contracts contribute improved credibility to the task of managing contractors and to the important aspect of contract administration.

A multiple prime project will not only utilize about the same number of trade contractors, but it is also likely that they will be the same trade contractors that would be employed as subcontractors if the project used the GC or D–B system. The difference is that the project team, through the CM, will coordinate trade contractor performance and administer trade contractor contracts instead of the GC or D-B contractor. Once this is understood, doubts about multiple contracting should diminish considerably.

11.6.1 Contractor Payments

An example of improved credibility is in the progress payment requirement of contract administration. On a single prime project, a single payment is due each pay period. Under a lump sum contract, the payment is determined by applying completion percentages to a schedule of values that covers all the work on the project, whether it is done by the prime contractor or the prime contractor's subcontractors.

The progress payment is made by the owner to the single prime contractor. The contractor then passes on individual payments to subcontractors who are entitled to a share of the progress payment based on the work they accomplished.

When the schedule of values was typically developed by the single prime contractor, every effort was made to front-load—to inflate the value for work done early in the project by equally deflating the cost of work to be done toward the end of the project. For example, funds earmarked for final landscaping in the schedule of values might be transferred to clearing and grubbing or demolition.

This billing tactic improves the prime contractor's cash flow but has the opposite effect on the owner's cash flow and equity. Although the prime contractor will not receive a greater total payment than called for in the contract, partial payments are received before they are earned.

This seems harmless unless you consider this scenario. If the contractor defaults on the performance bond, the surety is not responsible for overpayment made to the contractor prior to default. This means the owner would suffer financial loss.

Payments to contractors under multiple contracts are based on individual schedules of value from each contractor. Payments cannot be moved from one contractor's schedule to another, and front-loading is virtually eliminated.

Increased credibility is also influenced by where owner payments to contractors finally end up. With a single prime payment, the owner has no way of telling when and if the prime contractor is paying his subcontractors properly and on time. With multiple contracts, the owner pays the individual contractors directly, eliminating the concern for liens being filed by a complete tier of contractors (on private sector projects) or claims against the labor and material bond on public projects.

11.6.2 Qualification of Bidding Contractors

Another example of how multiple contracts and CM contract management contribute to improved credibility is in the bidding process. On GC projects, the selection (and qualification) of trade contractors is left entirely to the general contractor. While this is traditional, it is not always in the owner's best interests from a logical point of view.

Trade contractors (subcontractors) build buildings. The general contractor usually undertakes about 15% of the construction with his own forces. Trade contractors undertake the remaining 85%. If trade contractors build buildings, they are the ones that should be screened as to their potential performance—it is their work that will determine the completed quality of the building.

Due to intense competition, general contractors tend to assemble their dollar proposals using the most economical trade contractor quotes they can find. Other than reflecting on their experience with trade contractors in the area of the project, general contractors do not formally qualify trade contractors. More often than not, low-quote trade contractors' prices will be used.

When using multiple contracts, bidding competition occurs at the trade contractor level; three or four contractors will be in competition with each other in 30 or 40 bidding work-scopes. Because they are bidding directly to the owner, they can be prequalified by the owner regarding experience, financial status, current work load, and references regarding their performance in areas such as safety and punch list cooperation.

Although contractors usually cannot be excluded from bidding on public projects, there is no restriction on obtaining qualification information that can be helpful in evaluating proposals, awarding contracts, and working with contractors during construction. On private sector projects, qualification information can be used to select bid lists.

11.7 THE CONTRACT MANAGEMENT PLAN

Although contract management appears to be complex, especially on projects using multiple contracts, it should be routine for experienced construction managers. The project participants, with the exception of the main characters (the CM, or the GC or D–B contractor) are the same regardless of which contracting system is used; only the contractual relationships have changed.

The contractability review, initiated at the brainstorming session, is the starting point for the contract management plan. At that session, the CM should extract information from the owner that will lead to determining the CM form that fits the owner's requirements. The CM must also extract firm information regarding the physical aspects of the project, the desired time schedule, how the owner expects occupancy to occur, and the contracting climate in the area where the project is to be built.

Although it would be possible to decide the CM form at the brainstorming meeting, it is usually better to discuss the possibilities and ramifications with the owner and A/E, let them absorb the data, ask questions, and make the decision at a later date. The choice of contracting strategy is one of the most important decisions that will be made on the project, and it should be made with ample forethought.

If the CM was chosen by the owner without a preconceived idea of which contracting system(s) to use, the options available to the owner are all those listed in Figure 11.1, including general contracting and design–build contracting. The CM should understand that the productive definition of CM is innovative contracting and that the optimum contracting strategy—the one that effectively and economically provides as many of the owner's needs as possible—should be installed on the project.

Experience has shown that on certain projects there is a place for all three systems simultaneously. If such an arrangement benefits the owner, it should be used. However, more often than not, the owner has predetermined which system will be used. If CM is chosen, then only the form and variation probably remains to be determined. A comprehensive contracting strategy proposed for an actual project that includes the use of all three systems is described in Appendix C.

11.7.1 Formulating the Contracting Plan

Once the contracting strategy has been decided upon, the construction manager can formulate a contracting plan that incorporates as many of the owner's requirements as possible, taking into consideration owner restrictions and priorities, construction industry practices, and location-of-the-project contracting practicality.

Public sector projects most always require open competitive bidding procedures. Private sector projects do not. Consequently, some of the features a public owner would like to incorporate (such as restricting the bidders list by qualification, negotiation of dollar proposals, and a closed bidding format) are not available.

Private sector projects, without public financing involved, can be whatever the owner wishes, at least within acceptable construction industry practices and location-of-the-project contracting practicality. This contracting "freedom" in the private sector often leads to contracting plans that look good on paper but do not work very well in practice.

11.7.2 Bidding

An example of private sector "freedom" is the private bid opening. Experience has shown that contractors prefer to bid projects where the work-scope is definitive and their proposals are opened and read with bidding contractors in attendance. They are well aware of the bid shopping potential when work-scopes are vague and their proposals are opened privately (see Appendix F).

For some reason, private sector owners believe that clever tactics such as these will create a better project by eventually extracting a lower construction cost through the process of negotiations (post-bid shopping). By privately vying one contractor's price against another's, a low-bid auction is created after dollar proposals have been revealed.

What owners fail to realize is that contractors are also clever, much more so than owners presumably suspect. The construction industry is a contractor's arena, and construction contracting is their business. Contractors have no intention of negotiating themselves into an agreement where they will lose money. If they do negotiate a lower price than the work-scope demands, they will make up the difference in the way they perform during construction—much to the regret of the owner.

The CM contracting system facilitates competitive bidding that extracts the lowest price without bid shopping. What it offers as routine in the public sector should be adopted as routine in the private sector. CM provides finely defined work-scopes for all bidding contractors and an opportunity to have a dollar proposal opened and read in the presence of the competition. The apparent low bidder is promptly identified.

Bidding procedures on CM projects should follow the same format regardless of whether the project is public or private sector. If a private owner requires project cost confidentiality, only those contractors submitting proposals should be at the opening. When multiple bidding and contracting is used, which it is more often than not, only bidders for the work-scopes being opened should be present. This limits cost information to portions of the project. Chapter 22, Multiple Bidding and Contracting, expands on this procedure.

11.7.3 Schedules of Values

It has been mentioned that the Schedule of Values under multiple contracting essentially eliminates contractor front-loading for payment purposes. In addition, Schedules of Values can aid in keeping the site free of construction debris for reasons of safety and convenience, and make housekeeping and final clean-up more readily enforceable.

Each contractor's Schedule of Values should have two clean-up line items, one for day-to-day housekeeping and the other for final clean-up as a punch list condition. Clean-up is a contract requirement for which contractors must allocate cost. By providing clean-up line items, the contractor can only include dollars in a progress payment request when and if clean-up is performed. Work under these line items is verified the same as work under other line items is verified.

When using multiple contracts, the nature of the work of each contractor influences the amount of money to set aside for housekeeping and final clean-up. The mason could have a high housekeeping cost and a comparatively low final clean-up cost. The curtain-wall contractor could have a comparatively low housekeeping cost and a high final clean-up cost, assuming window cleaning was part of the curtain-wall contractors work-scope. A typical CM Schedule of Values including incentive items is shown in Figure 11.2.

11.7.4 Punch Lists

Punch list payment reserves can also be selectively handled as part of the schedule of values. Note the punch list item in Figure 11.2. On multiple contract projects, each contractor will have a unique completion date and a different potential for punch list liability. Instead of including a single retained percentage for all contractors, the liability potential can be determined from the same perspective that clean-up and housekeeping potential was determined.

For example: a pile driving contractor that has each pile inspected and approved as it is driven and is paid for clean-up under housekeeping and final clean-up line items

Sequence Work Item	Out Building BD 04000		Original Contract	$112,712.00
Work Item Number	**Work Item Description**	**Percent Complete**	**Allocated Cost**	**Cost To Date**
04001	General Conditions	90.0	$ 2,100.00	$ 1,090.00
04002	Site Work	75.0	$ 2,450.00	$ 1,838.00
04003	Concrete Work	35.0	$ 12,000.00	$ 4,200.00
04004	Masonry	90.0	$ 18,000.00	$16,200.00
04005	Metals	50.0	$ 12,900.00	$ 6,450.00
04006	Carpentry	80.0	$ 21,762.00	$17,409.00
04007	Thermal/Moisture Protection	80.0	$ 3,900.00	$ 3,120.00
04008	Doors, Windows and Glass	50.0	$ 12,200.00	$ 6,100.00
04009	Finishes	0.0	$ 7,800.00	$ 0.00
04010	Specialties	0.0	$ 4,350.00	$ 0.00
04011	Miscellaneous	70.0	$ 450.00	$ 360.00
04012	*Housekeeping*	50.0	$ 2,000.00	$ 1,000.00
04013	*Clean-up*	0.0	$ 1,500.00	$ 0.00
04014	*Punch List*	0.0	$ 11,300.00	$ 0.00
C0040	C0-4-3771-1	0.0	$ 570.00	$ 0.00
	Totals:	46.3	$113,282.00	$52,207.00

FIGURE 11.2 Typical Schedule of Values.

should not have any punch list liability when the pile driving work-scope is complete. The punch list line item in the pile driving schedule of values should be zero.

A masonry contractor has a potential punch list liability with regard to both quality and quantity of the work after the masonry contractor declares the work-scope complete. The amount listed in the punch list line item of the mason's Schedule of Values should be determined as a function of risk management. What is the punch list liability for a mason contractor based on the requirements of the defined work-scope?

The advantage of using Schedules of Values as a strategic approach to contract management is apparent. Contractors appreciate whatever contracting concessions they receive, especially those involving progress payments. Trade contractors have long been on the defensive regarding contract provisions, especially blanket provisions that fail to take into account the contractors' unique position in the project delivery process and that put them in situations they should not have to endure. (See Appendix F.)

Schedule of Value amounts, based on the type of work being performed, are consistent with the higher level of management CMs are expected to provide. Instead of conveniently lumping all trade contractors under one classification and burdening them without due consideration, contract management demonstrates cooperation to contractors without undue exposure to the owner. It has a positive impact on contractor performance.

11.8 CONTRACT MANAGEMENT IN THE FIELD

The use of construction management represents a dramatic change in the way traditional (GC) field operations are perceived and conducted. The effectiveness of CM field personnel does not depend on intimidation of subcontractors and the handling of individual craftsmen but rather on an ability to induce contractors to perform cooperatively. Three common goals should be kept in view—safety, quality, and timely completion—toward an end that produces profits for participating contractors.

On GC projects the same goals are kept in focus, but the handling of the contractor's own work-force is often an overriding concern. More often than not, more time is spent striving to efficiently use the GC's employees and equipment to produce profits, than in the efficient performance of subcontractors where profits have already been assured by contract. On typical building projects, where 80 or 90% of the project is usually subcontracted, this situation sometimes poses a priority problem. The problem does not exist on projects that employ broker general contractors (where 100% of the work is subcontracted). All supervisory efforts can be directed to subcontractor coordination, if in fact supervision is provided.

However, there are very significant on-site differences between GC and CM projects that should be accepted. Supervisors on GC projects are in the service of the GC, are advocates of the GC, keep records that benefit the GC, and are oriented toward profit-making for the GC.

Supervisors on CM projects are in the service of the owner, are advocates of the owner, keep records that benefit the owner, and are not directly involved in profit-making from construction operations.

The forms and variations of CM notwithstanding, CM field personnel should have no construction workers to direct. This is certain when using ACM and should be no different when using other forms and variations of CM if dual contracts are properly used (as pointed out in Chapter 6, Construction Management Under Dual Services Agreements). Planning, coordination of trade contractors, record keeping, and reporting are prime CM responsibilities.

Planning focuses on the efficient use of trade contractor employees and equipment. There are no hiring or firing distractions. The principal effort is to maximize efficiency—to see that the project is adequately staffed and completed in the shortest time under the circumstances that prevail. By achieving these goals, all project participants should profit to their own satisfaction.

11.8.1 Single Field CMs

On many CM projects, only one full-time field person is needed on-site. There is a full complement of resource and support people in the home-office with the necessary expertise to assist in carrying out the field CM's responsibilities.

Planning, the coordination of contractors, record keeping and reporting fully occupies the field person's day, and assistance in scheduling, expediting, estimating and resource planning is necessary to get the job done.

Field CMs are the eyes, ears, and voice of the project team, especially of the Level 2 CM team member. This close connection adds credibility and expedience to on-site problem solving. When a problem arises that is beyond the authority of the Level 3 team members, Level 2 team members are readily available to help.

Upper level team members can rely on the authenticity of information provided from the field and expeditiously find solutions and relay them to the field for execution.

When using the GC system, this simple reliance is not always possible. There is a potential for conflict of interest rather than a trusting allegiance between the GC field person and the A/E. Under these conditions, information provided by the GC is considered less than credible and usually must be verified by a time-consuming A/E site visit.

Full-time, on-site CM residence should begin shortly before the initial contractor arrives on the site. The CM field person and his/her Level 2 supervisor establishes and staffs the CM office facility and installs the communication protocol with the CM home office and with the other team members.

A computer link should be installed between the CM home office and the field office to minimize hard copy files at the site and provide immediate access to project information. This last suggestion significantly changes the traditional job description for field supervisory personnel.

To effectively manage contracts/contractors and perform contract administration duties, the CM field person should be available on-site as much as possible during construction and should work at least the same hours that contractors do. In the event of a required absence for whatever reason, the team and contractors should be given a means of contacting him/her or a temporary field person should be assigned to the site. Contractors should never be without access to someone from the team that has authority at the site.

1. Short term scheduling
2. Chairing progress meetings
3. Conducting contractor start-up interviews
4. Reporting contractor activities
5. Expediting materials and equipment
6. Recording site and weather conditions
7. Recording contractor progress
8. Verifying progress payment quantities
9. Expediting and logging shop drawings
10. Recording as-built information
11. Resolving design discrepancies
12. Observing force account work
13. Posting legal notices
14. Attending contractor safety meetings
15. Providing construction support items
16. Receiving/storing owner purchases
17. Checking site security
18. Supervising other on-site CM persons
19. Obtaining specification clarifications
20. Coordinating contractors

FIGURE 11.3 Typical contract management duties of a field CM.

11.8.2 Site Person's Duties

The routine contract management duties of the CM site person on an ACM project include but are not limited to the list in Figure 11.3.

11.8.3 Quality Inspection Exemption

It is important to note that construction inspection is not listed as a CM site person's contract management duty. Day-to-day inspection of contractors' work is usually not required by the terms of standard ACM contracts and only becomes a CM responsibility when specifically called for in the CM's contract with the owner. When inspection is required by contract, inspection procedures should be included in the quality management plan.

11.8.4 Field CM Planning

The responsibilities of the CM site person should be short-range. Time does not permit adequate study and evaluation to effectively cope with long-range conditions. Long-range responsibility is the assignment of the Level 2 CM Person with the advice of home office resource and support persons. Credible and comprehensive communications with the field person provide the necessary input for proper long-term evaluation at this management tier. All the abilities of the CM field person(s) must be directed to the tasks immediately at hand.

The CM site person's relationships with contractors can only be productive if the respect and confidence of the contractors is attained. While it is difficult to be both critic and leader at the same time, the site person must find a way to balance both functions on a day-to-day basis. When it can be shown that the contractors will benefit from a well-managed project, this balance can be achieved.

11.9 CONTRACT COMPLETION

One of the most persistent contracting problems is getting contractors to completely finish their work, an inherent problem that the CM system cannot completely eliminate.

Toward the end of a GC project, there will be work still to be done and problems with completed work that require correction. The work is mostly subcontractor or trade contractor work, not work undertaken by the GC's own forces. This remaining work seems insignificant to contractors but is significant to the owner, and it must be completed by the terms of the contract for construction.

To help protect against nonperformance, the owner retains funds from the prime contractor's entitlement as an inducement for completion. Standard contracts generally specify that the owner is entitled to retain the estimated cost of the remaining work after the date of substantial completion is certified. However, on GC projects, partially as a result of front-loading, the retained funds are entirely owed to subcontractors.

Unfortunately, subcontractors whose work does not appear as punch list items are held hostage by the retained amount at the same percentage as those whose work does appear (unless, of course, the GC took an uncommon step and made equitable payments to subcontractors in the progress payment prior to the establishment of the retained amount established for the punch list).

To compound the problem, the GC has already collected profit and overhead on the collective value of the punch list items when the owner paid inflated Schedule-of-Value billings at the start of the project. Additionally, the cost to some subcontractors to eliminate construction deficiencies and complete unfinished work exceeds the amount retained by the owner to pay contractors for compliance. Without a financial stake in the completion of punch list items, GC's are not anxious to force the issue with their subcontractors.

It becomes a matter of simple economics. With little financial incentive to finish the work, an attractive alternative for some of the subcontractors and the GC is to accept a reduction in the contract amount for the incomplete punch list items. While this may solve the GC's and deficient subcontractors' problems, it does not solve the owner's problem. The owner is left to her/his own initiative.

When the CM format utilizes multiple prime contracts, an opportunity to minimize the contract completion dilemma arises. On GC projects, subcontractors (trade contractors whose work constitutes a major portion of the punch list on a building) can only be urged toward completion by applying leverage to the general contractor. Under a CM multiple prime contract, format subcontractors (now prime contractors) can be dealt with directly and selectively by the owner through the CM. This is accomplished with the punch list procedure and the contractor's Schedule of Values.

11.9.1 Multiple Contract Punch Lists

With CM multiple bidding, each contractor will reach completion individually. This contrasts with GC contracting where, from the owner's perspective, subcontractor contract completion coincides with GC contract completion. Although some CM multiple contracts will extend to project completion, many will be completed earlier.

The use of multiple contracts requires and permits approaches to contract administration that are different than when using a single contract (contract completion in particular). While it is no problem to certify the dates of contract substantial completion individually, it must be decisively determined exactly when a contract is acceptably completed so that retainages can be released.

Multiple contract completions require new concepts and changes. Traditional documentation previously used to initiate and certify that substantial completion had to be modified. Several traditional document clauses have been put to work that previously were seldom called upon under the GC format.

One such clause permits an owner to take over performance of a contractor's work if the contractor is negligent in the performance of duties. While an owner might think twice before applying that clause to a general contractor—doing so could cause a bigger problem than the one at hand—an owner can prudently apply the clause to trade contractors without fear of escalating the problem. Threatening to remove a contractor has a positive effect toward completing his portion of the project.

Figure 11.4 is an example of the content of a provision that will improve punch list completion from slow-finishing contractors.

To facilitate punch listing, quality conformance should be observed whenever the A/E is on-site. The A/E should review the work of contractors on a continuing basis and not wait for a request for substantial completion certification to generate a punch list. When on-site for some primary purpose, the A/E should keep an eye out for work items that obviously require correction.

The punch list should be a collaboration between the A/E and CM. The CM should address work that is incomplete; the A/E should address the work that is not properly done. By continuous observation, a provocatively long punch list can be avoided when a contractor files for substantial completion.

> If the contractor neglects or fails to carry out the work of the contract in accordance with the requirements of the contract, the Owner may provide written notice to the contractor that the owner will do the remaining work with resources other than those of the contractor.
>
> To avert owner action pursuant to such notice, the contractor shall, no later than the end of the second work day after receipt of the owner's notice, provide a detailed schedule of the work to be completed that is acceptable to the owner. The owner will show acceptance by endorsement.
>
> If the contractor fails to provide a schedule that is acceptable to the owner or fails to meet the schedule that is acceptable to the owner, the owner will notify the contractor that the contractor shall cease operation on the project and that the outstanding work will be done with other resources.
>
> In such case a change order shall be issued to the contractor deducting the cost of correcting any deficiencies and the cost of completing the work, including the cost of the Architect's and the Construction Manager's additional services made necessary by such default, neglect or failure.

FIGURE 11.4 Owner's right to do remaining work.

Experience shows that some A/E's are disinclined toward the continuous quality assessment concept. Their expressed concern is for increased liability more than for the additional effort required during on-site visits. However, the concept is spreading, mainly as a result of the cooperative allegiance of the CM site person and the contractual accessibility to trade contractors.

11.9.2 GC Substantial Completion

When reviewing the GC system, we find that a certificate of substantial completion is issued to the general contractor when the owner can occupy the facility, in whole or in part, for its intended purpose. In the case of a building, this could mean partial occupancy (even though there are items that need correction or are incomplete, the remaining work is so minor that it can be accomplished while the building is beneficially occupied by the owner).

When the GC assumes the project has reached the point of substantial completion, he/she fills out a request form, compiles a list of work still to be completed and corrected, and submits these to the A/E for review and owner acceptance. The A/E reviews the list and adds items the contractor has overlooked. If the A/E determines the list only includes work that will not interfere with the owner's use of the building, the A/E accepts the GC's requested date as the official date of substantial completion. The A/E then forwards the certificate to the owner for concurrence and signature.

The significance of the date of substantial completion is that it is the ending date for the time requirements of the contract for construction, the starting date for many of the guarantees and warranties provided by the contractor, and it lowers the amount of the retainage held by the owner.

11.9.3 CM Substantial Completion

The major contrast between GC substantial completion and CM multiple-contract substantial completion is that the A/E cannot measure multiple-contractor substantial completion by the occupancy criteria. The foundation contractor or steel erector will reach substantial completion long before the project can be occupied by the owner under any circumstances. Therefore, CM multiple prime projects require a different approach to establishing contractor substantial completion.

A date of substantial completion for multiple contracts is applied for individually by each contractor, but the owner-occupancy criterion is not necessarily used as the deciding factor. A date of substantial completion may be requested by a multiple contractor after approximately 95% of the contractor's work is completed. The request and a list of work still to be completed or corrected is forwarded to the CM for review. The CM physically verifies and if necessary adds to the list of incomplete items; the request and amended list are then forwarded to the A/E. The A/E physically verifies and if necessary adds to the list.

In place of the owner-use or occupancy criteria used on a GC project, the A/E and CM must determine if the contractor's work is at the point where subsequent contractors can perform their work without being inconvenienced.

If the A/E deems that only minor work and rework remain, and completion is to the point where the work of interfacing contractors will not be adversely affected, the request is forwarded to the owner with a recommendation for approval and signature. If the request is rejected, the contractor may resubmit after completing the specific deficiencies noted in the submittal returned by the CM.

11.9.4 Completion Motivation

As a means of urging contractors to complete their obligations on the punch list, two contract provisions should be included: (1) a punch list line item of a sufficient amount is included in the contractor's Schedule of Values, and (2) a notice is sent advising the contractor that the owner will finish the uncompleted work if the contractor fails to make significant and sustained progress on punch list items.

Experience has shown that these steps have a positive impact on completion performance. There is enough money retained to motivate the contractor to complete the work, and no contractor wants to have his surety notified of nonperformance. Needless to say, the second device should not be installed as an empty threat. It must be demonstrated through overt preparations that replacement of the contractor is a viable alternative for the owner.

There will be situations where it would be inappropriate or unwise to remove a contractor. Completion of the work may be stymied by back-ordered equipment or material, or the contractor is too involved in the construction to be considered for replacement. When this occurs, there is little else to do but work the retainage to its ultimate advantage and encourage the contractor to completion.

11.9.5 Occupancy Pressure

Another strategy for approaching timely contract completion is to have an obvious and valid occupancy date. Contractors will respond positively to must-occupy situations, but on projects that are running far ahead of schedule or where completion is not viewed as essential, contractors become indifferent in their performance. Fast-moving, well-managed projects, running on an uncompromising schedule, have the best chances of completing on time.

When a completion date change does occur, it is the construction manager's responsibility to mobilize contractors and the team to respond. Those experienced in construction operations will testify that progress momentum is difficult to speed up or slow down without suffering consequences.

If the project has been well managed and the completion date change is not perceived as the result of poor planning or misguidance, a completion date change should be able to be met with minimum consequence. By earning the respect of the contractors early in the project, the CM should have no trouble making the correction.

11.9.6 Contractor Scheduling

Good scheduling has a lot to do with how well contractors respond to construction coordination. Chapter 20, Schedule Management, covers the schedules recommended for CM projects. If the CM's scheduling efforts are realistic in terms of contractor accomplishment, construction coordination can be achieved more smoothly and successfully. Contractors tend to be defensive where scheduling is concerned and are quick to notice when the scheduling of interfaces breaks down even the slightest bit.

Managing contractors is much simpler when the CM's scheduling efforts produce results and predicted dates are met. They are more inclined to cooperate when the CM is perceived as a competent planner and scheduler. Instead of accepting the CM's performance dates as guesses, contractors begin to rely on the CM to assist them in their own planning as well. Once a position of respect such as this is gained, the level of contract management improves.

Exhibited proficiency in other management areas such as decision, quality, resource, and safety also increase the performance stature of the CM in the minds of contractors but none any more than schedule management.

The two schedules that significantly affect contract management during construction are the Short Term CAP and the completion schedules. The CM should make a maximum effort to ensure the credibility of these schedules to carry out contract management responsibilities.

CHAPTER 12

Decision Management

The CM contracting system places construction managers in an advisory position with an owner—a position of agency, with fiduciary responsibilities based on a standard of care. Although the construction manager's role is mostly advisory and the owner makes most of the decisions, the agency relationship places considerable responsibility on the construction manager to provide well-founded advice; advice that is consistently in the owner's best interests.

The advisory and decision-making domain of the CM differs considerably from that of the A/E, who is also an agent of the owner. A/E's decisions are oriented toward technical design, with shared responsibility with the CM for providing advice in project management and contract administration. CM decisions are oriented toward management and CM advice is oriented toward the practical aspects of design.

12.1 STATIC AND DYNAMIC DECISIONS

There are two types of decisions in the CM project-delivery process: static and dynamic.

Static decisions are those that can be made without concern for an immediate response. There is time for review, study, and approval before those affected by the decision will respond by action. They can be thought of as being made in a static or stationary environment, a pause during the progression of the project. Decisions collectively made by the owner, A/E, and CM are static decisions.

Dynamic decisions are the opposite; they will be acted upon immediately by those who are expected to react to them. There is only time for a cursory review and study of the issues, and approval may or may not be obtained before the decision is made. They are in effect autonomous decisions based on whatever policies or collaboration is at hand to provide guidance for the decision maker. Dynamic decisions are made to keep the project going without interruption from the time-consuming static decision-making process.

Static decisions are made by synthesizing the expertise and experience of all three team members. Static decisions are preferred in the CM system and should be used whenever possible.

Dynamic decisions can only reflect the expertise and experience of the team member(s) who makes them. They are a necessary part of the CM system but should be avoided when possible.

12.1.1 A/E Advice and Decisions

A/E design decisions are extensive. The owner makes the decisions that establish the project's broad design parameters, and the A/E makes the many design decisions that are required to convert parameters into detailed design requirements. All A/E design decisions are fully documented and legally sealed by the A/E's license before they are released by the owner to be used by contractors.

Design documents are exposed to review from the nontechnical perspective of the owner, the practical perspective of the CM, and the oversight perspective of the A/E firm. The documents, and consequently the design decisions of the A/E, are all subject to approval before they are responded to during bidding and construction.

A/E advice is provided throughout the project during both design and construction. A/E decisions, however, are concentrated during the design phase of the project when the A/E's expertise is germane.

A/E design decisions are both dynamic and static. Those that are made during the design documentation are necessarily dynamic; they reflect the exclusive expertise and experience of the A/E (skills which convinced the owner to hire the A/E in the first place). However, decisions to proceed with the various phases of design and to release the work of the A/E to contractors are static decisions, because they are subject to extensive team review, study, and approval. They must survive the tests of designability and constructability and the satisfaction of the owner before they are released.

12.1.2 CM Advice and Decisions

CM advice is mainly provided during design and, along with the expertise of the A/E and owner, helps to make static decisions. Most of the CM's decisions, however, are made during construction when the CM's expertise is germane and there is only time for a cursory review and study of the issues. Consequently, the major differences between A/E and CM decisions should be understood.

Too often the services of a CM are favorably compared with the services of an A/E, although as explained above, they are totally different in their context. The vast majority of A/E decisions are subject to pre-review, whereas the majority of CM decisions are not. Yet when judging the performance of a CM, a standard of care suitable for an A/E is most often the criteria used for dispute resolution. CM performance should be judged on a CM's standard of care, which remains in the development stage and hopefully will be established in the near future.

Therefore, CMs should realize that this inequity looms over their shoulders and should take additional steps to avoid making decisions that from their experience appear to be correct but could somehow prove otherwise. The following advice is recommended:

- When pressed for a decision, take advantage of all the time available to make that decision.
- If advice is available, make every effort to seek that advice and use it to advantage.
- Record every CM decision in appropriate detail.

12.1.3 Advice vs. Decisions

The charge of the CM is to advise and consult with the owner, A/E, and other members of the project team on all decisions pertaining to the cost, time, and quality of the project and its potential to cause excess business interruption to the owner.

Volunteering cost-saving advice to the A/E on design does not shift the responsibility for the design decision to the CM. Advising the owner to accept a qualified contractor as a bidder does not shift the responsibility for awarding the contract from the owner to the CM.

In the context of decision making, the charge of the CM is to make decisions that provide a proper and expedient course for the construction team and project to follow. The CM is not to exceed the authority stipulated in the owner-CM agreement and must fully consider the advice provided by the A/E and owner.

Ordering a contractor to continue work on a specific date or purchasing gravel for a road provided as a construction support item would be valid unilateral CM decisions; changing the water-cement ratio for a concrete slab pour or adding an item to the work-scope definition of a contractor would not.

12.1.4 Design Decision Responsibility

The responsibility for the project's technical design is singularly in the hands of the A/E. The owner and CM may influence the technical design but neither can force the A/E to change it. The A/E is professionally and legally obligated to existing statutes, codes, regulations, standards of practice and industry standards.

Advice on design in the context of designability and constructability is all that should be offered by the owner and the CM and all that needs be considered by the A/E.

The reality of a CM's very limited role in design and design decisions must overcome a major misconception which is held by owners, design professionals, the courts, and even some CMs. Contrary to a common belief, the CM has no technical design responsibility to the owner.

Liberal interpretations of owner-CM contract language referring to the CM's responsibility in the area of design (especially during pre-bid document reviews) have fueled this misconception. However, a literal interpretation of the qualifications required by law to provide design services, and the qualifications of a CM firm, clearly nullifies the possibility of the CM's involvement in technical design.

12.1.5 Decision Risk-Taking

Construction industry users and many of its observers believe that the project delivery process and the contracting aspect in particular is a gamble, that the industry is an arena for risk-takers who make risk-based decisions that sometimes produce winners, sometimes losers. It is commonly and wrongfully assumed that contractors expect to lose money on some projects and to make money on others.

This impression is understandable in the context of the GC contracting system as explained in Chapters 2 and 3. Many decisions made by GCs are dynamic decisions made under the pressure of time, especially those made during the bidding phase of a project. Decisions involving construction means, methods, and techniques often contain an element of risk but when they do, experience rather than impulse drives the decisions of the contractor. It is certain that contractors at every level never expect to lose money on a project.

The CM system, as demonstrated by its contracting structures, attempts to eliminate the impression and mitigate the reality of risk-taking in the project-delivery process. As explained in Chapter 19, Risk Management, risks are inherent to the project-delivery process and must be addressed. Risk decisions should be neither impulsive nor based strictly on odds. They should be anticipated, identified, evaluated, and disposed of by a static decision.

12.2 DECISION MANAGEMENT AREA OF KNOWLEDGE

The decision management area of knowledge encompasses the development and handling of the project and construction teams and the relationship of their respective members. This area of knowledge is the least technical but most important when providing CM services.

It is the CM's responsibility to consistently extract decisions from the team which are in the best interest of the owner without alienating any team members. Team members must approach decisions and make decisions cooperatively, respecting each other's project function, expertise, and operational capacities. Decisions which become contentious must have a prescribed path for resolution.

The construction manager must effectively establish each team member's involvement in the team as early as possible and guide the decision-making process throughout the project. The CM must have a clear understanding of the services being provided by team members and their business operation and style; induce a productive, low-profile system of decision-making checks and balances, generate synergistic team action on decisions, create a productive team management hierarchy, and make available a fair decision-making process when decisions are contended.

The decision management area of knowledge includes a generous understanding of business structures, organizations, practices, procedures, motivations, and philosophies, plus detailed insight into the design profession and contracting business operations. High-level communication skills and ethical standards are required, and a grasp of human resource management and alternative dispute resolution practices is important.

12.3 DECISIONS AND DECISION MAKING

The best decisions are derived from timely, credible input and discussed in an environment that ensures adequate checks and balances. The CM contracting structure provides this environment to a greater degree than either GC or D–B contracting.

By contract, the A/E and CM agree to serve the owner's best interests when decisions are made. All team members have a common goal with few if any self-serving distractions.

The CM structure permits the owner to be directly involved in the static decision-making process to whatever extent desired. This is a major (beneficial) revision to traditional contracting systems which give the owner unprecedented access to the project during design and construction. The expertise of the team members facilitates the decision-making process and ensures the owner that well-founded decisions can be made. To alleviate owner concern for too much involvement, the owner has the authority to transfer decision making, in whole or in part, to the singular or combined expertise of other team members.

Owners who do not wish to be immersed in static decision-making can contractually assign specific decision-making responsibilities to other team members. If this option is chosen, the owner must use discretion when making the assignment; it is important that the synergistic balance of the team not be compromised.

As stated in Chapter 4, The Agency Form of CM, team leadership must shift between team members depending on which team member's expertise is germane. This is a commitment made by each team member through the partnering process and fully supported by the owner.

Assigning owner responsibilities to either the A/E or CM could be interpreted as assigning team leadership. It is important that this does not occur—neither the A/E nor CM should be assigned leadership of the team under any circumstances.

12.3.1 Checks and Balances

By design, the CM team structure requires that the range of expertise of the A/E and CM overlap sufficiently to prevent unilateral decisions from becoming team decisions by default. The owner's involvement in decision-making precludes A/E-CM bilateral decisions from becoming team decisions. The experience of each team member, and their peer position in the team structure, produces a dependable checks and balances environment that essentially guarantees sound static decision-making.

The full benefit of checks and balances will not be realized, however, unless the relied-upon knowledge of each team member is well-honed and available when needed. For this reason, static decision-making should be a preplanned team activity whenever possible. Many decisions will have to be made without preplanning due to the relatively fast pace of the project and the vagaries of the project-delivery process.

However, the team (especially the A/E and CM) should constantly anticipate major decisions and schedule their discussion well in advance. The combined experience of the A/E and CM should produce a decision-making agenda that minimizes dynamic decisions, and static decisions which have to be made without sufficient preparation.

12.4 THE DECISION-MAKING HIERARCHY

The project team's hierarchy is explained in Chapter 4. This structure facilitates decision making by assigning responsibility for decisions of certain types and values to three defined team levels (see Figure 12.1).

To function effectively, it is important that the team's decision-making hierarchy be established early in the project. The CM's and A/E's executive level and management level persons are usually designated prior to the brainstorming session, and it is assumed that the owner's executive and management level selections will be made by the brainstorming session or shortly thereafter.

Although it is common for an owner to have the CM and A/E identify their executive and administrative level persons prior to signing contracts with either, it is not always possible to identify the A/E's and CM's administration persons until design is

Level	Function Area	Decision Responsibility	Monetary Responsibility
Executive Principal of the Firm	Policy-making for the Team and Team members.	The highest Team decision level; the next step is Dispute Resolution. Makes all policy decisions.	Limited only by Company policy. Can amend budget; approve extra expenditures.
Management Representative in responsible charge of the project	Convert Team and Team member policies to procedures. Lead Team representative.	Intermediate Team decision level. Generates all project procedures. Makes/Approves all project decisions. Attends all Team meetings where decisions are made.	Limited by Company and Team Policy. Usually set at a maximum figure by the Executive Level. This figure can replace but not supplement the figure set for the Administrative Level.
Administrative Field Representative	Execute Team and Team member procedures.	Day-to-day decision level during construction. Passes Team decisions to the Management Level.	Limited by Company and Team Policy. Set at a maximum figure less than set for the Management Level.

FIGURE 12.1 The team's decision-making hierarchy.

underway, due to the comparatively long design phase time. However, their involvement in the project need not commence until design is almost complete.

The advantage in identifying the Level 1 and 2 project team members early is access to their ideas and input at the organizational meeting when the responsibility chart is developed. The Level 1 and 2 persons from all team members should be involved in that meeting because the chart deals extensively with the decision-making process.

Establishing the decision-making hierarchy need not be postponed until team members are identified. The hierarchy can be established academically if necessary.

12.4.1 Establishing Decision Responsibilities

Decision responsibilities are on the agenda of all early program management meetings until all foreseeable decision-making responsibilities have been agreeably assigned.

As the project moves ahead, it is probable that responsibility assignments will be added or altered. Consequently, the decision-management plan should include recording decision-making assignments as well as making or changing assignments. The plan should become a part of the CM project manual for future reference.

Decision-making responsibilities should be specific and as clear as possible. This is especially true regarding monetary limits and the kinds of decisions made. There must be a clear understanding between team members as to level-appropriate discussions and which decisions should be passed to the next level. The goal should be to make all decisions at the lowest possible level without violating rules of hierarchy.

12.4.2 Recording Decisions

The number of decisions that will be made throughout the course of the project is enormous. There will be far more dynamic decisions than static decisions. Because it will be impossible to formally record all dynamic decisions, recording must be left to the prerogative of the individual(s) that makes them. However, all static decisions should be documented as part of the appropriate meeting minutes. Particular attention should be paid to decisions that impact the progress of the project and the interaction of the project team.

The number of meetings on a CM project is substantial; the team concept and the requirements of static decision-making form the basis. Most meetings require that all members of at least one hierarchy level be in attendance. Consequently, decisions made at meetings will automatically be recorded in the minutes of that meeting and be made available to team members at that level.

If a team member feels that decisions made at one level should be passed on to another level(s), it is the responsibility of the team member to see that the information is passed on to the other level.

12.4.3 Exceptions

Although all three parties should be involved in all static decisions, bilateral decisions can be made if the third party would have no apparent reason to disagree with the decision and there is no need for the decision safety net of checks and balances. Bilateral static decisions might develop during a telephone conversation or at an informal meeting. They require a special effort to inform the missing party of the bilateral static decision and the conditions under which it was made. One of the team members involved in the decision should be designated as the informant.

Bilateral decisions made via correspondence present no difficulty if the information management plan requires that copies of each team member's project correspondence be forwarded to the other two members. In this instance, the copy alerts the third party to the bilateral decision.

12.4.4 Decisions and Schedule Adherence

Timely decisions are vital if the project schedule is to be maintained. A firm commitment must be made by each team member to adhere to the decision schedule. It must be accepted that time lost during the course of the project cannot be recovered without the owner incurring costs in some form. When static decisions are due, they must be provided unless the reason for the delay is in the best interest of the owner.

Activities in the early part of the program schedule are decision oriented. It is during this phase of the project when an enormous amount of information must be exchanged and numerous important static decisions must be made by the team. A special dedication to commitment and promptness must be made by all.

The CM must inspire a sense of decision-making urgency in the team and then support and maintain it by example. As the designated scheduler, the CM should be seen as a fair-handed facilitator, one who maintains a sharp focus on decision-making.

The CM should never lose sight of each team member's professional commitments beyond the project at hand or the effort expended by those who provide checks and balances to static decisions. To this end, the CM must persistently draw each team member into the decision-scheduling process.

12.5 THE DECISION MANAGEMENT PLAN

The decision management plan is neither lengthy nor involved, but it is necessary both as a reminder to team members of the importance of the decision-making process and a guide to that process.

The decision management plan should stipulate the types of decisions that can be made and a suitable definition of each decision for common understanding by team members.

A statement defining the difference between advice and decisions should be made. A simple procedure for placing static decisions on the agenda for program schedule and project meetings should be stated. A static-decision file should be established and a procedure written to record decisions. The criteria for bilateral static decision-making should be clearly stated and a procedure established for recording these decisions.

A chart that establishes the decision-making hierarchy (similar to Figure 12.1) should be created.

CHAPTER 13

Information Management

The CM contracting structure and its assignment of contractual responsibilities generates an unprecedented amount of project-related information. Its synergistic strategy for delivering a project mandates a flow of information between project team members. This combination could inundate participants with dispensable information and prove counterproductive, unless controlled.

Concomitantly, the integrity of the CM system is founded in openness, a timely exposure to information that facilitates the checks and balances of effective decision making. Chapter 12, Decision Management, dealt with the decision-making process, emphasizing a need for team action rather than individual action and promoting synergism and discouraging unilateral decisions. This chapter identifies information that is specifically necessary to manage a CM project successfully.

The owner, A/E, and CM, have the exact same objectives: (1) to design the project to meet the owner's needs, (2) to construct the project in conformance with the contract documents, and (3) to accomplish both tasks with minimum interruption to the owner's business activities. However, they may not have the same approach to reaching those objectives without a predetermined plan, especially in the area of information management.

13.1 THE INFORMATION MANAGEMENT PLAN

Sharing information is essential to a team's performance. However, if the team is large and dispersed, it is not practical for all information to reach all team members concomitantly, nor is it possible for every team member to productively use all the information available. Consequently, decisions should be made early in the project regarding the distribution of project information.

As with all other management areas, information management procedures should be documented in a management plan and incorporated into the CM project manual. The information management plan should clearly state: (1) the information to be gathered, (2) who is responsible for gathering it, (3) in what form will it to be presented, (4) how it will be disseminated, (5) who is to receive it, and (6) how, where, and for how long will it be filed and stored.

13.1.1 The Need-To-Know Philosophy

The large volume of information produced by the team concept must be handled prudently but expeditiously. Productive team actions are dependent on complete and timely awareness of the facts and data available for the team decisions that must be made.

Ideally, all three team members should be involved in every project decision. Bilateral decisions are more acceptable than unilateral decisions but neither is desirable. Bilateral and unilateral decisions will be necessary under the operational conditions that prevail, and when they are, consideration should be given to the team member(s) who could not participate by passing on the decision and its rationale as soon as possible.

To achieve the level of communication that produces consistently good team decisions, the information available from all sources must be harnessed. The decision management plan should prioritize decisions categorically and guidelines should be set for decisions which may be made without full team involvement. This is not a complete solution because situations are bound to arise that will not fit the guidelines, but it is a positive step.

A need-to-know philosophy best serves the management of information in the CM team structure. This philosophy can only be attained by common commitment to a way of conducting project business (in the current vernacular, partnering between team members).

Need-to-know departs from more common information philosophies which flood those in the loop with all available information, letting each recipient determine its importance. This is common practice in businesses of all types. It provides an illusion of control while burdening managers with data, much of which is not relevant to immediate concerns.

Certain concepts of total quality management (TQM) provide an analogy that fits perfectly.

As in TQM, each team member's information is received as a customer, analyzed as a processor, and forwarded on to the next party as a supplier. As a customer, the team member wants the information received to be concise, accurate, germane, and in useable form. As a processor, the team member is expected to refine and appropriately comment on the information; and as a supplier, the team member is expected to package the information to the satisfaction of its recipient.

Need-to-know expects suppliers of information to discreetly determine importance and urgency. Recipients should be confident that suppliers only forward accurate information and do so on a timely basis. Suppliers, committed to team allegiance and responsive to the mores of team participation, select need-to-know recipients prudently. Those team members responsible or qualified to act, communicate, make the decision, document it, and inform those who were out of the loop as soon as possible.

The need-to-know information philosophy works effectively on CM projects because of the common goal of the team and the fact that all decision-makers are peers. Neither the A/E nor CM have a financial stake in the project to influence their decisions but do have their reputation to lose by providing inaccurate or untimely information. The checks and balances umbrella that protects against irresponsible actions provides additional assurance of information and decision credibility.

The information management plan should specifically address areas where all team members are involved in the decision-making process. All other information areas should rely on the need-to-know philosophy.

13.2 INFORMATION MANAGEMENT AREA OF KNOWLEDGE

The information management area of knowledge encompasses the collection, documentation, dissemination, safe keeping, and disposal of verbal and graphic project-related information. The team structure and the use of multiple contracts significantly increases the information available to the owner. The volume of information (generated for project accountability purposes and by team member participation in decision-making checks and balances) requires a multi-level, need-to-know reporting structure and an efficient information storage and retrieval system.

The construction manager must be able to properly and effectively manage the information generated by the project. The CM must be familiar with all forms and means of communications, especially computer-based information systems, and will recommend and install those most appropriate in all areas for each level of management. The CM must be proficient in setting agendas, chairing meetings, recording meeting minutes, presenting oral and written reports, and perceiving the need-to-know requirements of team members at each management level.

The information management area of knowledge requires high-level communication skills, including personal conversation, correspondence, technical writing, meeting leadership, note taking, meeting recording and reporting, report form design, knowledge of information management systems, business protocol, and computer systems, and a strong sense of ethics.

13.3 THE KNOWLEDGE CIRCLE

The flow of project information should facilitate the team's structure for decision management covered in Chapter 8, The CM Organization. In general, information dissemination should be lateral, between project team members, and vertical within a team member's organization. There will be exceptions for this practice and situations when the management plan dictates otherwise. However, the integrity of the information management network hinges on consistency and conformance to the need-to-know philosophy.

Figure 13.1 shows the lateral and vertical flow of project information for a typical project. Executives communicate with executives, administrators with administrators, and management level persons with counterpart management level persons. Vertical

Level	The CM	The A/E	The Owner
Executive	Executive	Executive	Executive
Management	Project Manager	Project Architect/Engineer	Management Representative
	Value Managers Value Engineers	Architects Engineers	Engineers Technicians
Administrative	Field CM	Field A/E	Field Representative

FIGURE 13.1 The lateral and vertical flow of information.

communication should be confined to each column. Ignoring this structure by taking shortcuts, either advertently or inadvertently, will progressively undermine team morale and cripple team performance.

An important rule of information management is to be certain that all information is available to all parties in a timely manner. If the decision-management plan exempts a team member from primary involvement in a decision, the information management plan should stipulate that copies of meeting minutes, reports, memos, or correspondence be forwarded to the excluded team member.

Surprises should be avoided under all circumstances. A comment such as "I wasn't aware of that" will quickly undermine the synergistic goals of the team (providing, of course, that it was a breakdown in communication that prompted the comment). If information is properly managed, surprises will seldom occur. It should be evident that information must be micromanaged, especially early in the project when the need-to-know philosophy has not yet taken hold.

13.4 INFORMATION CREDIBILITY

The checks and balances of the CM team structure should not be relied upon to establish the credibility of information. The prime purpose of checks and balances is to extract the highest level of performance from each team member, not to correct mistakes in information provided.

As a processor and supplier of information, each team member is expected to perform to the expectations of the other team members. Information is expected to be correctly and completely processed before supplying the information to the recipient.

The CM's, A/E's, and owner's resource people will process most of the information based on data and direction provided by the A/E's Project Architect/Engineer, the CM's Project Manager, or the owner's Management Representative. These three managers should be held accountable for the credibility and timeliness of the information provided to the team.

13.5 COMMUNICATIONS

In the early days of CM, communication was limited to meetings, the mail, telephone and telegrams; time consumed by communications was a sizeable handicap when dealing with information transfer. Today, with so many communication options available, it is only a matter of selecting the method best suited to accomplish the task at hand.

Meetings are by far the best way to communicate when discussions must take place. Teleconferencing is a convenient option but should only be used when the meeting agenda is brief and the subject matter fairly noncontroversial. Although it reduces time and eliminates travel, teleconferencing is more formal and tends to stifle the uninhibited flow of information that should take place during confrontations. Queries by team members prompted by checks and balances (that are so important to fundamental decision making) will be more candidly stated during face-to-face encounters.

While the telephone provides the quickest access to information, the FAX automatically provides a record of what transpires. The FAX should replace the telephone whenever possible and should be used to back-up telephone discussions by capsulization. In the latter case, a memo or letter will also suffice.

Information transfer options should be chosen wisely to ensure timely delivery and minimize duplication of effort. The information management plan can only provide guidelines in this area; therefore, prudent individual choices must be relied upon.

A similar situation exists in scheduling. It is noted in Chapter 20, Schedule Management, that CPM scheduling is a computerized tool with a much greater capability than required for effective CM scheduling. It is noted that there is a tendency to overuse scheduling software simply because the capability is there.

The most prudent information transfer method should be used. There is a tendency to schedule the information processing so that the fastest and most expensive method of transfer must be used to meet deadlines. As with scheduling, available elaborate options seems to influence choice. While certainly not a material threat to the budget, the cost of last-minute transfer is an indication of inefficiency and a questionable attitude toward project costs. While this is a minor point, it is flagrantly contrary to CM philosophy.

13.5.1 Computer Communication

It is difficult to visualize a contemporary CM project without computers. They are essential to scheduling and a major convenience when budgeting or making progress payments. They also become indispensable in communication and record keeping. When properly used, computers can serve as the prime communication link between team members. However, as with scheduling, computers can be overused. While the ideal of having a computer on everyone's desk has merit, their use should be carefully controlled if efficiency is to be realized.

The project team should determine how computers will be used in communications. It is suggested that project communication remain private through the use of local area networks and modem transmissions. Project communications should not occur on the Internet unless it is necessary or deemed prudent.

Communications within the CM's organization should also be controlled. The job site should be linked with the main office to facilitate communications between management and administrative level persons. However, the information accessible to the field should be selective; the information accessible to the office should not be selective.

On projects that have a sizeable field-based staff, a direct link between the field and the offices of the A/E and owner is justified. On most projects where field-based personnel consist of one or two persons, it is not justified. The lines of communication established in the information management plan should determine direct link justification.

There is no single computer communication system that will serve all projects equally well. Each project should devise its own system based on the computer capacities of the team members and the type of communication that will take place. The system should be detailed in the information management plan. If sufficient forethought is put into the decision and the resultant plan is followed, the system will serve its intended purpose.

13.6 CORRESPONDENCE

Two suggestions could make the CM's correspondence involvements more efficient. The first facilitates identifying each project's correspondence; the second provides a convenient way for project managers to monitor outgoing correspondence from resource personnel. Both are convenient and economical.

Construction managers can use a color-code system to quickly identify paperwork connected with a specific project. Self-adhering tags or marker strokes can be applied to all internal and incoming paperwork and file folders. Each project is assigned a different color, making identification much simpler than reading multi-digit job numbers.

To monitor daily outgoing correspondence, a daily letter file can be created. This is done by producing an extra copy of each outgoing letter on unique colored paper. At the end of each day all copies are stapled together, marking the date noticeably on the first page and routing them to those in the need-to-know circle for company correspondence.

Correspondence monitoring is especially useful to Level 2 managers on a daily basis and to Level 1 executives on a spot-check basis. Notes and memoranda can be made directly on the letters because the file is not part of the formal record-keeping system. After the daily letter file has served its prime purpose, it can be kept in a chronological file for future reference or destroyed if considered dispensable.

There are more elaborate systems that can replace the two simple ones suggested here (such as creating computer files rather than using hard copies) but the portability and convenience of a hard copy, daily letter file makes the suggested system difficult to beat.

13.7 FORMAL REPORTS AND RECORDS

The information management plan should identify the major reports to be provided but should refer to each separate management plan for reporting details. The major ongoing reports and records are listed below. There will be periodic, secondary, and back-up reports as well as those listed.

Budget Management

Monthly Financial Report	The primary project report containing the following documentation.
Application Summary	A month-by-month listing of project payments made by the owner to date.
Dispersement Report	A listing of disbursements made by the owner for the current period.
Schedule of Values Report	A summary of schedules of values updated to the current pay period.

Billing Summary	A current status report on each contractor's dollar contract.
Budget Report	A budget summary showing costs to date and anticipated in the future.
Change Order Report	A summary of Change Orders to date, including contingency transfers.
Purchase Order Report	A listing of Purchase Orders issued by the owner for the current period.
Contract Management	
Daily Field Report	An objective report on the day's activities, including conditions and comments.
Decision Management	
Major Decision Log	A log of major decisions made by the team during the project.
Information Management	
Minutes of meetings	Minutes of signal meetings that take place during the project.
Material/Equipment Management	
Expediting Report	A report on the whereabouts of material and equipment purchased for the project.
Warranty/Guarantee Report	A listing of all warranties/guarantees with expiration dates.
Project Management	
Monthly Meeting Report	A report supporting the topics on the agenda for project monthly meetings.
Quality Management	
Punch List Report	A status report on contractor punch list items.
Resource Management	
Contractor Manning Report	A daily report identifying contractors on-site and their crew make-up and size.
Risk Management	
Insurance Report	A monthly update of the status of contractor's required insurance.
Safety Management	
Lost Time Accident Report	A monthly update on lost time accidents, including types and causes.

Schedule Management

Occupancy Schedule Report	A current update of the occupancy schedule and its interface with construction.
Milestone Schedule Report	A demand (daily, weekly, monthly) update on milestone schedule progress.
Program Schedule Report	A demand (daily, weekly, monthly) update on program schedule progress.
Short Term CAP Report	A periodic (milestone date) update on construction progress.

Value Management

Value Management Report	A monthly report on value management activities during design and construction.

The names or titles of the reports are not cast in stone, and the list is neither a minimum nor a maximum requirement. The budget management monthly financial report is a product of the team's financial management information system (FMIS) diagrammed in Figure 13.2 on page 196. It should be presented in a format acceptable to the owner, one that coincides with the owner's accounting requirements.

13.7.1 The FMIS System

A computerized FMIS can be designed using spreadsheet software. The sophistication depends on the computer capabilities of the construction manager. These general rules apply: (1) develop the FMIS program so that manual entry need only be done once for each item of data. This reduces the chance for error, and speeds up the process; (2) create files that cross-check each other wherever possible. This will flag errors if they occur; and (3) provide reports in the least complicated format so that effortless communication is facilitated.

13.7.2 Formatting Reports

Figures 13.3 through 13.9 show typical formats for the seven reports and summaries included in the Monthly Financial Report. It is suggested that the monthly financial report be mechanically bound so its credibility and permanence is assured. On many projects, this report provides the only source of accountability and is the owner's prime record for the project.

Figure 13.3 is an Application Summary format used on a private sector project. (The format as shown was prescribed by the owner.) Information pertaining to change orders and retainages are found on other reports in the monthly financial report.

On projects that utilize standard contract documents, the Application Summary page would be replaced with a standard application and certificate for payment form that is part of the family of contract documents used on the project. (The standard application and certificate for payment forms include information on retainages and change orders.)

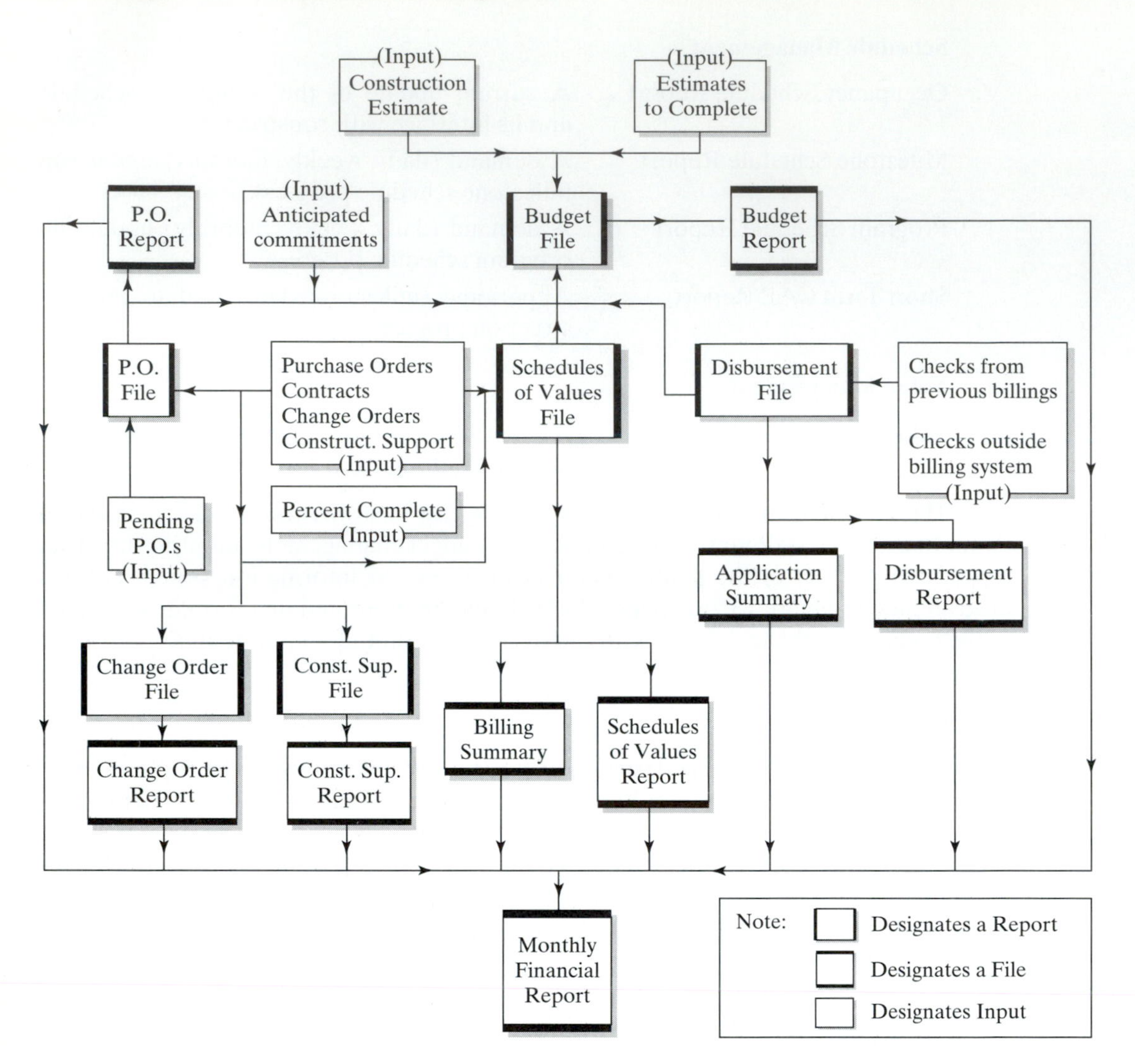

FIGURE 13.2 Diagram of a typical Financial Management Information System.

For added owner convenience, a dispersement report is sometimes included that lists the names of current progress payment recipients and the amount to which they are entitled. This page can be enhanced by including the mailing addresses of the recipients and the purchase order number for each contract. The goal is to separate the contractor progress payment process from the financial reporting process and make issuing progress payment checks as simple as possible for the owner. Back-up information can be found in the included reports.

The column format for the report in Figure 13.4 is optional. The example uses a column-stacking format to save report width. If width is not a problem, each heading can have a column of its own, and the page can be printed broadside (landscape). The last page of this report will show the totals for the project, and the total of the certified

Monthly Financial Report No. 07

Application Summary

Project No. ______ Project: ________________________
Owner: ________________________

Date: 07-01-96 Project Number: ______
Time: _______ Application No. 07 Page ____ of ____

Application No.	01	01-12-96	$ 378,944.69
Application No.	02	02-09-96	103,252.30
Application No.	03	03-15-96	103,531.03
Application No.	04	03-31-96	183,693.37
Application No.	05	05-10-96	310,990.28
Application No.	06	06-12-96	585,292.18
Application No.	07	07-19-96	170,152.45
Total Applications to Date:			$1,835,856.30

Payment Authorization:
Owner's Representative: ______________________ Date: __________
Architect's Representative: ______________________ Date: __________
Construction Manager's Representative: ______________________ Date: __________

FIGURE 13.3 Typical Application Summary.

net amount payable will be the amount of the current application on the application summary. In Figure 13.4:

Column (1) identifies the work-scope number.

Column (2) identifies the work-scope description and contractor.

Column (3) lists the estimated final budget amount including change orders and the balance remaining in the budgeted amount.

Column (4) shows the amount earned to date and the amount retained to date from that amount.

Column (5) lists the percent earned to date and percent to be retained.

Column (6) lists the previous amount billed, previous amount retained, and the amount certified for payment.

Column (7) identifies the current billed amount, current retained amount, and stored material.

Column (8) shows the current amount certified as payable to the contractor.

Column (9) lists the certified net payments made to the contractor to date.

Figure 13.5 is a format for a typical budget report similar to Figure 10.7 in Chapter 10, Budget Management.

Figure 13.6 is an example of a change order report or, more descriptively, a Change Order/Contingency Report. It is a recap of change orders appearing on other reports with added information for tracking contingency expenditures, transfers, and balances. This report is usually of considerable interest to owners. As noted in

Continued on page 201

Monthly Financial Report No. 07

Billing Summary

Project No. ______ Project: ________________________

Owner: ________________________

Date: 07-05-96 Project Number: ______

Time: _______ Billing Summary No. 07 Period Ending: 06-12-96 Page ____ of ____

Work-Scope No.	Work-Scope Description and Contractor	Est. Final Budget Amount - - - Balance Remaining	To Date: Completed - - - Retained	Percent Complete - - - Percent Retained	Previous: Billed - - - Retained - - - Certified	Current: Billed - - - Retained - - - Stored Materials	Current Certified Net Amount Payable	Certified Net Payments To Date
99	CM FEE	431,600.00	252,984.75	58%	239,048.75	13,936.00		
	T. D. & H. Limited		0.00	0%	0.00	0.00	13,936.00	252,984.75
	424 Smith Avenue	178,615.25			239,048.75	0.00		
	Evermore, MI 49931							
07	CONCRETE	200,822.18	177,647.18	88.4%	176,734.00	913.18		
	Johnson Contractors, Inc.		8,882.36	5.0%	8,836.70	45.66	867.52	168,764.82
	624 W. Cross St.	23,175.00			167, 897.30	0.00		
	Evermore, MI 49931							
60	CHAIN LINK FENCE	7,100.00	7,100.00	100.0%	7,100.00	0.00		
	Moran Fence Co.		0.00	0%	0.00	0.00	0.00	7,100.00
	222 E. 35th St.	0.00			7,100.00	0.00		
	Evermore, MI 49931							
65	BITUMINOUS PAVING	122,623.00	0.00	0%	0.00	0.00		
	Carter's Blacktop Inc.		0.00	10%	0.00	0.00	0.00	0.00
	1648 W. 10th St.	122,623.00			0.00	0.00		
	Evermore, MI 49931							
(1)	(2)	(3)	(4)	(5)	(6)	(7)	(8)	(9)

FIGURE 13.4 Typical Billing Summary.

Monthly Financial Report No. 07

Budget Report

Project No. ______ Project: ____________________

Owner: ____________________

Date: 07-05-96 Project Number: ______

Time: ______ Billing Summary No. 07 Period Ending: 06-12-96 Page ___ of ___

Work Scope No.	Work-Scope Description	Previous Cost to Date	Current Cost to Date	Cost for the Period	Estimated Cost to Complete	Estimated Final Cost	Budgeted Amount	Variance (+/-)	Adjusted Budgeted Amount
99	CM FEE	239,048.75	252,984.75	13,936.00	178,615.25	431,600.00	431,600.00	0	431,600.00
07	CONCRETE	176,734.00	177,647.18	913.18	23,175.00	200,822.18	200,822.18	0	200,822.18
60	FENCE	7,100.00	7,100.00	0.00	0.00	7,100.00	7,100.00	0	7,100.00
65	BITUMINOUS PAVING	0.00	0.00	0.00	122,623.00	122,623.00	122,623.00	0	122,623.00
		X	X	X	X	X	X		X
TOTAL THIS PAGE:		422,882.75	431,731.93	14,849.18	324,413.25	762,145.18	762,145.18	0	762,145.18

FIGURE 13.5 Typical Budget Report.

Monthly Financial Report No. 07

Change Order Report

Project No. ______ Project: ________________________

Owner: ________________________

Date: 07-05-96 Project Number: ______

Time: _______ Change Order Report No. 07 Period Ending: 06-12-96 Page ___ of ___

Work Scope No.	Line-Item No.	C.O. Amount	Work-Scope Contingency Balance	Design Contingency Balance	Interface Contingency Balance	Total C.O. Amount	Notes
			40,000	52,000	18,000		
06	0301	−1,020	X	53,020	X	−1,020	Moran Brothers
17	0302	+975	39,025	X	X	−45	Thompson, Graff Inc.
21	0302	+1,005	38,020	X	X	+960	Altman Inc.
06	0303	+1,650	X	X	16,350	+2,610	Moran Brothers
14	0304	+2,080	35,940	X	X	+4,690	Dorson Excavators, Inc.
21	0304	+650	35,290	X	X	+5,340	Rapid Steel, Inc.
22	0304	+842	34,448	X	X	+6,182	Regal Concrete Co.
						X	
C.O.s Report No. 01:						+6,182	
Totals to Date:			34,448	53,020	16,350	+6,182	
37	0305	+2,650	X	X	13,700	+2,650	Gordon Electrical Inc.
42	0306	+480	X	52,540	X	+3,130	Levin Structures Ltd.
						X	
C.O.s Report No. 02						+3,130	
Totals to Date:			34,448	52,540	13,700	+9,312	
28	0306	+14,280	20,160	X	X	+14,280	Martin Blacktop Co.
27	0306	+6,430	13,738	X	X	+20,710	Quigley, Stone & Sons
42	0307	+3,200	X	49,340	X	+23,910	Leven Structures Ltd.
						X	
Report No. 03:						+23,910	
Totals to Date:			13,738	49,340	13,700	+33,222	
Note (a)		X	+10,000		−10,000		Note (a) Transfer, 10,000: I.C. to A.C.
46	0322	+12,450	X	36,890	X	+12.450	Baker Ceilings, Inc.
						X	
Report No. 07:						+12,450	
Totals to Date:			23,783	36,890	3,700	+45,672	

FIGURE 13.6 An example Change Order/Contingency Report.

Chapter 10, contingencies are often a topic of controversy. Owners commonly question their justification and the necessity of their eventual depletion.

A brief description of each numbered Change Order (C.O.) should be available in the Change Order file by entering the C.O. number. On small projects the description of each C.O. could be included directly on the change order report.

Figures 13.7 and 13.8 provide detail and documentation for the Monthly Financial Report. Figure 13.7 shows a Typical Disbursement Report and Figure 13.8, a typical Schedule of Values Report.

A Purchase Order (P.O.) report is only required on projects where the owner's accounting system requires purchase orders be used to initiate every purchase including construction contracts and on projects that have a high incidence of direct owner purchases of equipment, furniture, and other items not installed by work-scope contractors. A P.O. file is unnecessary but helpful on other projects, especially for construction support items.

The P.O. file is an accumulated listing of purchase orders issued by the owner that can be sorted by number, vendor name and address, and date of issue. The P.O. Report lists the P.O.'s issued during the interval covered by the current monthly financial report. The file provides P.O. data to other reports as well as the Purchase Order Report. An example of a Purchase Order Report is shown in Figure 13.9.

The alpha-numeric designations under the work-scope column in Figure 13.9 identify work-scope accounts other than contracts with contractors for construction services. The "CS" identifies construction support items, "OF" identifies owner-purchased furnishings. "OQ" could be owner-purchased equipment; AE might be used for the A/E fee. Whatever coding system is decided upon should be logical and consistent, and the codes should be listed in the information management plan in the project's CM manual.

Work-scope coding consistency facilitates sorting project information for use in various reports and makes them less of a chore to produce. A logical, consistent coding system and well-developed reports provide checks and balances which substantiate the accuracy of monthly financial report information.

13.8 COLLATERAL INFORMATION

In obtaining and documenting the information necessary to administer a CM project, the contracting structure produces information that has the potential to raise the conceptual estimating performance of construction managers another notch.

The detailed project information and competitive work-scope cost data available to CMs on multiple-contract projects provides an unprecedented estimating resource. When diligently collected and appropriately stored in a data base, these data provide the best source for conceptual and construction estimating that the construction industry has produced (and all to the owner's benefit).

In the GC system, competitive subcontract cost information never becomes available to the A/E; consequently, this highly desirable information is not available when the owner is trying to establish a predesign budget or when the A/E is putting together a conceptual estimate.

Continued on page 203

Monthly Financial Report No. 07

Disbursement Report

Project No. ______ Project: ______________________
Owner: ______________________

Date: 07-05-96 Project Number: ______
Time: _______ Disbursement Report No. 07 Period Ending: 06-12-96 Page ____ of ____

Disburse To	Disbursement Amount	Work-Scope Item No.	Authorization	Notes
APPLICATION SUMMARY 07:				
T.D. & H. Limited 424 Smith Avenue Evermore, MI 49931	13,936.00	99	CTR 099	
Johnson Contractors 624 W. Cross St. Evermore, MI 49931	867.52	07	CTR 07	
Robins Mechanical 254 E. 12th St. Evermore, MI 49931	54,600.80	35	CTR 035	
(Etc.)	100,748.13 X			
APPLICATION 07 TOTAL:	$170,152.45			
Construction Support Report 07				
Porta-Jon, Inc. 422 E. 12th St. Evermore, MI 49931	24.00	CS02	PO 3457	05/31/96 to 06/28/96
Golden Light & Power Co. P.O. Box 4444 Golden, MI 49932	364.38	CS11	PO 2785	05/29/96 to 06/28/96
Contract Testing, Inc. 655 8th St. Micron, MI 49932	684.45	CS24	PO 0974	Concrete Cylinders
(Etc.)	3,145.80 X			
CONSTRUCTION SUPPORT 07 TOTAL:	$4,218.63			

FIGURE 13.7 An example Disbursement Report.

Monthly Financial Report No. 07

Schedule of Values Report

Project No. ______ Project: ________________________
Owner: ________________________

Date: 07-05-96 Project Number: _______
Time: ________ Schedule of Values Report No. 07 Period Ending: 06-12-96 Page ____ of ____

Item	Description	C.O. Amount	% Complete Prev.	% Complete Cur.	Budget Amount	Earned To Date	Previous Earnings	Earned For Period
15010102	Vinyl Tile		0	0	10,000	0	0	0
15010206	Ceramic Tile		0	0	12,000	0	0	0
15010303	Partitions		50	70	115,000	80,500	57,500	23,000
15010408	Acoustical Ceilings		15	15	16,800	2,520	2,520	0
15010508	Gyp Board Ceilings		0	25	7,800	1,950	0	1,950
15010603	Acoustical Walls		0	10	2,400	240	0	240
15010712	Housekeeping		0	0	9,217	4,500	3,000	1,500
15010823	Final Cleanup		0	0	3,000	0	0	0
15010930	Punch List		0	0	6,400	0	0	0
15011005	Decorative Channel		0	0	3,000	0	0	0
15031105	Decorative Channel	900	0	0	0	0	0	0
15031201	Skylight Liners	1,400	40	100	0	1,400	560	840
15031303	Relocate Bulkhead	138	50	100	0	138	66	72
15031403	Add Stud Line	119	100	100	0	119	119	0
15031502	Deduct Vinyl Tile	(300)	0	100	0	(300)	0	0
15031603	Add Bulkhead	400	100	100	0	400	400	0
		X			X	X	X	X
Work-Scope Div. Totals:		$2,657			$185,617	$91,467	$64,165	$27,302
C.O. Percent Variance:		+1.43%						
Work-Scope % Complete:						49%	34%	
Estimated Final Earnings:		(+2,657 + 185,617) =			$188,274			

FIGURE 13.8 An example Schedule of Values Report.

Subcontract costs are the exclusive property of GCs who cannot become part of the contracting system until after design is completed (unless of course the construction contract is negotiated). On public projects, negotiated contracts are a rare exception and not the rule. Consequently, the party that really needs this information—the A/E, who can use it to the owner's advantage—has no way of accessing it in the GC system.

To complete this point, in the design–build system, the D–B contractor has this information and can use it in conceptual estimating. However, the independent contractor status of D–B contractors permits them to use the information to their best advantage, not necessarily the owner's. Competitive bidding, under a preponderance of current public law, prevents negotiating construction contracts. This collateral benefit of the CM system strengthens performance under the system if construction managers take the steps required to harness it.

Monthly Financial Report No. 07

Purchase Order Report

Project No. ______ Project: ______________________
Owner: ______________________

Date: 07-05-96 Project Number: _______
Time: _______ Purchase Order Report No. 07 Period Ending: 06-12-96 Page ____ of ____

P.O. Number	Vendor Name and Address	Brief Description of Item	P.O. Amount	Work-Scope	Date Issued	Notes
PO3457	Porta-Jon, Inc. 422 E. 12th St. Evermore, MI 49931	Sanitary service 05/31 to 06/28/96	$24.00	CS02	12/25/95	$24.00/mo for 18 mos.; no price change.
PO2785	Golden Light & Power Co. P.O. Box 4444 Golden, MI 49932	Office power and light 05/29 to 06/28/96	$364.38	CS11	12/18/95	Office trailers, not temporary power. Industrial rate.
PO0974	Contract Testing Inc. 655 8th St. Micron, MI 49932	All materials testing required by specs.	$684.45	CS24	01/11/96	Includes all testing on and off site. See P.O. for pricing.
PO7873	Office Designs, Inc. 862 Bilmore Ave. Evermore, MI 49931	Furnishing for conference room #602	$9,780.00	OF04	06/05/96	Furnishings/ sketch #F-42097. Del 1296 No equipment.

FIGURE 13.9 An example Purchase Order Report.

When establishing the alpha-numeric system for identifying work-scope divisions, it was suggested that an eight-digit number be used. The last two digits of that system should be reserved for internal CM cost tracking. If the numbers from 1 to 99 are used to identify similar work-scopes on every project, and the work-scope bids are filed with the physical construction the bids represent, it will not be long before a superior data base of estimating information is available to the CM for use on future projects.

The raw data will need to be refined, by using cost indices representing geographic areas, dates, and project conditions to bring the data to a common base, but the information could be invaluable when checking or establishing conceptual estimates.

CHAPTER 14

Material/Equipment Management

Material/Equipment (M/E) refers to all manufactured items installed in the project plus processed and raw materials incorporated in the project. Equipment used to install and construct is included in Chapter 18, Resource Management.

The scope and complexity of M/E management varies widely on a project basis. However, even the smallest, simplest project will require M/E management, and many large, complex projects have M/E management requirements that literally dominate the construction manager's project responsibilities.

The M/E to be managed extends from owner-furnished M/E and direct purchase M/E to the M/E provided by trade contractors under their contract with the owner. Owner-furnished and direct purchase M/E management is readily achievable; trade contractor-furnished M/E management can only be accomplished through appropriate owner-contractor contract language and the full cooperation of the trade contractors.

Material/Equipment management provides four important results. It provides a means to: (1) reliably incorporate the owner's M/E requirements into design, (2) permit closer control of the project schedule, (3) extract competitive prices when acquiring M/E, and (4) provide practical follow-up on contractor and supplier M/E warranties and guarantees.

14.1 THE MATERIAL/EQUIPMENT AREA OF KNOWLEDGE

The material/equipment area of knowledge encompasses all activities relating to acquiring material and equipment from specification to installation and warranty. The CM format facilitates direct owner purchase of material and equipment for the project. The advantages (and disadvantages) of direct owner purchases must be evaluated and decisions on direct purchase items extracted from the team in a timely manner. The planning, specifying, bidding, acquisition, expediting, receiving, handling, and storing of direct purchases must conform to the owner's purchasing policies and accurately reflect the requirements of the project schedule.

The CM must be able to identify materials and equipment which are in short supply, have long-lead delivery times, or would provide an economic advantage to the owner for direct purchase items early in the design phase of the project. The CM arranges for their proper procurement, timely delivery, and physical disposition at the site. The CM must expedite owner purchases from their determination of need to delivery at the site, and must be familiar with procurement procedures and strategies and understand the material and equipment marketplace.

The material/equipment area of knowledge includes but is not limited to technical specifications, purchasing techniques, bidding and negotiations, transportation, expediting, inspection, materials handling, storage and warehousing, the Commercial Code, lien statutes, and material and equipment costs and values. Excellent communication skills are required.

14.2 OWNER-FURNISHED EQUIPMENT

On some projects, especially those in the industrial arena, the owner furnishes large and small equipment to be installed, by contractors or the owner's own forces, during construction or as part of the occupancy process. As the owner is the knowledgeable party, these purchases are specified and accomplished by the owner without assistance from the A/E or CM. However, in many cases, this equipment becomes an integral part of the project and must be properly and timely received, inspected, stored, and systematically incorporated into the project. Even if owner-furnished equipment is not part of the contracts for construction, there is an obligation on the part of the CM to see that it does not interfere with construction or the performing contractors. To this end, the CM must obtain delivery and disposition information on all owner-furnished equipment and incorporate consequential information into the project schedules.

In essence, equipment furnished by the owner must be treated by the CM no less attentively than all other material and equipment used on the project.

14.3 OWNER-PURCHASED MATERIAL/EQUIPMENT

Owner-purchased M/E are items that will be installed by contractors as part of their construction work-scopes but which for strategic reasons have been purchased directly by the owner. Owner purchases must be justified in the context of common construction industry practices to provide beneficial results to the owner.

There are only two valid reasons for direct owner purchases: (1) an item has a long delivery time (long-lead item) that requires early purchase to ensure on-time delivery for schedule reasons, or (2) the purchase will provide an economic advantage to the owner without contradicting common construction industry contracting practices at the location of the project.

14.3.1 Long-Lead Items

Long-lead items usually involve extensive fabrication after the order has been placed or time-consuming shop drawing requirements that prevent timely delivery. Non-stock items such as mechanical and electrical equipment requiring team approval and extensive fabrication fall into this category. (A boiler or special air-handling unit is an example of a long-lead item.)

Waiting for a contract to be awarded to a mechanical contractor would put delivery long into the future. After the contract is awarded, the contractor must place the order to a manufacturer and go through the shop drawing process before fabrication can begin by the manufacturer. Adding fabrication time to the delay-in-award time and the shop drawing processing time could cause avoidable construction delays.

The project team must identify long-lead items as early in the design phase of the project as possible and then arrange for timely prepurchase by the owner, usually after obtaining competitive bids from manufacturers or suppliers. The CM should canvass the sources of long-lead items to determine their position in and influence on the construction schedule.

The CM must also specifically assign receiving, care custody, control, and installation responsibilities of long-lead items and include them in contractor work-scope descriptions. Although desirable, assignment is not always made to contractors; there are times when receipt, care, custody, and control are preferably handled as a general condition line-item until the designated contractor is ready to proceed with the installation.

14.3.2 Economic Advantage Items

Some owners in the public sector are exempt from paying sales tax when purchasing materials. However, the owner must pay sales tax on materials provided as part of a construction contract: the owner's tax exemption cannot be passed on to the contractor who makes the purchase.

An economic advantage accrues when material is purchased by the owner and turned over to a contractor for installation. Not only is the amount of the sales tax saved, but open competitive bidding by suppliers usually results in additional savings to the owner. However, the separation of labor and material purchases must not contradict common construction industry contracting practices to be successful.

Most trade contractors on a project use their labor to modify and install raw or processed construction materials on the site. However, some construction materials are prefabricated and require only minor on-site modifications during installation. Raw or processed materials, because they are financially tied to labor productivity and profit, are not suitable for direct owner purchase. Although tied to profit, prefabricated materials are not as closely tied to productivity and are generally suitable for direct owner purchase.

An example of a qualified economic advantage item is structural steel. Steel erection contractors often do not fabricate *and* supply structural steel; steel fabricator contractors often do not erect structural steel. By directly purchasing fabricated steel and acquiring a separate erection contract, the owner saves the amount of the sales tax that an erector would have to pay if supply and erection were combined in a single contractor's work-scope. This arrangement obviously provides an economic advantage to the owner without contradicting construction industry practices.

Other materials which could suitably provide an economic advantage to the owner, after due consideration, are: doors, door frames, hardware, metal deck, open-web joists, casework, cabinetry, carpet, and reinforcing steel.

14.3.4 Competitive Bidding

It is highly recommended that all owner material/equipment purchases be accomplished through the competitive bidding process regardless of whether the project is public or private sector. To this end, the project team must develop technical specifications, coordinated work-scopes, delivery information and schedules, and bidding documents. Information on competitively bidding materials can be found in Chapter 22, Multiple Bidding and Contracting.

14.3.5 Care, Custody, and Control Assignment

Owner-purchased material and equipment must be received, inspected, properly stored, and eventually installed by contractors. The CM is responsible for assigning these responsibilities to contractors in very specific terms. The location of this information is in the work-scope descriptions written for each contract to be awarded.

14.3.6 Contractor-Furnished Material/Equipment

The vast majority of material/equipment will be purchased by the contractors who install it. The few items purchased or furnished by the owner can be tracked closely by the team from the first-hand information obtained from the various vendors. Because the owner usually has no direct access to information regarding trade contractor purchases, the status of contractor-furnished M/E is a concern that must be pointedly addressed in the owner-contractor contract documents.

14.4 M/E MANAGEMENT PLAN

An M/E Management Plan that is acceptable to all team members should be developed by the CM early in the design phase. It should be installed before the initial purchase for the project is made and maintained by the CM during the course of the project. It is terminated after the last M/E item has reached its warranty/guarantee date.

The complex information gathering and sharing, and the documentation requirements of material/equipment management, calls for a management plan to execute this responsibility. A CM's in-house M/E management plan can usually be modified to accommodate team needs. As a management plan, it should be included in the project manual.

The plan should have four main components: (1) a procedure for M/E selection guidance during the design phase, one that assures the availability of the materials and equipment specified for the project; (2) a coordination procedure for reviewing shop drawings, product data, and samples; (3) a material/equipment expediting system, one that accurately and conveniently tracks all purchases by the owner and the contractors from the P.O issuance to the installation of the material/equipment; and (4) a comprehensive installed material and equipment warranty/guarantee follow-up procedure.

The M/E selection procedure should interface with the Quality Management Plan (Chapter 17); the M/E expediting procedure should interface with the Schedule Management Plan (Chapter 20).

14.4.1 Material/Equipment Selection

The main purpose of the M/E selection plan is to exclude material and equipment from the contract specifications that are out of production, have delivery dates which are incompatible with the construction schedule, or are unfamiliar to the available contractors.

It is the CM's responsibility to identify material/equipment that fall in these three categories and bring the facts to the team's attention early in the design phase. To accomplish this, the CM must be familiar with the local contracting establishment, canvass the available suppliers of material and equipment, have a technical understanding of the material/equipment under consideration by the A/E, and provide cost information for alternatives.

As with all decisions pertaining to design, the CM can only assume an advisory role in the selection process; the A/E makes the final determination.

14.4.2 Shop Drawing, Product Data, and Sample Coordination

Figures 17.4a and 17.4b (Chapter 17) show the involvement of a CM in the shop drawing, product data, and sample preconstruction procedures. In this case, the CM is directly involved in the "approval" process and concerned with component comparability in systems situations. However, on projects where component comparability is not a major issue, the CM can delegate the "approval" role to interfacing contractors. The CM's role converts to one of facilitator and coordinator of the review process. The procedure selected should work for both quality management and M/E management without duplication.

Figure 14.1 shows a CM's alternative involvement where the CM delegates an "approval role" to interfacing contractors.

14.5 MATERIAL/EQUIPMENT EXPEDITING

A major function of the CM is to expedite all material/equipment incorporated into the project. The main expediting tool is an expediting report accessible at all times to all team members and contractors. The report should be updated on an information-available basis and founded on reliable information gathered by the CM.

A simple computerized spreadsheet can be used. The report should be construction-schedule oriented, referenced to contractor work-scopes, and crossreferenced to specific items under procurement. Each item should be assigned a mark-number; the contractor procuring the item should be numerically identified within the mark-number as well as the number of the division of work. Sorts should be possible by item number, contracted division of work number, and contractor number.

14.5.1 Expediting Report Column Headings

Typical headings for an expediting report are listed in Figure 14.2 on page 211. Each time the report is updated, the date of the update and the person who did the update should be noted on the form. The column headings should be customized on a project basis.

In use, the expediting report should immediately provide the status of any M/E item on the current date and include: the order status or date placed, promised delivery date, shop drawing submittal date, shop drawing status, shop drawing approval date, terms of delivery, forecasted delivery date, the received date, partial shipment information, back orders, and the condition of the item upon receipt.

Information for the report should be gathered daily by the CM. Owner-furnished and owner-purchased M/E information should be conveniently available due to the CM's direct involvement with the owner. However, information on contractor M/E purchases is voluntary (and unreliable) unless contract provisions require contractors to make this information available in specific terms.

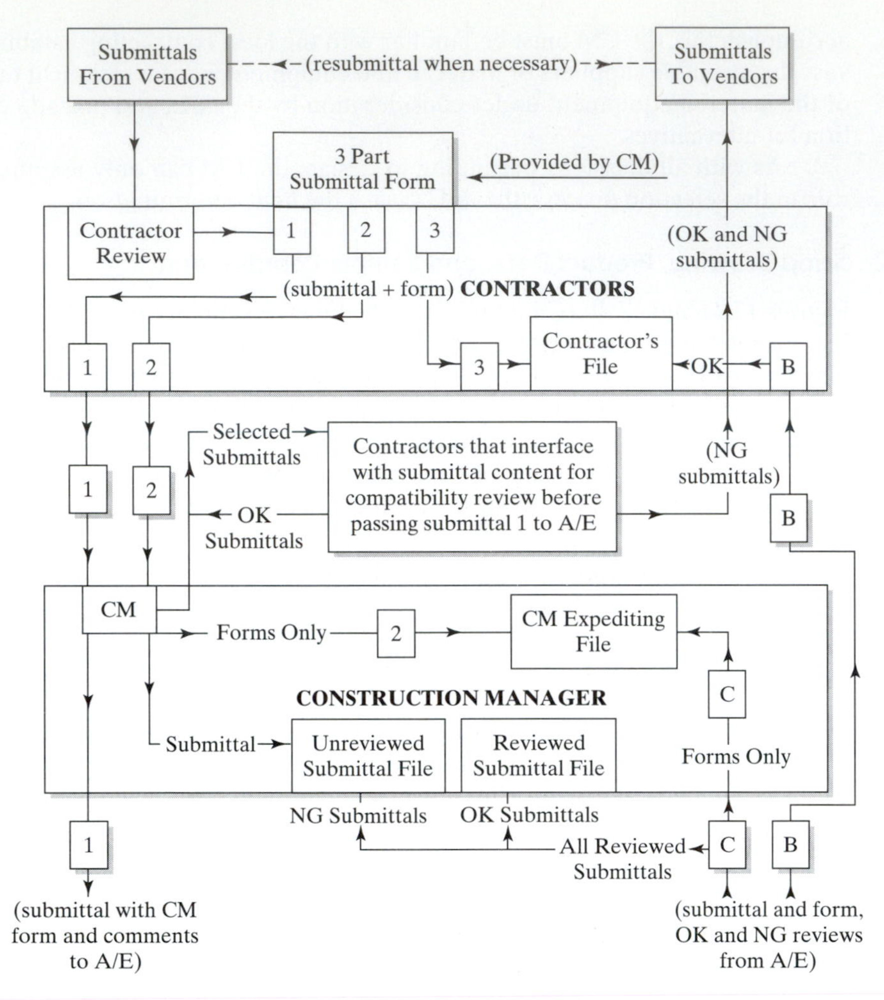

FIGURE 14.1 An alternate preconstruction submittal review procedure.

It is suggested that contractors be required to designate someone in their organization to provide expediting information on all purchases made by the contractor when requested by the CM's expediter. It adds credibility to the information received when contractors are required to forward copies of purchase orders (without pricing) and M/E receiving slips to the CM. The goal is to acquire accurate delivery status information on a current basis for all material to be incorporated into the project.

14.6 M/E WARRANTY AND GUARANTEE FOLLOW-UP

A standard provision of construction contracts (often overlooked, to the detriment of the owner) is the warranty/guarantee stipulation. Warranties and guarantees on labor, material, and equipment seldom get the attention they deserve unless a blatant problem develops. By then, the condition is usually troublesome to the owner and difficult to correct.

Expediting Report Column Headings	
• Item	Description of item
• Item Mark Number	Distinctive number for item
• Work-Scope Number	Installing contrator's W.S. #
• Quantity Required	Each, Tons, CYs, Lbs, Gals, etc.
• Furnished By	W.S.#, Owner, CM, etc.
• Required Delivery Date	According to Short Term CAP
• Status of Order	
• Ordered Date	Date order placed with supplier
• Shop Drawing Submittal Date	Date S.D. initially submitted
• Shop Drawing Approval Date	Date S.D. leaves A/E office
• Delivery Date	Date promised by supplier
• Delivery Method	Truck to site; rail at siding, etc.
• Unloading Party	Unloading contractor's W.S.#
• Custody Party	Party assigned custody at site
• Status of Delivery	
• Shipping Date	Date shipped by supplier
• Delivered Date	Date received at site, siding, etc.
• Condition Upon Delivery	Partial, damaged, O.K., etc.
• Comments	Problems, required action, sources

FIGURE 14.2 Typical column headings for an Expediting Report.

As part of the M/E Management Plan, the CM should establish a formal warranty/guarantee procedure that extracts the most from the contract provisions. The procedure should begin with the installation of the first item under warranty/guarantee and continue until all warranty/guarantee time periods expire, which with some exceptions is within a year after installation or use.

A tool which not only extracts the most from warranty/guarantee provisions but also serves the CM as an excellent client relations tools is a warranty/guarantee tickler file. In generic terms, a tickler file is one that contains time-related information and can be browsed periodically to flag impending events. From a convenience standpoint, the file should be computerized, but a hard copy file is equally effective.

The file (organized by month of expiration) should contain the list of all contracted warranties/guarantees with their start and expiration dates. The location of the actual warranties/guarantees should be noted along with the names, addresses, and contact information pertaining to the contractors and vendors who provided them. The tickler files should be maintained on a project basis.

At the first of each month, the next month's file should be reviewed to determine which warranties/guarantees are due to expire. Letters should be sent to the owner identifying the items and their expiration dates with an advisory to check each item and report any problems. It is assumed that a close inspection might uncover a condition that may otherwise develop into a problem. The CM should offer to appropriately assist the owner if necessary.

Besides the owner's benefit, warranty/guarantee tickler files provide three important benefits for the CM: (1) they demonstrate that the CM is looking after the owner's best interests, even after the project is completed, (2) they provide a legitimate reason for the CM to stay in close contact with owners for sales purposes, and (3) they provide insight into the performance of the material/equipment used in the project for future reference.

CHAPTER 15

Value Management

Value Management is the CM process that provides owners with optimum construction projects at minimum costs. Value management or VM is an extension of the familiar terms, *Value Engineering* (VE), *Value Analysis* (VA), *Life-Cycle Costing* (LCC), and constructability. VM uniquely includes designability and contractability as value influences and interjects budget and cost control, quality, and efficiency factors into value-management decisions.

Of all the services provided by a CM, value management is the one owners wanted most when they prompted the start-up of the CM contracting system in the 1960s. As stated in Chapter 2, The Reasons for a Third System of Contracting, owners suspected that they were not getting a fair return on capital they invested in construction projects. They felt they needed help, in addition to the A/E, to find their way through the maze of a construction project, an ally to champion their practical requirements and a manager to extract the best return from their capital investment. VM services should be geared to accomplish this.

The overall result of value management is impossible to authenticate; it is impossible to provide an owner with a complete, supported list of savings resulting from the VM effort. This is because VM is a continuing intuitive process as well as a periodic initiative, in many respects similar to quality management. Both depend heavily on team intuition and checks and balances to achieve results.

Under the circumstances, the CM should develop a VM plan that not only gets the job done but makes the owner aware that it is being accomplished. The plan should promote the team's intuitive VM efforts and document the periodic VM initiatives where obvious VM progress is made. The owner must be made aware of VM results.

15.1 VALUE MANAGEMENT AREA OF KNOWLEDGE

The value management area of knowledge encompasses a project's cost versus value issue. It has three value components: designability, constructability, and contractability. Designability relates value to overall project design. Constructability relates value to construction materials, details, means, methods, and techniques. Contractability relates value to contracting options, contractual assignments, and contracting procedures. The CM is expected to extract maximum value for the owner from the constructability and contractability options which are available.

In the area of designability, the CM must be capable of extracting the optimum overall design. In the area of constructability, the CM must be capable of securing the owner's prescribed level of value from the design and construction of the project. In the area of contractability, the CM must be capable of translating the owner's goals and the project's characteristics into a contracting structure that will extract the most value from the construction industry. The CM must consider the four major areas of owner concern (time, cost, quality, and business interruption) when recommending courses of action to the team.

The value management area of knowledge includes an understanding of contracting structures, formats, and procedures and the construction industry as a supplier of services. A thorough knowledge of construction materials and equipment, value engineering techniques, and life-cycle cost analysis is essential. Support knowledge in the areas of design, materials technology, estimating, scheduling, and procurement is important. Of major importance are communication skills and ethics.

15.2 VALUE MANAGEMENT DECISIONS

Figure 15.1, Value Management Decision Making, is a flow diagram showing how VM decisions can be made. The terminology used in the diagram requires explanation; some may differ from provincial usage and others are new. VE, VA, and LCC fit their common usage. The accepted scope of constructability has been reduced; the contracting aspects have their own term (contractability). The term designability has been introduced to manifest the CM's involvement in pragmatic design and to emphatically stress the CM's noninvolvement in technical design.

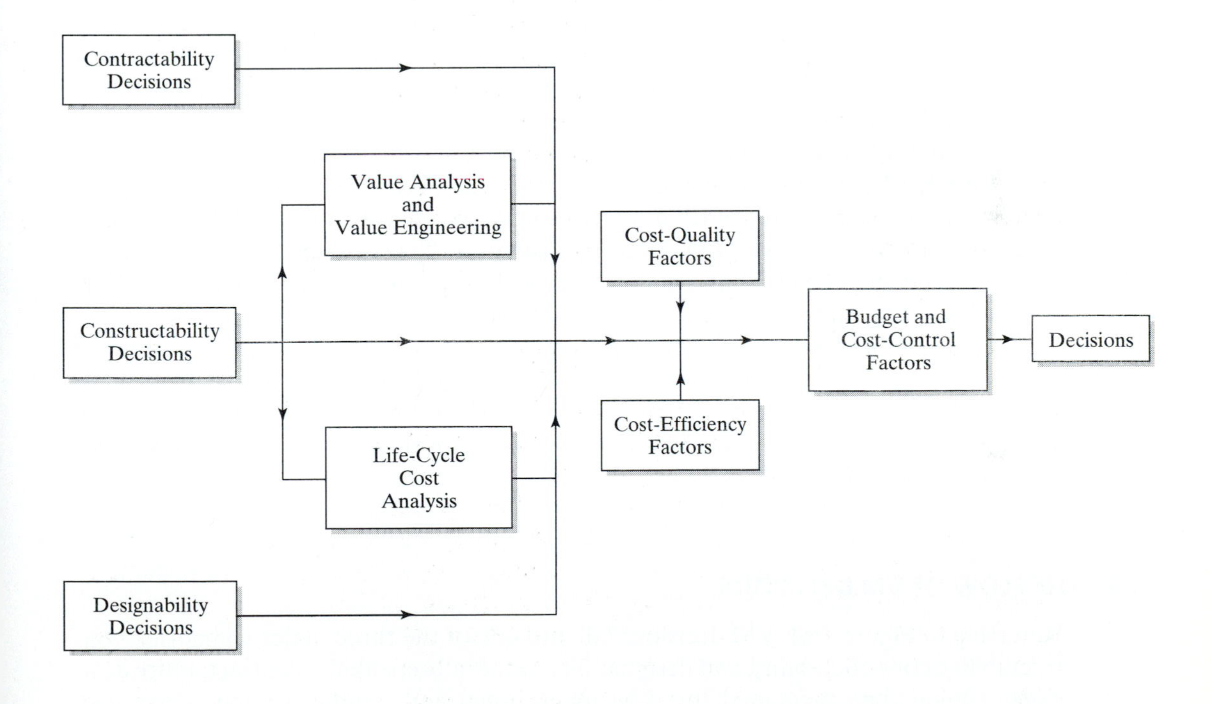

FIGURE 15.1 The flow of Value Management Decision Making.

Value Engineering refers to the practice of reducing project costs before, during, and after the design phase by actively seeking out equally performing but least expensive design alternatives.

Value Analysis is essentially after-the-fact value engineering. It is an objective review of a formulated design or design component to determine if there are ways to improve it to benefit the owner from a physical and economic perspective.

Life-Cycle Costing is comparing design alternatives on a cost-of-construction versus a cost-during-use basis. The results of LCC provide the required information for owner pay-now or pay-later decisions (i.e., should the construction budget be increased or is it more viable to pay higher operating/maintenance costs during occupancy or use).

Constructability is a pragmatic, value-based opinion of the construction features and details designed into the project. Constructability is rated high to low. Features and details with high ratings meet design criteria and can be economically constructed.

Contractability is a pragmatic, value-based opinion of the contractual distribution of the six required service elements of project delivery covered in Chapter 1, The Fundamentals of the Root Form of CM. Contractability is rated high to low. A high rating indicates an economical assignment of contractual responsibility to project participants.

Designability is a pragmatic, value-based opinion of the project's design and major design elements from the perspective of the owner's requirements and aesthetic preferences. Designability is rated high to low. Design and design elements with high designability ratings accurately and economically meet the owner's design requirements.

Budget and Cost Control as presented in Chapter 10, Budget Management, is an integral part of value management. One point of budget and cost control interjection is in LCC decisions where the possibility exists that construction costs could increase to save on future operating expense. The goal of all other VM efforts is to reduce cost.

Cost-Efficiency Factors influence the economic outcome of the project through exercising expedience. Cost-efficiency factors inject time, cost, and schedule consideration into the VM decision-making process. Examples of cost-efficiency queries are the best time to begin construction on a project, or the value of a delivery delay resulting from substituting one piece of equipment for another.

Cost-Quality Factors influence economic project outcome through their ultimate value to the owner. Practically every material or product has a cost-quality factor. Cost-quality factors inject owner preferences into the decision-making process. An example of a cost-quality query is whether to use imported marble for its prestige or local marble for its economy.

15.3 THE FLOW OF VM DECISIONS

Referring to Figure 15.1, VM decisions fall into one of the three major categories: contractability, constructability, and designability. Generally speaking, contractability decisions should be developed by CM operations and administration personnel, constructability or designability decisions by operations and resource personnel. The basis for this breakdown can be found in Chapter 8, The CM Organization, where the talent of the CM firm's personnel is described.

However, when dealing with something as important as VM decisions, all individuals in the CM organization with pertinent expertise should be involved in the development process regardless of departmental ties.

It is the operations management level person's responsibility to handle each VM decision appropriately to get the best results. Intuitive VM decision development might only require informal discussions between the management level and appropriate operations, resource, and administrative personnel. Intuitive VM discussions with the A/E could also be informal and exclude the owner if a decision is within the owner's project parameters. VM initiatives demand a more formal approach because of their size and complexity.

15.3.1 Constructability Decisions

Constructability decisions are based on construction experience. It is common knowledge that many design professionals do not acquire enough field experience to understand the problems connected with construction features and details. Mandatory field experience should perhaps be part of a designer's qualifications but, unfortunately, is not. Consequently, many design features are based on standards and instinct with little consideration for the difficulty and cost to construct them.

Construction managers, especially those with a construction and contracting background, have experience in the field and know the ease or difficulty connected with construction features. Additionally, many CM's have experience on varying project types (hospitals, warehouses, schools, commercial buildings, and civil works) and consequently have a wide range of constructability exposure.

On the other hand, architects tend to specialize in project types which limits their constructability exposure. It is common for a design solution successfully used on a warehouse or civil works to provide equal quality, purpose, and value, at less expense, than one proposed for a hospital or commercial building project.

15.3.2 Compound Solutions

Constructability decisions may or may not require VE, VA, or LCC analysis depending on their complexity but may also involve designability decisions.

For example, changing two 45° masonry corners of a building to one 90° corner will obviously save construction costs but will also increase floor and ceiling area, exterior and internal wall area, building volume, excavation volume, foundation length, and roof area. The loss will be an architectural feature that may affect the building's aesthetics.

If only one corner of the building is involved and the hypotenuse dimension (from 45° corner to 45° corner) was, say 14 feet, the change would not be worth VM consideration. However, if ten corners with hypotenuse dimensions of 14 feet are involved, considerable plus and minus life-cycle costs will be involved.

In the case of one corner, neither the cost quality nor the budget and cost-control factors would come into play. However, in the case of ten corners, the cost-quality factor (and especially the budget and cost-control factors) could be decisive.

15.3.3 A Constructability Quandary

The above example brings up a very important point for CMs to consider when performing VM services. The chamfered corners are features used by the architect to produce a structure that is distinctive and pleasing to the eye, one that enhances the landscape and conveys the image of the owner. The architect considers the corners to be products of a responsibility to the client and the profession to produce architecture as well as buildings.

The CM considers the corners expensive and unnecessary from a constructability perspective. Square corners are less expensive to construct. The CM's fiduciary responsibility to the owner mandates that the corners be proposed for a VM constructability study. There is an obvious conflict between aesthetics and cost. It is the CM's responsibility to see that it does not become a conflict between the CM and A/E.

The decision should be made by the owner, based on the aesthetic value of the building's appearance to the owner versus the cost to produce it. This particular decision must be the result of discussion between the owner and A/E. The CM is not qualified to comment on architectural aesthetics and should not enter into discussion relative to them; the CM should simply provide estimates of cost and budget information until a decision is reached by the owner.

15.3.4 Designability Decisions

Designability decisions are the most sensitive of all value management decisions which the CM must extract from the team. It is impossible to establish one guideline for success because each design decision has its own set of circumstances that control the decision-making process. Additionally, there is an inherent polarity among team members when dealing with decisions predicated on the technical and aesthetic expertise of one team member and subject to the review/approval of others.

The A/E has design expertise, owners and CMs essentially do not. However, owners have physical requirements and preferences that must be incorporated into the their project design and can tell if they are included or not. CMs should have sufficient experience to critique the design on the basis of value and the stated needs of the owner.

Consequently, the team should always approach designability decisions with maximum discretion and an unwavering dedication to the project's goals. When presenting their input, the CM should always approach designability discussions with maximum respect for the owner's and A/E's expertise. If the owner and A/E are at odds on a point, the CM should use extreme care when siding with one or the other. If the CM and A/E are at odds on a point, the CM should take the lead in using tact and logic to effect resolution and never force a concession without a good reason.

15.3.5 The Designability Example

The constructability example used above is also a good designability example. In addition to a question of construction cost, there is a question of additional interior space on every floor at every chamfered corner. Comparing the more aesthetic design with the lesser aesthetic design not only relates to construction cost but also to the space needs of the owner. Either a constructability or a designability decision would be valid in this case, and both should be developed.

Electrical and especially mechanical systems are common areas for designability concern. Current instrument technology provides many cost-saving options based on life-cycle costs but can also overcontrol heating and cooling systems beyond the owner requirements. Most CM's will agree that electrical and mechanical systems provide fruitful areas for designability reviews and should be closely reviewed on every project where they comprise a major portion of the construction budget.

15.3.6 Contractability Decisions

The success of contractability decisions depends on the CM's mastery of the contract management area of knowledge (Chapter 11) and of the owner's contracting requirements and limitations. It could prove to the owner's detriment for the CM to proceed in the contractability area without the necessary expertise.

The primary contractability decision is made when the owner selects either the CM, GC, or D–B contracting system. Once the system is selected, there are more contractability decisions that must be made. Of the three systems, CM provides more contracting options than the GC and D–B systems combined.

As discussed in Chapter 5, CM System Forms and Variations, the first step in constructability decision-making is to determine the best CM form and variation for the project, one that is completely compatible with the owner's purchasing philosophy as well as her/his contracting requirements and limitations.

From this point, contractability decisions lead to the development of the project's contracting strategy and plan. Guidance for the decisions is extracted from the other eleven areas of CM knowledge.

15.3.7 Contractability Limitations

Every contracting option is not available to all CM variations and formats. Construction industry protocols, local area practice, and common sense must enter into the contractability decisions made for each project.

For example, arbitrary bid packaging on multiple-contract CM projects might nullify the cost advantages of competitive bidding. Fast tracking should not be used unless confirmed as necessary. Qualifying contractors works well in the private sector but may be limited in the public sector. Certain dual-service combinations can only succeed in the private sector. The mandatory use of proprietary contract documents could limit contracting options.

To put certain contractability options into effect requires extensive modifications to standard and proprietary contract documents. Although the CM is not an attorney, the need exists to express terms and conditions in contract language for review by the owner's attorney.

The required knowledge of contracts and the construction industry and the potential consequences of a factious contracting plan elevates contractability to a high-level responsibility. Not all construction managers have the expertise to develop an optimum contracting plan.

15.4 THE TIMING OF VM DECISIONS

To determine the contracting system to use and facilitate the selection of contract documents, the primary contractability decision should be made by the owner before hiring anyone for the project. If the owner needs input to these decisions, help can be provided by a consultant, preferably one that will *not* be part of the project after it is underway. This exclusion helps the owner to make an objective decision.

The choice of CM form and variation should be made before selecting a CM; not all CM firms are capable of providing all forms and variations. Contrarily, a CM is probably the best source for helping the owner with this decision, providing the CM's partiality does not impede an objective decision.

All other contractability decisions can be made by the team during the design phase as the project develops and more information is provided on the schedule, building systems, and available contracting resources. Designability and constructability decisions cannot be made until the team is fully functioning in the design phase. Figure 15.6, Potential Savings vs. Cost of Change, promotes the case for early VM decisions. When reviewing completed drawings and specifications, experienced value managers have no difficulty locating features that have the potential for VM improvement. However, VM activity should not be deferred until drawings and specifications are complete. Good VM is a progressive application that finds improvement potential as design is in progress by working closely with the A/E as design develops. In other words, value engineering is preferred to value analysis in cases where constructability is an issue.

As early as the brainstorming and organizational meetings, CM personnel should take copious notes pertaining to design when developing budget numbers or informally discussing project features with the A/E and owner. Although it will be too early to be objective on casual design comments it is not too early to form an opinion of the A/E's design inclinations and philosophies.

When discussions of schematic or preliminary design begin, it is essential that operations and resource persons stay current on potential VM issues. One way to do this is for them to visit the A/E office as much as is practical when design gets underway and establish a productive rapport with their counterparts in the A/E organization. If this is impractical, formal and informal VM meetings should be scheduled as soon as possible at the convenience of the A/E. In either case, communication must be established on both a one-on-one and team basis to accomplish the CM's commitment to VM.

It is not unusual for some A/Es to be less than enthusiastic when CM persons show an active interest in the project's design, even though the A/E is aware of the CM's obligation to the owner in this area. The CM must accept the fact that the A/E is the knowledgeable design team member and was specifically hired by the owner to perform design services. The CM's role is to provide checks and balances to A/E design decisions from a VM perspective and to influence design in a spirit of cooperation.

When confronted with a nonnegotiable A/E design preference, the CM must weigh the VM advantage to the owner against the potential consequences to future team efforts and decide on the appropriate course of action. VM services are the most productive and most challenging of all CM services to perform. Resource Management, covered in Chapter 18, expands on team member relationships.

15.5 VALUE ENGINEERING

Value engineering is used to conclude complex constructability issues. Its purpose is to optimize the material/equipment/systems aspects of a project to conform to the owner's cost-quality objectives. VE is accomplished through the interaction of the resource members of the project team. It is a continuous process with established check-points during design and on-demand involvement during construction.

The CM's ongoing awareness of A/E output during design influences appropriate changes before design solidifies. VE is not a parallel design effort by the CM, nor are CM VE proposals dictatorial in any way. VE is the product of a team effort which must be performed at a professional level. Economics should not be the exclusive criteria for VE decisions, and the A/E should always be acknowledged as the team member responsible for design.

Cost quality and cost efficiency factors can influence VE decisions in the final analysis. Cost-quality factors address the budget; cost-efficiency factors, the schedule. It is the responsibility of the CM to clearly state the effect that either will have on the project. It is the responsibility of the A/E to properly equate design alternatives. It is the responsibility of the owner to select an alternative, and the responsibility of the team to support the owner's decision.

15.5.1 Contractor VE

The value engineering effort by the A/E and CM during the design phase should fully optimize the material, equipment, and systems included in the contract documents. However, further optimization could be provided by trade contractors. To capture the benefits of this possibility, the design in the contract documents should be opened up to further value engineering by awarded work-scope contractors.

Provisions in the supplementary and special conditions should encourage contractors to review the documents and suggest changes that would provide the same function and quality as the design, but at less cost. The suggestions should be accompanied by documentation supporting the design equivalence and the cost savings. Documentation and savings must account for changes in the work of other contractors as a result of the change. The CM should provide an example of the submittal format for the contractor to follow.

Suggestions would be reviewed by the team and, if accepted, be developed as a contract change to the contractors involved. Reward to the submitting contractor would be a percentage of the savings to the owner after the cost of modifying the contract documents by the A/E is taken into account. The percentage of the savings passed on to the contractor should be in the range of 50% to 60%. If a contractor VE program is to be included, it should be clearly defined in the contract documents with the shared savings amount specifically stated.

15.6 LIFE-CYCLE COSTING

If the useful life of a structure could be accurately predicted, the theoretical intent of life-cycle costing would be to match the life of the construction materials and equipment used in its construction with the useful life of the structure. By doing so, the owner would have received the predicted return on investment, the salvage value of the structure at the end of its useful life would be zero, and all components would fail simultaneously. Obviously this scenario is not possible due to the varying durability of construction components, many of which can survive indefinitely. Therefore, the practical intent of life-cycle costing is to produce a project that is the most economical to construct, operate, and maintain for a definite period of useful ownership with the understanding that there will be a salvage value, however small.

Over a 40 year period, the operating and maintenance costs of a typical building consume approximately 30% of the total dollars spent by the owner. Finance costs consume about 45%, construction costs about 20%; land, fees and other costs about 5%. Assuming the typical building is of average design and construction, for every dollar spent on construction, $1.50 is spent on operating and maintenance costs and $2.25 is consumed by financing.

Assuming the building is aesthetically and functionally acceptable to the owner and contains the minimum required space, the only way to reduce construction costs would be to use less expensive material/equipment. By doing so, finance costs would decrease but operating and maintenance costs would increase; or as an alternate, the productive life of the building would be shortened.

Many projects (especially in the private sector) are justified on the basis of a required return on an investment. The owner is interested in the income from the building or structure over a period of years. Other projects are justified on the basis of a resale soon after the project is completed. Obviously, operating and maintenance costs are of greater concern to the long-term owner than the short-term owner.

LCC decisions are based on "pay now or pay later." In many instances, by investing more construction dollars, an owner can reduce operating and maintenance costs during building ownership. Simply put, if a decrease in O & M dollars is greater than a resulting construction cost, the "pay now" option has merit, providing the construction budget can accommodate the increased cost.

The most receptive LCC items are those subject to maintenance, repair, and replacement during use (such as mechanical and electrical equipment); those that require maintenance alone (such as interior and exterior surfaces, doors and windows); those that affect fuel consumption (such as building envelops and HVAC systems and components); and those that reduce costs of facility operations (such as vertical/horizontal transportation and security systems).

To get the most from LCC , the owner's team members should have the authority to increase the construction budget to effect future savings. If that authority is not available, the results of LCC will be limited but not eliminated; LCC studies often uncover items with similar initial costs that can economically outperform items summarily selected by the A/E during the design process. CMs have a responsibility to identify those items with LCC potential, during formal and informal constructability reviews, and perform LCC analyses for a determination.

15.6.1 Life-Cycle Cost Analysis

An LCC analysis is not difficult to perform once the required data are in hand. It is a simple application of the time value of money that can be accomplished using financial interest formulas or tables. For some LCC studies, it is the gathering of authentic performance and cost information that is difficult.

For example, data for an LCC study on a building envelop and fuel consumption is available and dependable as published in handbooks. Data on competing mechanical equipment must be obtained from manufacturers who could be biased toward the performance and longevity of their products. To extract credibility in the latter case, project team members must have sufficient experience with the items to confirm the accuracy of the information provided.

LCC analysis will not be taught here—there are many sources that accomplish this in proper detail—but it may help those apprehensive about the technique to briefly outline an LCC procedure. This information is offered because owners sometimes comment that value management and especially LCC are services that CM firms claim to provide but show little evidence of delivery.

15.6.2 An Example of LCC Analysis

Let us assume that the major components of three different HVAC systems are being considered by the A/E. The following information has been gathered from the equipment manufacturers and verified as sufficiently credible by the A/E and CM resource members of the project team.

Description of Cost	*Alt. #1*	*Alt. #2*	*Alt.#3*
A. *Initial costs*			
1. Equipment purchase	816,000.	536,000.	738,000.
2. Interface costs	120,000.	100,000.	160,000.
3. Other related costs	64,000.	64,000.	2,000.
B. *Replacement costs*			
1. Cost of replacement	10,000.	200,000.	20,000.
2. Replacement interval	16 years	8 years	10 years
C. *Estimated Annual costs*			
1. Maintenance/Repair	25,000.	20,000.	16,000.
2. Operation	30,000.	35,000.	25,000.

The amortization period is 20 years and the interest rate is 10%. Anticipated salvage value at the end of 20 years for all options is 0.

Requirement: Calculate the Present Worth or the Annual Cost of owning and operating each system over the 20 year period. Determine which of the three is the LCC economic choice.

Procedure: To compare the alternates, a dollar value must be determined for each, based on the time-value of money. The dollar value can be expressed and compared as the *present worth* of the owning/operating/maintenance/replacement costs, or the *annual cost* of the owning/operating/maintenance/replacement costs.

The *present worth* of each alternate is the sum of the present worth of the annual operating/maintenance costs and the present worth of the replacement costs, calculated at year 0, plus the known construction costs. The alternative with the lowest present worth would be the economic choice.

The *annual cost* of each alternate is the sum of the annual cost of the equipment purchased/installed and the annual cost of the replacements, plus the annual operating and maintenance costs. The alternative with the lowest cost per year would be the economic choice.

Figure 15.2 on page 223 is a diagram of the costs of the three alternatives over the 20 year period. (C = construction costs; R = replacement costs; O/M = Operating and Maintenance costs; P/A = present worth factor @ 10% for *n* years; P/F = single payment factor @ 10% for *n* years.)

Alternate #3 has the lowest present worth even though the construction costs are $200,000 higher than alternate #2. If the owner can afford to add $200,000 to the construction budget, it will reduce the operating and maintenance budget during the life of the building. Pay now or pay later.

The factors used in the example can be found in any published interest tables under the P/A and P/F columns for *n* years at 10% interest.

If the annual cost calculation were used instead of the present worth calculation, the replacement costs would be converted to present worth, then converted to annual costs. The construction costs would be converted to annual costs. Both would then be added to the annual operating/maintenance costs to arrive at the annual costs.

15.7 BUDGET AND COST-CONTROL DECISIONS

Budget and Cost Control is a prime responsibility of the CM. Budget control is applied to the bottom line of the budget and controls the transfer of funds from line item to line item. Cost control is applied to each line item that comprises the budget and controls the use of funds within each line item. They are interdependent but envisioned and managed separately.

Budget control can be tested by assorted influences between the budget's establishment at the conceptual phase and the final payment to the last contractor. Influences such as differing site conditions, labor disputes, labor shortages, inflation, escalation, unresponsive bidding, errors and omissions, changes, and material shortages challenge the bottom line of the project budget.

Potential budget influences were anticipated when determining the bottom line of the budget by establishing contingencies to cover them if they arise. (These contingencies were addressed in Chapter 10, Budget Management.) The function of budget control is to manage the contingencies so they are used for their intended purpose or redistributed in the owner's best interests as a team VM decision.

Cost control can be tested by assorted influences on work-scope items during the course of design. Influences such as the following challenge the costs of budget line-items: insufficient construction details, dimension errors, closed specifications, misinterpretation of design intent or owner requirements, quantity survey errors; outdated labor, material and production rates; and obsolete products and material shortages.

Cost-control influences were anticipated when budget line items were developed from the conceptual budget, and funds were allotted to the design contingency to

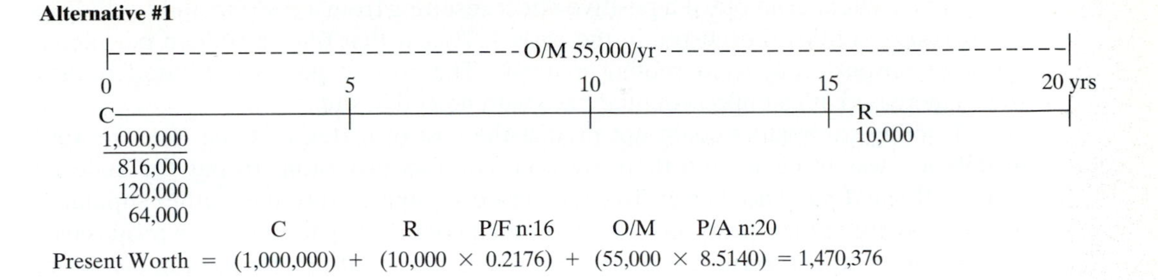

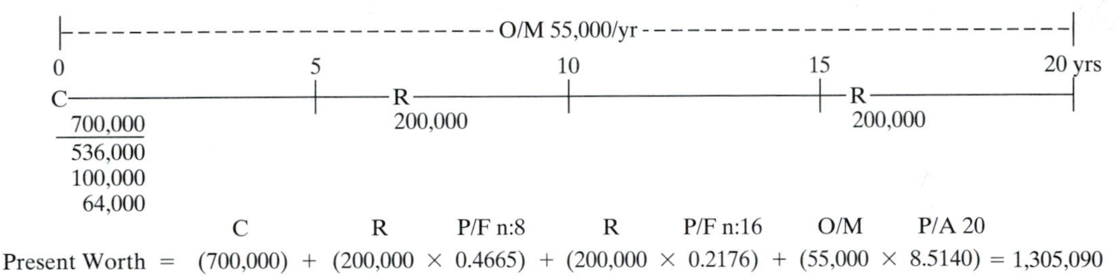

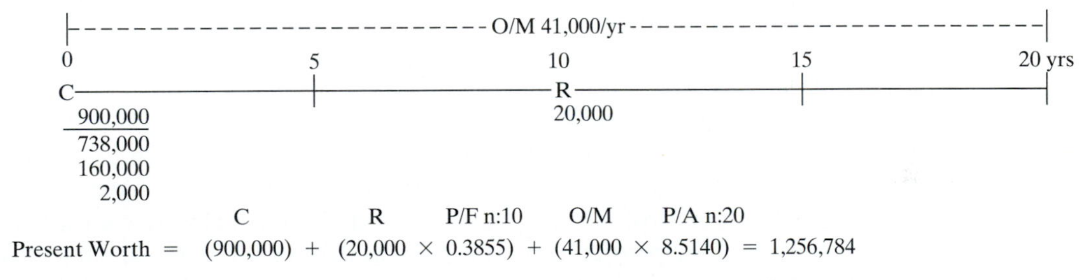

FIGURE 15.2 Diagram of the three life-cycle alternatives.

cover misinterpretation of design intent or owner requirements. However, the initial line items were broad-scope allocations of budgeted funds based on design concepts until design was specifically defined in the contract documents.

The function of cost control is to accurately allocate costs from one line item to another to reflect the work-scopes as they emerge, and to maintain line-item costs within the owner's parameters as design is defined and current contracting and construction information becomes available. When a line-item cost is exceeded, additional constructability, designability, and contractability studies must be made to bring the line item under control.

An excellent example of a positive effect resulting from a contractability study of a cost-control problem occurred in the early 1970s. At that time, a sudden petroleum shortage seriously impacted roofing material. The cost of petroleum-based roofing sheets increased at an unprecedented pace with no end in site.

Roofing contractors could not predict the cost of material at the time the roof installation was to begin, and there were no contract provisions to pay for material stored off-site if purchased early. To compensate for uncertainty, they either stipulated a cost escalator in their proposal or added a high contingency to lump sum proposals.

In order to get sensible roofing proposals during that period, roofing bids were taken as early as possible based on present material prices. The awarded contractor was permitted to order the material immediately and place it in storage. Roofing contractors were singled out in the bidding documents to be paid for materials stored off-site as long as evidence was provided that it was paid for and stored and insured in a bonded warehouse. Roofing contractor's proposals included the predictable costs of storage.

This unique approach, designed to eliminate the crippling effects of unknown escalation costs used by CMs in the 1970s, has become a standard contract provision in the construction industry today.

15.7.1 VM Consideration Example

An example of the complexity of value management decisions can often be found in the selection of a building's environment system. Decisions must be made on fuel, heating medium, distribution systems, energy conservation and system controls. These choices affect the structural system, electrical system, roof system, building envelope and interior materials and finishes.

In each choice, concern should be given to the construction, operating, replacement and maintenance costs, and to delivery dates and installation procedures that affect the construction schedule and breakdown of construction contracts.

Developing an optimum building design with proper consideration for designability, constructability and contractability (with due consideration for cost efficiency, cost quality, budget, and cost control factors) requires a well-planned, closely coordinated effort by the members of the project team. It is obvious that a competent, amiable professional-level match-up of A/E, owner, and CM personnel is required to satisfactorily accomplish this common but very complex task in the owner's best interests.

15.7.2 VM Credibility

Construction managers can positively impact VM decisions, especially if the CM firm has an extensive construction contracting background. Most VM decisions find credence in the information and instincts of a practicing constructor and contractor rather than a consultant that functions off-site. Continuing involvement in the day-to-day activities of competitive bidding, purchasing, contact with consultants and trade contractors and the work force on the construction site provides an intimate experience from which credible construction and contracting-based decisions can be made.

This is not to say that every CM should currently be part of an active construction or contracting firm, because over the years some CM firms have completely transitioned from their roots in contracting and construction to prosper as consultant-based construction managers. As pointed out in Chapter 9, The CM Body of Knowledge, as long as the CM firm has the necessary attributes and stays current with construction industry practices, the CM should be considered qualified.

15.8 THE VM OPTION OF FAST-TRACKING

Probably the most consequential VM decision is whether or not to fast-track a project. Fast-tracking has many ramifications, and a decision to use this contracting expedient should only be selected after a thorough analysis of the owner's needs. Considerations including the required occupancy date, climatic conditions, labor and material markets, and the unique demands on team members all need to be studied if fast-tracking is a potential option.

Fast-tracking should be considered one of several contractability options and a contracting tool that has a specific purpose. When properly utilized, fast-tracking shortens on-site construction time by allowing portions of the design and construction to proceed synchronously. However, fast-tracking should not be used unless it can solve a completion date problem.

Unlike the name implies, fast-tracking does not accelerate either design or construction. The same amount of time is devoted to design and construction as on a non-fast-tracked project. The acceleration is the result of overlapping early design and early construction activities.

Fast-tracking is a form of phased construction. Each phase is a work-scope or group of work-scopes designed, bid, and constructed as a unit or phase, separate from the rest of the project. Subsequent interfacing phases are handled in the same manner until the entire project is under construction. The result is a series of design–bid–build phases rather than the conventional single design–bid–build procedure that is mandatory in the GC system (and customarily used in the CM system).

It should be noted that the CM system, unlike the GC and D–B systems, has a contracting structure that facilitates competitive bidding of each fast-tracked phase. Fast-tracked phases on GC and D–B projects cannot be competitively bid; they must be negotiated. Consequently, fast-tracking is a viable option on most projects in the public sector when the CM system is used. This is one of the least recognized yet most outstanding features of CM.

15.8.1 Fast-Tracked Phases

Determining what to include in each fast-tracked phase is a project management/contract management decision. The work-scope(s) of each phase usually conform to the typical construction sequence of the total project. However, consideration should be given to work-scopes that can be designed early and do not physically interface too closely with each other.

On a typical building project, the phases that might fit this criteria are mass excavation, site utilities, building excavation, site drainage, partial roads and parking, partial landscaping, temporary structures, foundations, and ordering long-lead items.

Whatever the inclusions, the work included in the phase is selected for its expedient affect on total project completion and convenience to the design process.

15.8.2 Managing Fast-Track Projects

When fast-tracking a project, more demands are made on project management, budget management, and risk management than on single-phase projects. Contracts are being awarded and owner dollars are being expended on construction before the total cost of the project has been established by competitive bids. To protect budget credibility, the CM must establish construction level estimates for the remaining work-scopes of the project without construction-level design definition; this requires exceptional experience and expertise.

Fast-tracking also requires nontraditional design sequencing and additional design coordination on the part of the A/E. The structural system of a building is logically designed from the top down and constructed from the bottom up. Foundation design is based on building loads resulting from decisions made as late as the design development phase of a conventional project. Foundations and structural steel are good candidates for fast-tracking; consequently, assumptions must be made in order to bid the foundation work-scope early and at least get an order into a mill rolling for structural steel.

A building's utility requirements are usually determined after design has proceeded sufficiently to closely determine incoming power, water and gas, and outgoing loads for sewage and storm drainage. Site utilities are good candidates for fast-tracking; consequently utility requirements must be determined sooner than the A/E would like, and estimated loads must replace accumulated design data.

The early design of mass excavation, building excavation, site drainage, roads and parking, and landscaping work-scopes are not as problematic as foundations, site utilities, and structural steel. These work-scopes can be readily developed after the building has been sited and its grades established. They are excellent candidates for single or phased work-scope packages when fast-tracking.

To ensure design adequacy (especially in foundations, structural steel, and site utilities) overdesign is common. Consequently, the costs of these work-scopes will be higher than if they were designed in normal sequence. The extent and nature of overdesign is a risk-management decision to preclude costlier physical changes after construction of the work-scopes is completed. For example, site power could be split in two; an early work-scope for conduit, pull boxes, and manholes, and a later one for pulling conductors rather than installing direct burial cable that is safely oversized in a single early package. The cost of the alternatives would be the deciding factor.

To accommodate design and construction coordination problems inherent when design is out of sequence, changes must be made quickly in the field to compensate for design oversights. To accomplish this efficiently and economically, a good working relationship between the A/E and CM must exist. Although subsequent design is expected to conform to the physical configuration of work-scopes under construction, there will be times when field changes will have to be made. Both the A/E and CM should be on the lookout for these instances and proceed to correct them as quickly and economically as possible. The owner must be made aware of these potentials before deciding to fast-track and must also understand that the added expense of fast-tracking is a trade-off for an earlier completion date.

15.8.3 The Decision to Fast-Track

The decision to use fast-tracking should be one of the earliest decisions made. It should be prompted by the CM, endorsed by the A/E, and decided upon by the owner. The time, cost, and business interruption factors of the project should be given due consideration.

Decisive information is found in the program schedule covered in Chapter 20, Schedule Management, and the budget in Chapter 10, Budget Management. The schedule shows the owner's occupancy requirements and the estimated time to be consumed by design, construction, and move-in activities. The budget will reveal if there are enough contingency dollars to cover the potential increase in costs inherent to fast-tracking. With this information, the team can make a risk-management decision on the practicality of fast-tracking. The climatic conditions anticipated during construction, the building type (single or multiple story) the structural options, wall bearing, steel frame, pre-cast or cast-in-place concrete frame are the important factors in addition to schedule and budget.

Figure 15.3 shows how fast-tracking a school meets the owner's move-in date of August 1 when weather conditions are severe enough to force a shutdown. If the school was constructed when winter weather is not a factor, fast-tracking should not be considered. However, when early occupancy produces income for the owner (as in the case of a mall or condominium), the fast-tracking option would be an income-producing expedient and should be investigated on its economic merits, regardless of the climatic conditions at the site.

15.8.4 Contemporary Bidding

The fast-track example in Figure 15.3 depicts a project that is divided into four bidding phases; site, foundations, long-lead items, and all other. This is sometimes called "controlled" fast-tracking because it limits the bidding of work-scopes to the design time-frame of a single-phase project; all the fast-tracked phases are under contract between the start of design, March 1, and the end of design, July 31.

On large projects with construction durations of two years or more, it is often desirable to spread bidding over the construction span of the project rather than the design span. This is referred to as *contemporary bidding* or *just-in-time bidding*. The intent is to eliminate the guesswork inherent to long-lead bidding in favor of realistic pricing for work-scopes such as roofing, painting, landscaping, roads, parking, etc., that will not be start on-site for many months.

From the owner's perspective, it would seem that there is one arguable disadvantage to contemporary bidding: the chance of eliminating a contractor who underescalated a controlled fast-track bid in favor of a contractor who submitted a contemporary bid using current market prices.

Owners should be advised to pay current prices for contracted services rather than long-lead prices. The rationale is that although prices may rise and fall over the construction period, contemporary bids will ultimately provide an overall economy to

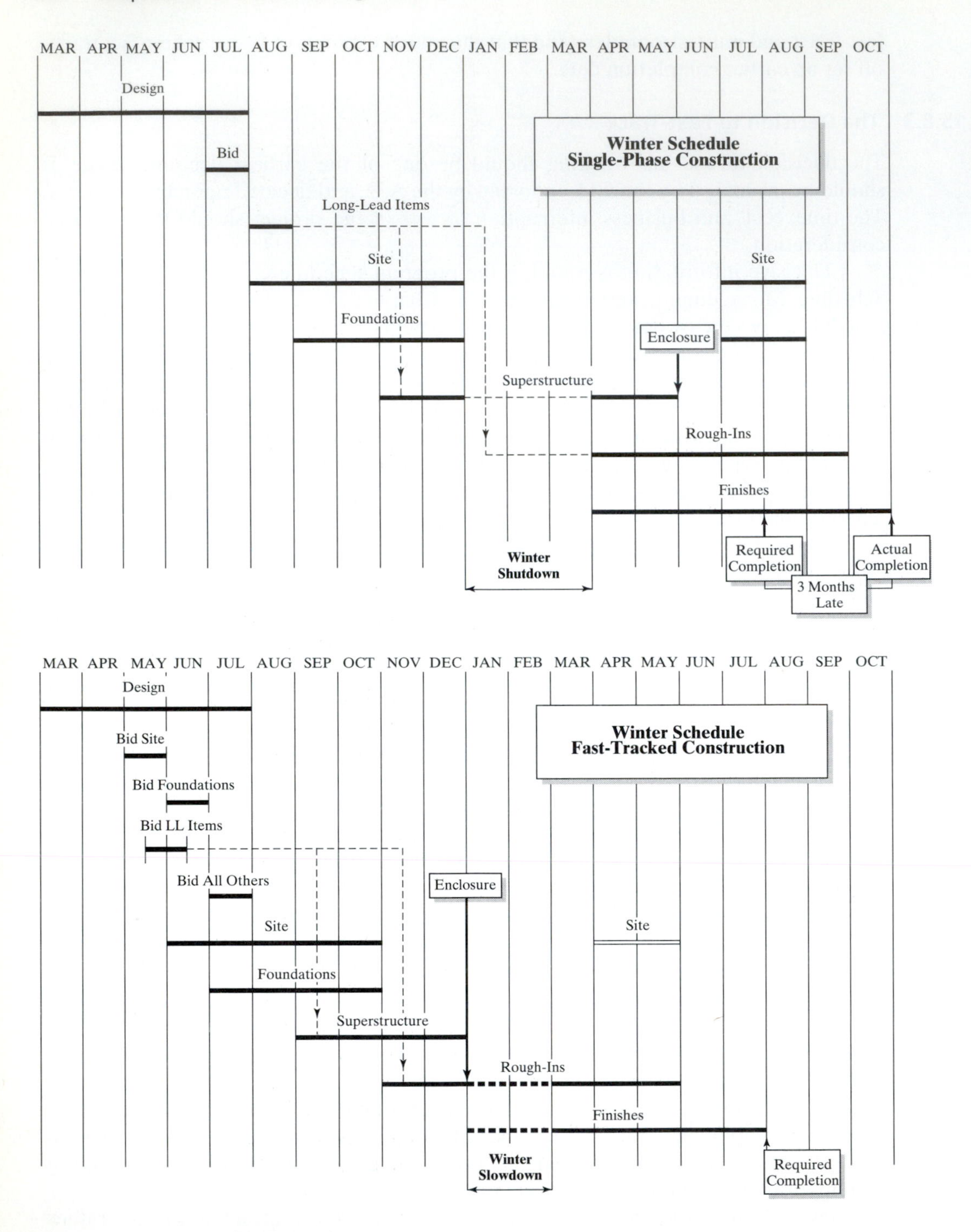

FIGURE 15.3 Comparison of single-phase and fast-track schedules.

the owner. The small saving that the owner accrues from an underescalated long-lead bid from a trade contractor is certainly not considered small to the contractor who made the miscalculation.

While the owner may interpret the error as the spoils of the bidding process, an economically troubled contractor often communicates his distress to the project in the quality and timeliness of his work. This adversely affects the work of interfacing contractors and disrupts the flow of the project in general. Unfortunately, there is little a CM can do to quickly reverse a situation like this.

Of course the greater possibility exists that long-lead bids will be overescalated. Contractors are prone to protect bid pricing in this manner in spite of the pressure of competition. When this occurs, the owner is the unwitting loser because the situation is never revealed.

Chapter 10, Budget Management, points out that an escalation contingency is appropriate on projects that extend over a long period of time. Fast-tracked projects using contemporary bidding should be protected under this contingency to prevent straining the budget. The CM and A/E should be well-qualified to determine what the escalation contingency should be.

15.8.5 The Impact of Fast-Tracking on A/E Fees

On a typical, single-phased project such as an elementary school, the program schedule will have about 100 activities. On a phased project, each additional phase adds about 20 activities. As a result, the program schedule for a two-phase project will have 120 activities.

A/E time and effort is only required in about 30% of the added activities, which are mostly routine; contract document reviews, advertisements for bids, distribution of bid documents, receiving bids, and pre-bid meetings. Nonetheless, more A/E time and effort is required.

Experience has shown that an equitable fee adjustment for added A/E time and effort will result if the number of phases less one, is multiplied by 30% of the added activities, and that amount multiplied by the dollar value of the phased portion of the project. The phased portion is the remaining amount after the dollar value of the largest phase is subtracted from the total dollar value of the project.

On the fast-tracked project in Figure 15.3, four phases are used. Consequently the program schedule would increase 60 activities ($[4-1] \times 20$); 20 of which ($60 \div 3$) will require additional A/E time. A/E activities have increased 20% ($20 \div 100$).

The budget for the project is $6 million and the budget for the largest phase is $4 million. Therefore, $2 million dollars is considered the value of the phased portion.

Assuming the A/E's quoted fee is 5% of the project's single-phase construction cost, the fee increase for a phased project would equal the 20% increase in activities times the A/E's percentage fee times the calculated phased amount ($0.05 \times 0.2 \times$ $2 million) or $20,000. The A/E's total fee for the project would equal 5.33% of the project's single-phase budget.

The calculation above is only one way to rationalize an A/E's request for a higher fee on fast-tracked and phased projects. There are others. However, this rationale, which has been successfully used in the past, can serve as a guideline for CMs and owners.

15.9 CM VM RESOURCES

To accomplish the goals and reach the full potential of VM, an adequate resource staff is essential. Figure 15.4 shows the expertise necessary to interact with the A/E and provide the information for VM decision making. The chart should not be misinterpreted; it indicates the expertise required, not the persons required.

It is probable that the value managers would be engineers and that the civil engineer was competent in structural engineering as well as site civil requirements. The administrative and scheduling functions of the chief value manager is normally assigned to one of the value managers.

Taking all of probabilities into consideration, the minimum VM resource staff would consist of four persons instead of the ten implied by the chart.

15.9.1 A/E-CM VM Interaction

At its best, VM is a continuous process with several checkpoints along the way to formally culminate VM decision making. To coordinate with the design process, checkpoint meetings coincide with the contractual phases of A/E participation. Meetings are scheduled near the end of the schematic and design development phases and at a point between 50% and 80% completion of the contract document phase.

The schematic and design development VM meetings can take advantage of the contractual pauses during design to verify the budget and petition the owner to con-

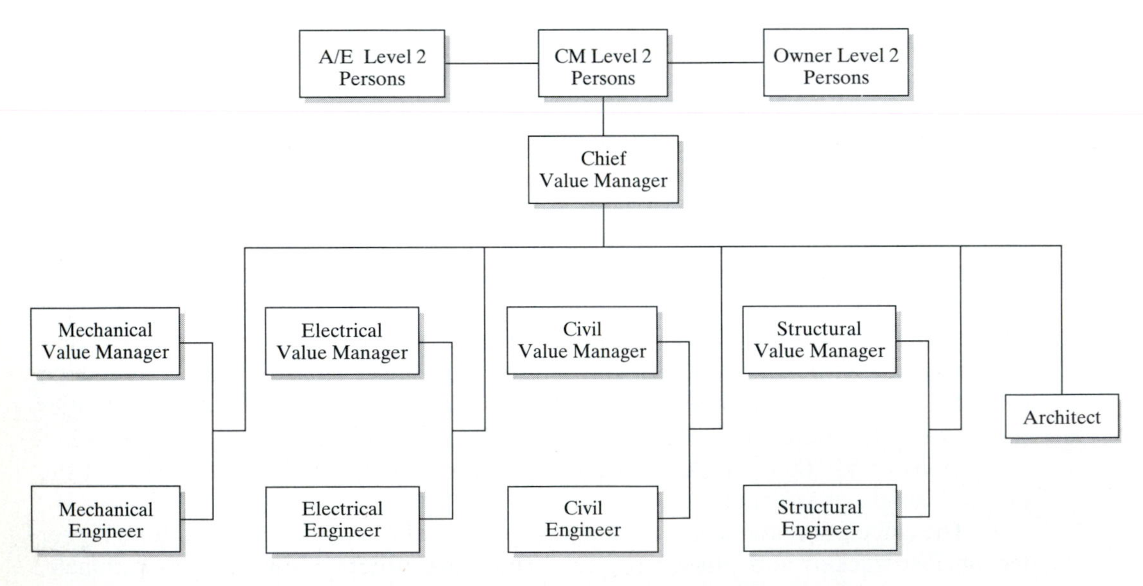

FIGURE 15.4 Typical value management resource team structure.

tinue on to the next phase of design. The meeting during the contract document phase should occur when architectural features are essentially final but electrical and mechanical features are still accessible for final VE and LCC analysis.

The reason for the timing of the contract document VM meeting is that electrical and especially mechanical systems have a tendency to "grow" when drawings and technical specifications are in the final stages of development, even though extra efforts were made in earlier phases to preclude this.

15.9.2 VM Records

The groundwork for VM discussions at checkpoint meetings should have been laid during the more casual interactions between VM team members prior to the meetings. Some VM decisions simply need the owner's concurrence; others require the owner's approval. Many VM decisions do not require the involvement of the owner at all—they are technical decisions agreed to by the A/E and CM based on the merits of available alternatives, any one of which would be acceptable to the owner.

However, it is suggested that the CM and A/E cooperatively expose the owner to as many VM decisions made as possible. VM is the one service that owners claim is promised by the A/E and CM but not delivered to the level of their expectations. Unfortunately, the owner is not privy to the continuous VM dialogue between the A/E and CM that produces most of the beneficial owner-oriented decisions.

To dispel this concern and produce a valuable record for future projects, a VM log should be kept and made available to the owner at VM meetings. While the VM log has owner-relations value, it is also a perfect way to disseminate VM solutions throughout the A/E and CM organizations for reference on other projects. Figure 15.5 is a suggested format for a VM log.

Value Management Log

A/E Proj#: A0741 CM Proj#: 16423 Proj Name: Carrington Clinic Owner: Oldburg Investment, Inc.

VM Prop No.	VM Basis	Value Management Proposal	Prop Initiator	Prop Reference	Cost Increase (+)	Cost Decrease (−)	Running Total (+) (−)	Date
26	Design	Pigmented block to burnished block	A/E	RBK	5,000	0	− 94,500	072895
27	VE	Change cooling tower, 400T to 360T	CM	ESL	0	3,600	− 90,900	072895
28	Design	Delete 2nd elevator, keep shaft	Owner	DEM	0	55,000	−145,000	081195
[1]	[2]	[3]	[4]	[5]	[6]	[7]	[8]	[9]

Column [1] Sequential number, significant proposals
[2] Const., Contr., Design, LCC, VE
[3] Brief description of proposal
[4] A/E, CM, Owner, Contr.
[5] Source (person's initials)
[6] Added cost
[7] Subtracted cost
[8] Accumulated cost
[9] Date accepted

FIGURE 15.5 A typical Value Management Log.

15.10 THE VALUE MANAGEMENT PLAN

A preamble to the value management plan should state that the design, construction, and contracting of the project can and will be optimized through the ongoing synergistic efforts of the project's team members. The VM concept of optimization is too important to the owner to forego this reminder.

Statements should be included that briefly define the three principal areas of VM (designability, constructability, and contractability) as visualized and accepted by the project team. Similar to a partnering agreement, each executive and management level member of the project team member should commit to cooperation in pursuit of VM opportunities.

Consequently, the value management plan is slightly different from the other management plans. It serves as a commitment to participation as well as a plan for team member interaction.

15.10.1 Team Participation

The owner, A/E, and CM should identify the VM/VE personnel that will be used on the project, list their credentials, and stipulate the areas of their expertise. The team VM chain of command should be diagrammed to establish VM communication links and maintain control of the process.

The timing and format for formal VM meetings should be established and provisions made to insert dates in the program schedule. Ground rules for informal VM meetings should be stipulated, and a means of conveying the decisions made at these meetings should be stated.

The format and use of the VM log should be outlined and the depository for accepted VM proposal backup data designated.

15.10.2 CM Internal Participation

Unless a project team member has other thoughts on participation, the VM organization structure in Figure 15.4 should work internally for the CM.

The most difficult problem will probably be scheduling internal VM team persons into the VM effort. The internal CM Master Program Schedule referred to in Chapter 20, Schedule Management, provides the best opportunity to allocate sufficient time to accomplish this. The chief value manager should be a person with better-than-average planning ability in order to have the right people at the right place at the right time.

There may be times when the firm's workload will not permit timely scheduling of VM personnel for a project. Under these circumstances, it is suggested that temporary VM help be enlisted to keep the VM process moving. Some mature CM firms have retired employees who are willing to help out temporarily or have connections with persons who can be hired on a fee basis.

It is tempting to do a less-than-thorough job, realizing that the owner and the A/E probably will not know the difference because the CM is the initiator of most VM efforts. This choice, however, is really not available to the CM under the terms of the contract, the CM's agency relationship with the owner, and the ethics of professionals who provide VM services. It is better to delay the project than to ignore a VM opportunity.

15.11 VM TIME-COST RELATIONSHIPS

The inescapable consequence of VM is change in the drawings and specifications or both. Changes resulting from VM efforts save money for the owner but cost money to effect. In general, the earlier a change is made, the less money it will cost to make.

The relationship between the savings produced by a change and the cost expended to make the change is not direct. The type, size, and timing of the change variably influences the relationship between the two costs. The CM must make this fact very clear to the owner, and both the CM and A/E must provide change-related services based on this premise.

Figure 15.6 illustrates the relationship between the cost of a change and the owner's cost return from a change. The major influence on the change cost is the timing when the change is made.

As design proceeds from phase to phase, contract documents are progressively approved by the owner; therefore, changes that affect previously-approved documentation will be billed by the A/E to the owner as additional services. Changes made within a design phase that do not affect previously-approved documentation are accomplished without added billing. Changes that require alteration to work already in

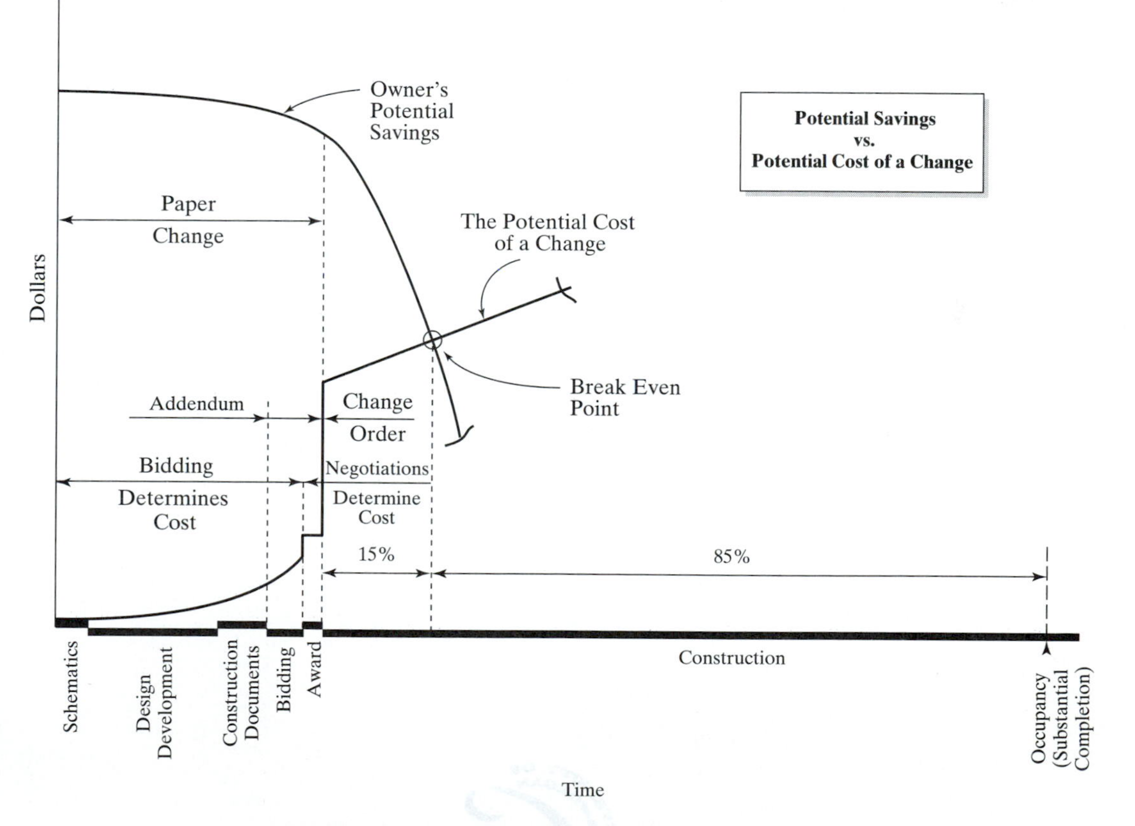

FIGURE 15.6 The cost of a change vs. the owner's cost return.

place are more expensive than those that do not require alteration, and the cost of work determined by competitive bidding will always be lower than if negotiated.

To determine either the net savings or the net added cost to the owner as the result of a change, any cost of amending documents must be added to any negative cost impact which results from derivation of the contractor's price for work included in the change. Consequently, there is a point during the construction of the project where VM changes will have a negative value.

Figure 15.6 is a generalization and does not take life-cycle cost into consideration. If the change has a positive impact on life-cycle cost savings, the break-even point could be moved to the right (later into the construction phase than shown).

Changes made between the start of design and the start of construction can be referred to as *paper changes*; they only affect contract documents. During the schematic phase, changes require little more than an eraser and a restart, a normal approach to schematic design. During the contract documents phase, redrafting of drawings and rewriting of specifications will probably be required.

Changes made after distribution of bidding documents, but before bids are received, can be included by addenda and their cost derived from competitive bidding. Changes made by addenda issued after bids are received, but before awards are made, can usually be favorably negotiated. However, changes made after awards are made must go through the change-order process; their cost is usually determined through difficult negotiations where contractors normally win.

The value part of value management activity, especially in the areas of designability and constructability, must be extracted early.

CHAPTER 16

Project Management

Project Management determines and executes the processes that are required to maneuver a project from start to completion. It is based on an initial plan of action that is modified as new information becomes known.

The Project Management Plan, prescribing when and how project management actions are to be performed, is incorporated into the CM Project Manual. It includes team member performance commitments and responsibilities and serves as the guide for orderly project progress.

16.1 EMPHASIZING FUNDAMENTAL PRACTICES

An important point is that the procedures under the heading of Project Management are so fundamental that some readers will view them as either unnecessary or capable of being accomplished with less concern. This attitude has for years been a problem in traditional construction industry practices. It is a major reason for embracing the CM system in the first place and accounts for its acceptance by owners. Good management cannot be achieved with a broad brush. It can only be realized by checking off the fundamentals on a step-by-step basis without ignoring those that are untimely or do not seem important.

Surprise can destroy any cooperative relationship, especially a team, and CM represents the ultimate team relationship in the construction industry. It is the CM's managerial responsibility to develop a CM format that is so fundamental that its chances of failure are minimal and to "dot every i and cross every t" to avoid surprises.

16.2 PROJECT MANAGEMENT AREA OF KNOWLEDGE

The project management area of knowledge encompasses all of the operations aspects of project delivery, including determining, formulating, developing, installing, coordinating and administering the necessary elements from the beginning of design to the termination of warranty and guarantee periods. The CM has the responsibility to make the selected CM process work, to coordinate the efforts of the team and the performing contractors in achieving their common goal.

The CM must provide discreet leadership and expertise in carrying out responsibilities, and effectively organize the other five required elements—design, contracting, construction, construction coordination and contract administration—into a functioning management format. This fully develops the team's resource potential and prudently orchestrates participation throughout the project.

Project management requires a thorough understanding of the design process, the contracting process, the construction industry, and all forms and variations of the construction management system; how the CM process works, what the required activities are and what is required of each activity and each team member. A good grasp of communications skills and ethics is necessary.

16.2.1 Major Project Management Components

The major components of project management are :

Brainstorming Session
Responsibility Matrix Chart
Program Schedule
Management Options
Project Management Plans
CM Project Manual
Exit Meetings
Other Meetings

16.3 BRAINSTORMING SESSION

The brainstorming session introduced in Chapter 7, ACM Procedures, sets the stage for future team interaction. The session should be scheduled as soon as possible after the owner has selected the CM and A/E. It is not necessary that contracts are signed; an owner's letter of intent to the CM and the A/E is appropriate.

The session is essentially the forerunner of partnering sessions used in the general contracting system—a chance to get to know who will be working with who, to surface unique problems, to generate a positive and cooperative attitude, and establish a common goal for the participants. The difference in CM and GC partnering is that no adversarial relationships are represented at this meeting. Traditionally considered adversaries by owners and A/Es, contractors will not get involved in the project for several months to come. At that time, preconstruction meetings will serve the partnering purpose.

It should be remembered that contact between team member organizations has been minimal to this point. The CM's and A/E's salespersons and perhaps Level 1 Managers are the ones who interfaced with the owner's executives when negotiating CM and A/E services. As Chapter 24, Acquiring CM Services, points out, it is recommended that more interfacing occur before an owner selects a construction management firm, and it is best if the CM and A/E firms have an opportunity to get to know each other before either is selected by the owner. (Unfortunately, this rarely happens.)

The brainstorming session is important because first impressions are important. Once a project gets under way, there is little time to change negative first impressions. The meeting should be structured to bring out the best in all attendees.

It should be assumed that the session can be accomplished in a day; to take longer is usually counterproductive. All participants should be well prepared to explain their involvement and discuss their assignments in detail with their counterparts. The CM is responsible for organizing the meeting.

16.3.1 Arranging the Session

The CM's Level 1 executive outlines the structure, purpose, and content of the session to the owner and A/E; they collectively select a time and location for the session and generate an agenda that will achieve the meeting's purpose. Along with the goals of the session, the agenda is provided to all participants with sufficient lead time to prepare for the session. They are advised that they will not only learn about the project in general but will interface with their team counterparts.

In terms of participants, the size of the meeting depends on the size of the project but more so on the extent of the owner's involvements. The interactions of the CM and A/E will not vary much from project to project but the owner's interaction will. No matter how small the project or how uninvolved the owner will be, it is important for the A/E and CM personnel to interface in the presence of the owner.

If the owner is a large cooperation or bureaucracy, chances are that many people in the owner's structure will be involved to represent their departmental interests. If the project is a manufacturing facility, the production and the plant engineering departments will have to provide input. The accounting department will have their own requirements for project billings and payments. If direct purchases of material and equipment will be made, the owner's purchasing department will be involved.

A major interface between the owner and CM could be computerized reporting of financial transactions. If the owner has a system in place, the CM has the obligation to format reporting information in a compatible way. Computerization is the trend in financial reporting, and the CM's financial management control system must have sufficient flexibility to meet reporting requirements.

In the early days of CM, the U.S. Corps of Engineers developed a computerized management control system of their own and provided it to their contractors on disks to ensure standardization between projects. Regardless of the quality of the contractor's resident system, the Corps' system had to be installed and used.

The brainstorming session is the time and place for all involved parties to interface. The owner's purchasing and accounting persons will meet with the CM's financial management control coordinator and computer programmer to find ways to integrate the CM financial reporting with the owner's accounting system. The owner's production and plant engineers will meet with their A/E and CM counterparts to strategize ways to ensure that their needs will flow smoothly into the design.

16.3.2 Meeting Location

Because the session will require several persons from each of the team member's organizations, it would be convenient for one of them to hold the meeting at their place of business. It is suggested that if the owner's personnel will be greatly involved that the meeting be held at the owner's location. (If the owner will not have many people involved, the meeting can be held at the A/E's location.)

If geography is a major factor, the session could be held at any location that is acceptable to all. Sometimes a neutral site is advantageous because it promotes the undivided attention of the attendees. On the other hand, meeting at a team member's location (especially the owner's) helps in the familiarization process and could add to the purpose of the meeting.

16.3.3 Brainstorming Session Agenda

The session agenda will vary from project to project; however, a basic format and list of items to cover is evident on every project. As the knowledgeable party but equal team member, the CM's Level 1 Manager should act as a facilitator rather than as chairperson.

The session's agenda should be complete, well structured, and informative. This is the first of many team meetings. It is an ideal opportunity for the CM to demonstrate strong management ability and a good opportunity for the other team members to assess that ability.

It is useful to have the meeting start in a formal mode and progress toward informality. This will ensure that all essential, formative project information is presented to and understood by all attendees. The team will function as a matrix organization, and each element must perform within the established criteria. As the agenda moves ahead, a break is provided and the break-out sessions begin. This is when attendees will get to know each other and begin to feel comfortable at the meeting. A break and lunch provide time for pleasantries and voluntary interaction. Experience has shown that the meeting will lighten up before the day is over. A typical brainstorming meeting agenda is shown in Figure 16.1.

On more involved projects, the break-out sessions could extend into the evening or next day. On less involved projects, break-out sessions could end after or even before lunch. Regardless, everything required to achieve the purposes of the brainstorming session should be planned for and executed.

16.3.4 Formulating Management Plans

The goal of having CM persons leading break-out sessions is to guide their groups in interface decisions and preliminary procedures for the areas in which they will function. The CM should review the preliminary procedures and send the results to the group members for their input and recommendations. This exchange should be repeated until the procedures are satisfactory to the group. The CM representative should then send it to the Level 2 Person for approval at the administrative and policy level. Once approved, the procedure is added to the CM project manual under its proper heading.

Experience has shown that the formulation of management plans requires about two to three weeks, depending mainly on the complexity of the owner's involvement. During that time, copies of the plans in progress will be used to assist decision making at the Organizational meeting.

16.4 THE RESPONSIBILITY CHART MATRIX

The sole purpose of the Organizational Meeting introduced in Chapter 7, ACM Procedures, is to produce the Responsibility Matrix Chart. The meeting should be attended by the Level 1 and 2 Managers from the owner, A/E, and CM. Due to the content of the meeting, it is appropriate for the owner's legal advisor to be present (specific activities relating to the owner-A/E and owner-CM contracts will be discussed and may be minutely changed).

Brainstorming Session Agenda

Location:	Acme Corporation, 1223 Pace Street, Hartlow, MI Conference Room 201
Date:	February 12, 1995
Time:	8:30am
8:30 – 9:15	Greeting (Owner's Level 1 Person) Indepth project overview Introduction of A/E and CM Level 1 Persons
9:15 – 9:30	A/E team member introductions (A/E Level 1 Person) Overview of the A/E firm/organization Indepth overview of A/E team organization
9:30 – 9:45	CM team member introductions (CM Level 1 Person) Overview of the CM firm/organization Indepth overview of CM team organization
9:45 – 10:15	Meeting purpose/goals (CM Level 1 Person) Meeting structure and timing Break-out sessions, participants/locations/leaders
10:15 – 10:45	Coffee and Conversation
10:45 – 12:00	Break-out sessions (at set locations)
12:00 – 13:00	Lunch (Oak Room, Acme Corporation cafeteria)
13:00 – 16:00	Break-out sessions continue (at set locations)
16:00 – 17:00	Plenary session (Level 1 Person) Questions/answers Meeting Summary The next steps Adjourn

Note: Some attendees will find it necessary to interface with other attendees at more than one break-out session location. Break-out session leaders must arrange to make this possible.

FIGURE 16.1 A typical Brainstorming Session Agenda.

The CM contracting structure joins the owner, A/E, and CM in a three-party team. Contracts between the owner and A/E and the owner and CM spell out the responsibilities of each to the owner. However, no agreement exists, formal or otherwise, that coordinates the responsibilities of the A/E and the CM. There is a document yet to be developed that conclusively prescribes the interactions of the three-party team, before construction contracts are awarded.

In support of this, the GC system also has a three-party team. It includes the owner, A/E, and GC. There is a contract between the owner and A/E and the owner and GC. To specify coordination of services between the three, the General Conditions of the Contract for Construction, a very lengthy document, is included as part of the owner-GC contract and is included by reference in the owner-A/E agreement.

The CM contracting structure also includes a general conditions document, but it is only included as part of owner-contractor contracts. A considerable amount of important team interaction goes on during design, bidding, and award that requires responsibility clarification in a similar way.

Experience has repeatedly demonstrated that problems resulting from the question, "who does what, the owner, A/E, or CM?" frequently occur on CM projects. There is no document that specifically spells out the interacting roles of the A/E and CM throughout the project, especially during the very important preconstruction phases. The responsibility matrix is an important document; it allows the project to proceed with less discussion and more direction regarding "who does what."

Every CM project should have a responsibility chart, regardless of the familiarity that develops as a CM team works together time after time. Just as a GC project would not proceed without a General Conditions document, a CM project should not proceed without a responsibility chart or another document that accomplishes the same thing.

16.4.1 Creating the Chart

Creating the responsibility chart takes time as much as a full day on projects where team members are working together for the first time (a little less, if the owner has previously worked with the CM, somewhat less if the owner has worked with the A/E on a CM project, and much less if the A/E and CM have previously worked together). Obviously the time required depends on the previous CM interaction of the team members and their knowledge of the CM system.

An important collateral benefit will accrue to the owner if the project is the owner's first CM experience. It is not common for owners to understand the step-by-step process of project delivery. Most owners are first-time users of the construction industry, and few build more than one project in their careers. Few owners have an accurate intimate knowledge of the construction industry and construction contracting, yet they invest large sums of money when they construct a facility.

As the responsibility chart is created at the organizational meeting, mostly by the interaction of the A/E and CM, the owner is exposed to the entire CM process in great detail. When the chart is completed, the owner should understand all the contract language and exactly what the A/E and CM will do in his/her behalf. The owner, as well as the A/E and CM, will also know the role he/she is to play and the importance of that role to the success of the project.

As stated in Chapter 7, ACM Procedures, "the facilitator should never use a completed responsibility chart from another project as a basis for modification." This should be a hard and fast rule. The CM's Level 1 Manager should be the facilitator although, depending on CM experience, it would also be appropriate for either the owner or A/E to act in that capacity.

Copies of a blank matrix form and a list of activities are called for. Figure 16.2 is a form that can be used as a final or preliminary matrix layout. It is a basic form which covers the activities of the owner, A/E, CM, and contractors. The contractor column is not necessary but is a helpful reference when assigning contractor responsibilities in documents.

Figure 16.3 is a final matrix form that can be used if it is important to differentiate the responsibilities of the team's three management levels. A single "contractor" column could be added to this form for the same purpose it serves in the final/preliminary matrix in Figure 16.2.

ACTIVITY	OWNER	A/E	CM	CONTRACTORS

FIGURE 16.2 Final/Preliminary Responsibility Matrix Form.

Both Figures 16.2 and 16.3 are examples of forms which could be used. It is up to the team to design a form that accomplishes the best end result.

The list of activities to which team member responsibilities are to be assigned should be listed chronologically rather than alphabetically. A project phase sequence is easier to follow and more educational to team members not familiar with the CM system or who have not previously worked together on a CM project. (Once the chart is completed in a chronological sequence, it can be converted to an alphabetical sequence for reference use.)

During the meeting, the activities are listed singularly or in appropriate groups down the left-hand column of the matrix form. Each activity is discussed by the team until it is fully understood and the responsibility for that activity is assigned. A descriptive word or phrase is then entered in each team member's column to describe the team member's specific responsibility in that activity. The process is repeated for each activity or group of activities.

	OWNER MANAGER LEVEL			A/E MANAGER LEVEL			CM MANAGER LEVEL		
ACTIVITY	1ST	2ND	3RD	1ST	2ND	3RD	1ST	2ND	3RD

FIGURE 16.3 Final Responsibility Matrix Form.

Many responsibilities will be determined by the intent of the wording in each team member's contract with the owner, producing a confirmation rather than a determination action. However, construction management creates team member situations that are not clearly defined and can only be settled by disclosure and debate.

For example, the A/E accepts that it is his/her responsibility to produce and distribute bidding documents to contractors, and the owner accepts that it is his/her responsibility to pay for production and distribution. The owner-A/E contract states that. However, if the owner and A/E are new to the CM process, it is unlikely that they are aware that the number of document sets to be produced, distributed, and paid for could be ten times the number used on a similar GC project. This could very well be the case on an ACM project using multiple contracts.

If the owner and A/E are not made aware of the extent of this responsibility, it can cause seriously disruptive problems when it is time to produce, distribute, and pay for 250 sets of bidding documents. The in-house facilities of the A/E may not be adequate to produce the sets, and the owner would question the large budget item covering bidding document costs. Everything should be done by the team to preclude a "why didn't I know this before?" query from a team member.

It should be made abundantly clear in the chart that the A/E always retains the responsibility for design, and the CM's presence does not mitigate this responsibility. It is easy to get the impression from the owner's contracts that the CM is responsible for checking design integrity and performing alternate designs. These misconceptions, if they exist, should be clarified in the chart.

Completing the chart should not be deterred by semantics. Experience has shown this to be a real possibility as responsibilities are discussed. Word-meaning controversy usually stems from concern for the assignment more than the way the word or phrase might be construed. It is natural for team members to try to minimize their contractual obligations if given the chance.

Many activities brought up for responsibility assignment on a CM project have been neglected or ignored on traditional projects (e.g., value management). How often does the A/E on a GC project conscientiously search for alternate design solutions or less-costly construction details? The product, design, or construction detail which has been established in the A/E's technical file is the one usually used in subsequent contract documents. The CM system enforces value management and the organizational meeting details it in the responsibility chart.

It is best to keep the action words and phrases as simple as possible and, if necessary, establish the meaning of words that may be unclear with referenced notations. The chart will be used by people other than those who produce it, and a single interpretation of responsibilities should be clearly established.

On occasion, splitting hairs during discussion might make it necessary to establish a list of action words to be used and provide their definition within the context of the chart. Experience has shown this to be a clarifying factor because the team's meaning rather than the dictionaries meaning will be less controversial.

A representative activity list is provided in Figure 16.4 as a primer. It should not be considered a complete or standard list; the activity list used should be the product of the team members.

Typical Responsibility Chart Activity List

Addenda
Advertisement for Bids
Alternates, Bidding
Award of Contracts
Bid Documents
Bid Openings
Bidders' Lists
Bidder Qualification
Borings, Soil
Budget, Conceptual
Budget, Interim
Budget, Construction
Budget, Contigencies
Cash Flow
Certificate of Substantial Completion
Change Orders
Change Order Requests
Clean-up, Site
Clean-up, Final
Contractability Studies
Constructability Studies
Construction Means, Methods, Techniques
Construction Support Items
Contract Documents
Contractor Coordination
Contractor Call-backs
Contracts, Contractors'
Design Approvals
Design Information
Document Review
Drawings, As-Built
Drawings, Record
Drawings, Working
Estimating, Conceptual
Estimating, Construction
Expediting, Owner Purchases
Field Layout
Field Reporting
Field Testing, Quality Control
Insurances, Contractors'
Insurances, Owner's
Letters of Intent
Meeting, Monthly Project
Meeting, Pre-Bid
Meetings, Pre-Construction
Meetings, Post-Bid
Meeting, Program Management
Meeting, Team
Meetings, Weekly Progress
Meetings, Workshop
Notice to Proceed
Pay Requests, Contractors'
Permits, Building
Plan Reviews, Agency
Prevailing Wage, Interviews
Prevailing Wage, Reports
Project Manual, CM
Project Team
Proposal Forms
Punch List
Purchase Orders, Owner's
Quality Control Program
Safety
Sanitary Facilities
Schedule Enforcement
Scheduling
Security, Site/Building
Site Control
Specifications, Division I
Specifications, Front End
Specifications, Instructions to Bidders
Specifications, Outline
Specifications, Special/Supplemental
Specifications, Technical
Submittals, Samples
Submittals, Shop Drawings
Surety Bonds
Survey, Initial Layout
Surveys, Property
Testing, Field
Value Engineering
Waivers of Lien
Work-Scope Descriptions

FIGURE 16.4 Typical Responsibility Chart Activity List.

For all intents and purposes, creating the responsibility chart is the team's first test of cooperative decision-making and a precursor of the CM team's relationship.

The list of action words in Figure 16.5 is provided as an example. As with the activity list, the team should come up with their own words that describe action responsibility. The potential problem with semantics becomes apparent due to the similar meanings and double interpretation of some words.

Acknowledge	Endorse	Prepare
Advise	Enforce	Present
Approve	Evaluate	Provide
Arrange	Execute	Procure
Assemble	Expedite	Produce
Assist	Facilitate	Purchase
Attend	File	Receive
Award	Head	Recommend
Check	Include	Record
Comment	Issue	Request
Complete	Member	Respond
Comply	Monitor	Responsible
Conduct	Observe	Review
Conform	Obtain	Sign
Coordinate	Organize	Specify
Determine	Originate	Submit
Distribute	Participate	Update
Draft	Pay	Utilize

FIGURE 16.5 A sample list of action words.

16.4.2 An Example of a Completed Chart

The examples of charts in Figures 16.6a and 16.6b have been organized by phase in chronological order rather than alphabetically. As previously mentioned, this is the recommended approach because it allows the activities to be discussed in their natural sequence. It also makes the owner's learning process easier and more understandable. Only two phases are shown—Design and Bid/Award—and only some of the many activities inherent to each phase have been demonstrated.

The responsibility personnel headings used in Figure 16.6a and 16.6b are unique for the particular project they represent. The owner has vested responsibility for both second and third level management in one person, and there is no breakdown shown for either the A/E or CM. This would be appropriate for a small project (probably one smaller than the optimum project referenced in Chapter 8, The CM Organization).

16.4.3 The Responsibility Chart's Depository

The completed responsibility chart should be included as part of the CM Project Manual and referenced by the team when anticipating the next steps in the project. When team members know beforehand exactly who is responsible for each part of a collective activity, the activity planning goes much more smoothly. The responsibility chart facilitates this beneficial part of the management process.

It was stated that the organizational meeting and a GC partnering session were similar but a difference between the two is the absence of adversarial positions. Another much more significant difference can result if the A/E and CM agree to incorporate the responsibility chart into their contracts with the owner. In GC partnering, the partnering agreement is not contractual between parties; it is an informal expression of good intentions. In CM, the responsibility chart is an explicit expression of cooperation developed on the basis of good faith.

Responsibility Chart

Page____ / ____

DESIGN PHASE	OWNER LEVEL 1	OWNER LEVEL 2/3	CM	A/E	CONTRACTOR
14. Specifications, Technical	Review/ Comment/ Approve	Review/ Comment/ Advise	Revise/ Comment/ Advise	Provide	X
15. Alternates, Bidding	Approve as Req'd	Recommend	Analyze/ Recommend	Recommend/ Prepare	X
16. Value Engineering and Value Management	Approve as Req'd	Review/ Recommend	Provide	Assist/ Review	X
17. Estimating, Conceptual	Approve as Req'd	Review/ Recommend	Provide	Assist/ Review	X
18. Plan Reviews, Agencies	Monitor	Monitor	Monitor	Facilitate/ Responsible	X
19. Permits (Not Assigned in Bid Documents)	Pay	X	Arrange for and Obtain	Assist/ Consult w/ Agencies	X
20. Insurance, Builders Risk	Provide	Recommend	Assist/ Recommend	Includes in Specifications	X

FIGURE 16.6a An example Responsibility Chart (Design Phase activities).

BIDDING/AWARD PHASE	OWNER LEVEL 1	OWNER LEVEL 2/3	CM	A/E	CONTRACTOR
10. Bid Openings	Attend as Req'd	Attend as Req'd	Organize/ Conduct	Attend	X
11. Meetings, Post-Bid	Attend as Req'd	Attend as Req'd	Organize/ Conduct	Attend	Attend
12. Letter of Intent, To Award	Approve/ Sign	Recommend	Prepare/ Issue	Recommend/ Approve	Respond
13. Surety Bonds, P/L&M	Review/ File	Advise	Review & Approve	Review & Advise	Provide
14. Insurance, Contractors	Specify and Approve	Advise	Advise/ Monitor, File	Include Requirements in Bid Doc's	Provide
15. Notice to Proceed	Approve/ Sign	Advise	Prepare/ Issue	Recommend/ Approves	Respond
16. Award of Contracts	Award	Advise	Recommend	Recommend	Receive

FIGURE 16.6b An example Responsibility Chart (Bidding/Award activities).

An alternative to including the chart in the A/E and CM contracts with the owner is to simply have it signed by each Level 1 Manager as an additional agreement or operating by-law.

16.5 THE PROGRAM SCHEDULE

The program schedule is a management tool that expedites and tracks a project from start of design to occupancy. The schedule is covered in detail in Schedule Management, Chapter 20. However, its unique function in the management of CM projects compels additional coverage here.

The traditional GC contracting system separates design and construction as two distinct segments of the project-delivery process. This is because design, the prime function the A/E is hired to perform, virtually ends when a contract is awarded, and construction, the function the general contractor is hired to perform, begins when design is complete. Project-long continuity is only attained by the A/E's responsibility for contract administration.

The CM system defines design and construction as a continuum, a part of the design-to-occupancy process. Continuity is achieved through the construction manager whose role and responsibilities are equally prominent during design and construction and whose presence de-emphasizes the traditional switch from design to construction. The entire project is the concern; no one phase is more important than another.

CM program management takes the position that time lost at any point in the project delivery sequence is lost time. The design process is subject to similar time commitments as construction. Owner performance is scheduled no differently than the performance of the A/E, CM, and contractors. The management of construction program participants is based on commitments to perform and the collective monitoring of performance progress by the team.

16.5.1 Program Management Meetings

The first program management meeting should be held as soon as possible after the organizational meeting, and attended by the Level 1 and 2 Managers from the owner, A/E, and CM. The purpose of the meeting is: (1) to confirm the owner's time requirements, (2) to broad-scope the activities from the date of the meeting to the completion of occupancy, (3) to narrow-scope the sequence of activities which occur in the immediate future, and (4) to obtain scope and time performance commitments from team members for the broad- and narrow-scoped activities.

Between the owner-A/E and owner-CM agreements and the responsibility chart, team member commitments to program activities have been established. However, it is highly unlikely that either the A/E-owner or CM-owner agreements contain activity time-performance clauses, other than the CM's commitment to an owner-occupancy date. This means that activity and team member commitments dates must still be established.

Although there are two schools of thought with regard to how to schedule activity performance, experience has proven one to be more productive. The less-productive method reflects an overall misconception of a CM as a dominating dictator rather than competent leader. Although a CM should be proactive with regard to project and team activities, the "carrot" has proven much more productive than the "stick" when getting results on CM projects.

The program schedule uses team member cooperation and the milestone concept to motivate time performance. The milestone concept states: *If you have to be at a certain point at a certain time, and progress is dependent on time consumed by intermediate activities, the surest way to meet the overall time commitment is to establish and meet each intermediate activity time commitment.*

This concept is not new to construction schedulers but is to traditional contract requirements (especially in the GC system, where the contractor agrees to a date of substantial completion but is not contractually required to establish intermediate dates, from which potential completion can be more accurately measured). A typical response by a GC to an owner's inquiry concerning lack of progress is, "Not to worry, we intend to speed things up." When milestone dates are a contract requirement, progress can be assessed with certainty at each juncture.

16.5.2 Team Member Commitments

The team does not have a single or permanent team leader, only leaders pro tem. The team member whose expertise is most germane at a specific meeting or time period is looked upon for leadership by the other team members. The owner is the titular team leader by virtue of the fact that the she/he hired the A/E and CM to provide services. However, the owner must accept the fact that his/her expertise is limited and must agree to alternate leadership with the A/E and CM when their expertise is germane.

The CM assumes the leadership role for program management meetings. It is the CM's responsibility to schedule, plan, chair, and record these meeting for the team. The CM should not come to the meeting with preconceived activity durations. However, when planning the meeting, the CM should know what has to be done to reach a milestone as well as the scheduled time to reach it.

The CM should ask for performance commitments from the A/E and owner, preliminarily accept the responses, and then attempt to improve the commitments, by reviewing the common project goals of the team. If team consensus indicates that a milestone(s) date should be changed, the CM should be prepared to provide valid consequences.

Experience has shown that parties are more apt to stick to commitments made themselves rather than those imposed by others. This reality should be accepted by the team and held as a tenet throughout the project.

16.5.3 Scheduling Design

Architects and engineers have been able to keep rigid time commitments out of their contracts with owners, on the arguable premise that design should not be hampered by time constraints. The premise has merit, but extended design-time durations are not always caused by a search for a better design. They are often caused by an overextended work load or plain procrastination.

When an A/E firm works on a GC project, the owner is the only party to which the A/E is accountable. Most owners are not sufficiently knowledgeable in either the design or project-delivery process to know if the A/E is doing everything possible to expedite design, or if the A/E is taking more time than needed. Without external accountability, A/E firms have the opportunity to maneuver their workload to fit their own requirements. This is a common business practice in the United States and one that is prevalent in professional and nonprofessional services.

The three-party team structure of the CM system provides external accountability to each team member's performance. The design, construction, and contracting expertise of the A/E and CM overlap enough to provide checks and balances to each of their decisions and activities, and the requirement that all decisions be collective brings all team members under the checks and balances umbrella.

All program schedule activities involve either the team or individual team members, and without exception each activity should be scheduled just as vigorously as the other.

A representative list of program schedule activities is shown in Figure 16.7 on page 249. The list is not complete; each project will have its own listing. However, activities on the list that are bold would be appropriate for every CM project.

As the list in Figure 16.7 indicates, not all activities are meetings. The CM team concept requires more meetings than the GC or D–B systems; however, not all meetings must be face to face. Conference calls or written exchanges are both acceptable substitutes. The goals of meetings are information exchange and consensus decisions; the format used to achieve those goals is not important.

The inventory of activities and actions that must be accomplished to move the project from start to finish is important. The list is a checklist, a simple management tool that keeps team members focused and on-task. The program schedule becomes a definitive plan of action by sequencing and assigning durations to the list of activities and actions.

16.6 MANAGEMENT OPTIONS

The major management decision is whether to use a strong office or field staffing arrangement. The difference between the two was covered in detail in Chapter 8, The CM Organization, but the timing of this decision was not. There are two opportunities to make the decision. The best time is before the CM firms submit their proposals for services to the owner. The other opportunity is after the CM is selected and project-management decisions are being made.

A construction manager is in the best position to make this decision, but the owner must make it if it is made prior to receiving CM proposals. Owners who are familiar with CM should have no difficulty handling this dilemma, but first-time users of the CM system would unless they sought expert advice beforehand.

The agenda for the first program management meeting should include the discussion and resolution of the team organization, particularly where team members will be located and how they will interact.

On optimum-size projects, manned essentially by one full-time Field CM (Level 3 Person), the owner and A/E should make their Level 3 Persons very accessible for

Brainstorming Session
Brick Selection
Budget, Conceptual
Budget, Construction
Building Permit(s), Issue
Building Permit(s), Application
Commissioning
Construction, Completion
Construction, Start
Contract Award(s)
Contractability Decisions
Contractability Review
Design Phase, Design Development
Design Approval(s), Owner
Design Meetings, Owner
Design Phase, Schematics
Design Meeting, Architectural
Design Phase, Contact Document
Design Meeting, Electrical
Design Meeting, Mechanical
Design Meeting, Site Civil
Exit Conference
Final Document Review
Finalize Owner Finances
Fire Marshal Inspection
Fire Marshal Review
Instructions for Bidders
Meeting, Quality Plan
Meeting, Value Management Plan
Meeting, Monthly Progress
Meeting, Decision Management Plan
Meeting, Risk Management Plan
Meeting, Information Plan
Meeting, Resource Management Plan
Meeting, Post-Bid
Meeting, Material Management Plan
Meeting, Pre-bid
Meeting, Contract Management Plan
Meetings, Weekly Progress
Meetings, Workshop
Meetings, Team
Meetings, Program Management
Occupancy Approval
Organizational Meeting
Owner Approval(s)
Owner's Board Approval(s)
Phase 2, Document Review
Phase 1, Document Review
Program Management Meeting
Proposal Forms
Purchase(s), Long-Lead Items
Schedule, Detail Construction
Schedule, Milestone
Schedule, Occupancy
Schedule, Short Term CAP
Specs, Outline Site Civil
Specs, Outline Architectural
Specs, Outline Mechanical
Specs, Outline Electrical
Work-Scope Descriptions, Write
Work-Scope Descriptions, Index

FIGURE 16.7 A representative List of Program Schedule Activities.

efficient team interaction. Modern communication techniques make access quicker and more credible. Consequently, their use will influence where the owner's and A/E's Level 3 Persons should be physically located.

As projects increase in size and complexity and time becomes a factor, owners and A/Es should consider locating their Level 3 Persons on-site. If strong field staffing is required, all team members must respond to expanded field personnel requirements.

16.6.1 Team Interaction

The CM team concept changes the philosophy of field operations which traditionally exists in the GC system. The CM team is dedicated to cooperation. Team members in a legal agency relationship have a common goal—the successful completion of the project and the owner's best interest. The A/E's and CM's stake in the project is reputation and the financial rewards derived from future projects. There is very little standing in the way of full cooperation between team members.

On a GC project, similar organizational circumstances can only be obtained from a partnering agreement—a voluntary commitment to cooperation without an agency foundation which may or may not be consistently honored. Experience has shown that some general contractors honor the spirit and commitment of partnering agreements more consistently than others (in the public sector, the apparent low bidder can be one or the other). Additionally, the GC's stake in the project is profit as well as performance. Under certain circumstances, cooperation with the owner and the A/E may be sacrificial and difficult to accept.

A significant difference between a GC and CM project is site management. On a CM project, the person in charge of the site is the Field person, a team member whose information can be relied upon by the A/E and owner as being credible and accurate. When technical problems arise in the field, and they often do, there can be an expeditious exchange of information between Level 2 team managers as to which timely decisions can be made and forwarded to the field. Efficient project management depends on the exchange of credible information, timely discussions, and the providing of timely solutions.

The person in charge of the GC project site is the GC's project manager or superintendent. When a problem arises, the GC contacts the A/E for a solution. A partnering agreement notwithstanding, the GC's status as an independent contractor provides no incentive for the A/E to accept the details of the problem as stated by the GC. The A/E's inclination is to visit the site to confirm the conditions before arriving at a solution. Because the A/E is the owner's agent in regard to the design, the A/E must check the problem's validity before providing a solution.

There are other examples where the CM team concept facilitates valid information exchange, expedites decisions, and enhances the management function. The assurance of credible team interaction under all circumstances is a major factor when selecting the management option for the project.

16.7 THE PROJECT MANAGEMENT PLAN

The program schedule is the graphic version of the project management plan. Once in place and functioning, the program schedule directs the efforts of the team in step-by-step order from project start to finish. Most of the activities in the program schedule have management plans of their own which in turn have procedures of their own.

The program management meetings determine when the various management plans must be started and completed, and the program schedule determines when the activity on which they will be used will occur. What remains is to designate the attendees at the various management plan meetings, designate the lead team members, and schedule the dates of plan completions. A major action is to specify the scheduling format for the program schedule and prescribe its update frequency. Further information on this important schedule can be found in Chapter 20, Schedule Management.

The team's Level 1 Persons should provide the leadership during the first stages of the project management plan development, at least until there is team consensus that leadership can be passed on to Level 2 Persons.

16.8 THE CM PROJECT MANUAL

The CM project manual is the team's depository for the owner-A/E agreement, the owner-CM agreement, other owner-consultant agreements, the responsibility chart, the twelve management plans, and the procedures developed to execute each plan.

The more CM experience a firm has, the easier it is to assemble a CM project manual, because each new manual is based on plans and procedures developed and used on previous projects. Management plan and procedure development was covered in Chapter 7, ACM Procedures. It is a major undertaking for new construction management firms but not a problem for firms which have several CM projects to their credit.

The physical manual should be a loose-leaf binder(s) to facilitate updating. At least nine copies should be produced and maintained; one for each of the three level managers of the owner, A/E, and CM. There will be frequent manual updating, and it is important that the nine official copies be identical and contain the latest information. If more copies are needed by a team member, they should be produced and maintained by that team member from her/his own official copy.

The information management plan should include procedures for producing and updating the manual. Accurate and timely management plan updating is important, especially in the early stages of a project while team decisions and procedures are fluid. CM manual updates should be handled just as formally as contract document changes.

Experience has shown that five to ten CM projects must be completed before an entrant CM firm produces a complete CM project manual, and that the project manual is one of the most neglected part of a CM's services, regardless of how many projects the firm has completed. Experience has also shown that the quality of management services on a project is directly proportional to the quality of the CM project manual produced for the project.

The only way to form an appreciation for the CM Project Manual is to understand that good management requires that everyone involved in the project know exactly what to do and when to do it, and have the assurance that everyone else is equally informed and dedicated.

A construction manager's fee is for management as the last half of the title states, not just for consulting or advice on how to handle problems. Good management anticipates problems and has procedures in place to mitigate problems or prevent them. The CM project manual contains the solutions.

16.9 EXIT MEETINGS

Exit meetings should be shown as activities in the program schedule. They should coincide with the termination of various participants in the project. The purpose of the meetings is to determine what could be done differently on the next project to improve the results obtained on the current project.

The CM and A/E will probably benefit the most from exit meetings because the owner's selection process for the next project is (or should be) dependent on past performance references. A service organization must become more proficient from project to project. The A/E and CM should welcome critiques from contractors, the owner, and each other so that adjustments can be made for the future.

Entry firms have the most to gain. Veteran firms, however, regardless of how long they have been providing services, can use exit meetings to determine if some of the energy is going out of their performance or to determine how new personnel are affecting the organization's standard of performance. There is much to be gained from a well planned and conducted exit meeting.

16.9.1 Exit Meeting Formats

Exit meetings should not be collective; the CM should meet separately with the owner, the A/E, and selected contractors. The CM-owner exit meeting should be attended by the three level managers from both organizations and any others that may have constructive input. The A/E-CM exit meeting should be structured the same way.

CM-contractor exit meetings could be collective but on a selected basis; they could also include the A/E. Owner attendance at CM and A/E contractor exit meetings serves no useful purpose unless the owner intends to undertake another project in the near future.

Exit meetings should be located, planned, and conducted in a non-confrontational manner. They should be held on neutral ground in an informal atmosphere (a retreat location such as a restaurant or club). They should be conducted in a cooperative spirit, with the understanding that both parties will benefit from their experience of working together on the project. Not only will participants find ways to improve performance on the next project, they could discover why certain things they expected were not done and why certain things they did not expect were.

Exit meetings should be unstructured, and each party at the meeting should have the opportunity to bring up items for discussion. Enough background should be provided to generate recall and response.

At a CM-A/E exit meeting, the logical discussion leaders would be each firm's Level 2 Person. They will be most aware of the course of the project. However, the selection of a discussion leader should not discount the quality of relationships which existed during the course of the project. A Level 1 Person would be a second choice.

It is important to remind those at an exit meeting that a construction project is unique, that it is a manufacturing process, without weather protection, a static work force, or the benefit of prototype guidance. It produces a one-of-a-kind structure, built by virtual strangers who are quickly assembled for a relatively short period of time. Until the project begins, usually very little is known about the participants.

Each construction project, regardless of the contracting system used to deliver it, includes first-time working relationships to one degree or another. More than likely there will be more initial relationships than recurring relationships. Construction projects represent one of management's most difficult challenges.

One of the CM system's major potentials lies in the fact that its contracting structure provides the best opportunity to combine optimized design and equitable cost competition between trade contractors with an unprecedented level of management. This combination is designed to bring out the best in construction industry practitioners who are part of a CM project. Exit meetings should confirm this fact to one degree or another. If they do not, adjustments will most certainly have to be made.

16.10 OTHER MEETINGS

With very few exceptions, meetings should be well planned, have a predistributed written agenda, provide timely notification to all participants, and be conducted as expeditiously as possible without compromising the intent of the meeting.

As noted earlier, meetings facilitate the operation of the CM system; there will be many meetings of one type or another during the course of a project. As long as they are productive, their frequency should never be questioned. Those who feel that meetings should be held to a minimum in any cooperative undertaking do not understand the intense interactive process upon which the CM project-delivery structure relies.

The strength of the CM system is vested in the successful interaction of project participants with the same goal and similar interests but different expertise. The multitude of decisions which must be made to move a construction project from start to finish should take full advantage of each team member's expertise and will then receive the benefit of the all-important checks and balances provided by a CM team structure.

Meetings would not be necessary if all the expertise assembled for a CM project were vested in one party who could be relied on to make every decision in the owner's best interest. If this were possible, decisions would be unilateral and other parties would not be involved. However, if it were possible, a contracting system structured around the single party would produce maximum efficiency and provide far superior results than either of the three contracting systems currently available. Such a system is not available or possible in the construction industry.

The only way to approach this perfect level of efficiency (and the superior results which would accrue) is to assimilate a one-party system as closely as possible. In the CM system, this translates to collective, open, frequent, communication between the A/E, CM, and owner. If it were practical to have all parties physically together on a daily basis throughout the project, it would be closer to the ideal than CM practices currently provide.

As explained in Chapter 13, Information Management, all communication from one team member to another must be copied to the third team member. Face-to-face meetings are an expensive but necessary way to achieve this objective when major decisions are being pondered. However, if it is practical to meet by conference call, a computer network, or team circulated letter, one of these options should be chosen.

Meetings should be looked upon as essential, not an inconvenience.

CHAPTER 17

Quality Management

The CM contracting structure provides new opportunities for achieving owner-specified construction quality. These opportunities are collateral benefits and inherent to the system. They stem from the unique alignment of traditional project participants which CM produces.

The CM structure places the CM, the construction/contracting-oriented party, on the same ideological side of the project's well-being as the owner and design professional. Quality in the project is pursued through the common efforts of all project team members.

The CM system's attributes also provide a logical answer to a question which has persisted for a long time: "What is construction quality?" CM philosophy promotes quality as "conformance to plans and specifications," thus leaving little room for debate about high or low quality materials and workmanship during construction. If a quality debate occurs, it will be between the project team members before quality is documented by the A/E in the technical specifications.

In the GC system, it is common for contractors to interpret specifications as the maximum requirements they must meet, while A/Es and owners interpret the same words as the minimum requirements the contractor must meet. CM philosophy accepts that quality can be specified at various levels during design but, once specified, conformance dictates contractor performance.

The intent of CM Quality Management is to accurately pinpoint and record the owner's quality needs and decisions as early as possible, ensure that the owner's choices are specified, and extract conforming performances from constructors in the field. In so doing, the CM neither preempts the sole responsibility of an A/E to write technical specifications for CSI Divisions 2 through 16 nor relieves the contractors of responsibility to conform to specifications during construction.

17.1 THE QUALITY MANAGEMENT AREA OF KNOWLEDGE

The quality management area of knowledge encompasses all elements of CM project delivery that contribute to the quality of the end product. Quality is stipulated by the client, designed into the project by the A/E, reviewed by the team, and constructed into the project by contractors. During design, quality has varying levels from high to low. Once specified, quality must conform to the levels specified. Quality management is a continuing process originating with client decisions and ending with contractor conformance.

A CM must be capable of designing, installing and directing a quality management system that fits the needs of the project. The CM must have knowledge of construction materials and products, understand their use and capabilities, and the available means, methods, and techniques applicable to their installation.

The CM must know how to interpret contract drawings, technical specifications and shop drawings, understand field and laboratory testing procedures, and stay current on construction means, methods, and techniques and the latest procedures for assessing material and installation quality.

The quality management area of knowledge includes technical specification writing, materials testing and measurement procedures, product and material characteristics and capabilities, manufacturing tolerances, contractor installation capabilities, building codes, and design standards. This area also requires excellent communucation skills and performance ethics.

17.2 THE QUALITY MANAGEMENT NETWORK

Quality, like safety, can only rely on constant awareness for success. It is important that all project participants become part of a collective quality effort and that quality is the primary goal of daily project activities from the start of design to owner occupancy. To accomplish this, a bold new approach must be considered. The diagram in Figure 17.1 shows the quality conformance network.

The quality management network is structured around five major interdependent components. They bring each project participant into the network in a reasonable way and prescribe pragmatic procedures that interface quality awareness with daily activities throughout the project.

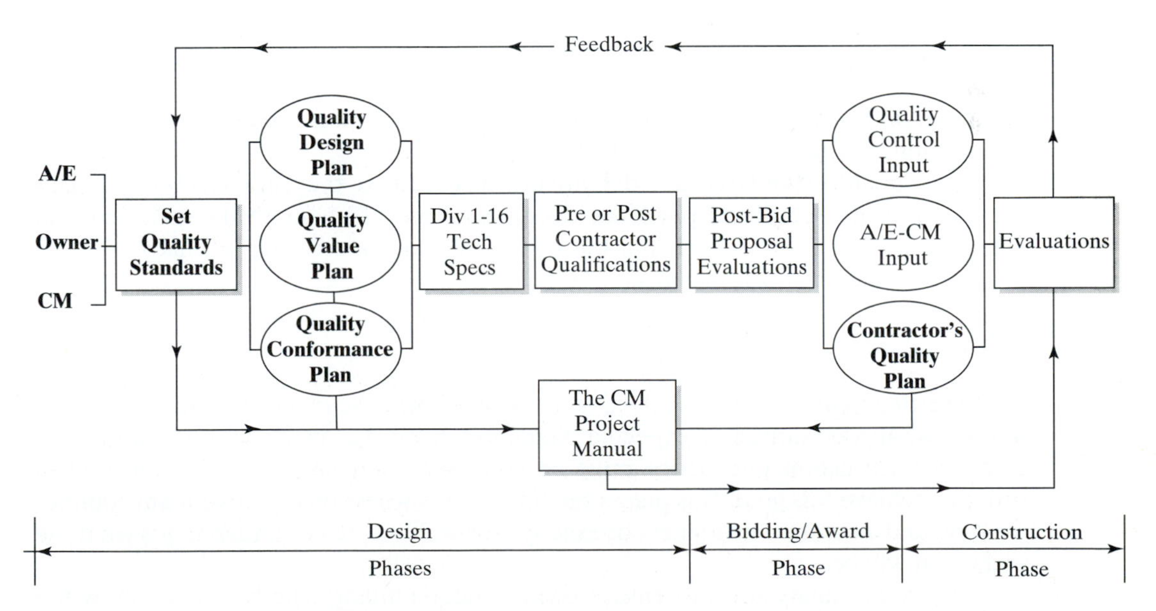

FIGURE 17.1 The Quality Management Flow Network.

As alluded to in the quality management area of knowledge, this action area demands the CM's communication skills and ability to organize the network and administer it in a productive manner. Quality is difficult for most owners to definitively prescribe and at times difficult for A/Es to retain at its prescribed level throughout design. The network components are designed to alleviate these problems as much as possible.

Although some of the components are directly applicable to GC and D–B contracting, the ACM form (with multiple contracts) is the structure that accommodates them all and has the potential to provide the full benefits from the quality management network.

17.3 MAJOR QUALITY MANAGEMENT COMPONENTS

The major components of quality management are the:

- Quality Standards proposed by the team
- Quality Design Plan proposed by the A/E
- Quality Value Plan proposed by the CM
- Quality Conformance Plan proposed by the team
- Contractor's Quality Plan proposed by each contractor.

17.4 THE QUALITY STANDARDS

Establishing and recording the owner's quality standards, as early in the project as possible, is the linchpin of productive quality management. The quality expected by the owner upon the completion of the project must be categorically extracted from the owner by the A/E and CM. What the A/E misses, the CM must pick-up. The establishment of quality standards is a team obligation; the maintaining of quality standard documentation is the responsibility of the CM.

The quality standards are definitively recorded as they are established. This record of owner decisions produces documentation which must be reviewed and amended as design progresses and the owner develops increasing intimacy with the project and construction elements.

17.4.1 The Goal of Quality Standards

When setting quality standards, the goal is to never hear the owner say that she/he did not get what was wanted or expected. Somehow, even the owner with the minimum level of construction knowledge must be constantly and progressively educated in order to achieve this goal. This places the burden of success on effective team communication and a constant awareness of exactly how well the owner understands what the end result will be.

There are many ways to ensure owner understanding. The best is to show the owner examples of design, materials, and equipment as they actually are in completed facilities and relate their reality to the specifications developed for his/her project. The pursuit for understanding need not be exhaustive but must be sufficient to convince the A/E and CM that the owner knows what to expect when construction is complete.

Architects, engineers, and contractors relate to specifications and drawings; they can visualize space, shape, appearance, quality and their related costs. They can usually communicate on alternatives without additional visual dependency. To use this level of communication with owners when discussing the attributes of the project would be a grave mistake. The A/E and CM must provide more.

17.4.2 Quality, Cost, and Value

When establishing the owner's quality standards, cost is usually a consideration in every decision. This is where value management interfaces with quality management. It is a very important interface that cannot be sidelined. The A/E and CM must constantly remember that cost and value are not the same. The ideal project would be one specified solely on the basis of value, not cost. However, it is highly unlikely that the ideal project will ever come along. Therefore quality standards must be established with a balance between the two; where value is too expensive, cost must provide a solution.

17.5 THE QUALITY DESIGN PLAN

The objective of the Quality Design Plan is to ensure that the owner's documented quality standards are first and foremost in the minds of everyone involved in project design.

To produce a project from feasibility through design to the contract document phase requires the efforts of the many individuals on the design team. Everyone responsible for decisions that could reflect the owner's preferences must be made aware of those preferences through documentation. This can be accomplished by distributing updated quality standards to all those involved.

However, the structure and interdependent operations of a design team present a communication problem that must be overcome. A plan must be formulated to bring all design interests together in a timely and effective manner. The plan should provide for design team and design review meetings on both a scheduled and as-needed basis. The project team determines when these meeting should be held and incorporates them into the program schedule as part of project management.

What is suggested here currently takes place to one degree or another on all projects. It is referred to as *design coordination*. The difference between the CM approach to this responsibility and what is commonly practiced is that a CM is available to inject pragmatic construction expertise and comparative cost information and add checks and balances to design decisions. Additionally, the design coordination responsibility of the A/E is emphasized and formalized, making it more effective from the owner's perspective.

The Quality Design Plan should be developed by the team under leadership of the CM, with principal input and strategy supplied by the A/E representing the interests all design team members. The plan should clearly state when and how the quality standards are to be updated and distributed to design team members, how the A/E intends to monitor design coordination, how the team will be involved in quality decisions, and how the Quality Design Plan will interface with other quality plans and the project management plan.

17.6 THE QUALITY VALUE PLAN

The goal of the Quality Value Plan is to ensure that owner values are incorporated into the project and, where applicable, that alternative designs be developed and evaluated for consideration by the owner.

Constructability, value engineering, and life-cycle costing are part of the CM's responsibility under the broader heading of Value Management (Chapter 15). The CM's prime construction expertise and supportive design expertise facilitates pragmatic design input and has the potential to improve the overall value of the project from the owner's perspective.

Project design is a moving target that expands as it proceeds—similar to a snowball rolling down an incline. Each time a design decision is made, it becomes part of the total design, complicating any alternative that surfaces in the future. With this in mind, the CM must formulate a Quality Value Plan that will discretely incorporate value assessment activities into the design phase which minimally interfere with the momentum of the A/Es design activities. A process of stop, redesign, and restart should be avoided.

17.6.1 Design Reviews

Periodic Project Design Reviews involving the owner, the A/E's design team, and the CM's value team are necessary. However, unless the reviews are numerous and close together, significant redesign could result, a situation which should be avoided if at all possible. It is important that each team know what the other is doing on a current, ongoing basis so their input can be integrated into the design in a timely manner.

The solution is to organize the A/E design team and CM value team into subteams to micromanage the specialty areas of design (mechanical, electrical, architectural, civil site, structural, etc.). If the project is of such a nature that an owner design team is also to be involved, it too should be similarly organized. With this matrix organizational structure, specialty design reviews by subteams can occur at a necessary frequency, thus maximizing the efficiency of the design review process and minimizing the extraneous involvement of others in project design team meetings.

To facilitate the absolute need to interface all subteam decisions on a current basis, a team hierarchy should be created to quickly pass on the information to the project design review team and other subteams that may be affected. Project Design Reviews should be scheduled by the project team based on overall design progress and the consequential impact of subteam decisions at any point during design.

17.6.2 Plan Development

The Quality Value Plan should be developed as early as possible by the CM with significant input from and the complete concurrence of the A/E. The brainstorming session covered in Chapter 16, Project Management, is the mandatory starting point for the plan's development. All A/E design team members and all CM value team members should be present at the session to meet and to find common ground for their important roles in the value management process.

As with all management plans, the owner should fully understand what the plan will accomplish and how it will be carried out. This is particularly true of the Quality Value Plan. It is an area of CM responsibility that owners often question and an area that is often remote from the owner's design/construction expertise.

It is common for owners to question if value management contributed to the success of the project or if the value management function was even provided. Value management has the potential to be the most prolific value-producing attribute of the CM contracting structure, and every effort should be made to apply it generously and to demonstrate its effectiveness to the owner during design.

17.6.3 Design and Value Team Representation

Figure 17.2 on page 260 graphically shows the quality and value expertise that can be effectively used to produce value-based quality in the constructed project. Teams and subteams can be formed from the structure to fit the cooperative requirements. When using subteams, one of the members of the subteam should be designated coordinator and be responsible for passing information to other subteams and the Level 2 Manager. A typical mechanical design subteam is shown in Figure 17.3 on page 261.

It should be understood that the boxes in Figures 17.2 and 17.3 do not imply that an individual must be provided for each of the designated areas of expertise. A person with more than one area of expertise can function in as many areas as his/her capabilities and capacities permit.

Figure 17.2 shows the owner as one with design and value expertise on-staff. This is the exception rather than the rule. However, the structure welcomes owner expertise if it is available. In any case, owners should provide quality/value representation at the Level 2 Manager position to validate the setting of quality standards and critique design as it develops.

The quality standards, quality design, and Quality Value Plans must be inserted in the CM project manual and working before the design phase moves ahead too far. The value decisions for architectural, structural, and site design should be made early in the preliminary or schematic phase. Value decisions for mechanical and electrical design should be made as soon as possible thereafter but no later than the early part of the design development phase or mid-part of the preliminary design phase.

The two plans should cover probable activities during bidding, award, and construction phases as well. The standards and the two plans that maintain the standards will facilitate technical changes that are requested or required after final design or the contract document phase is complete.

17.6.4 The Design/Value Notebook

As previously mentioned, whether or not the CM is responsibly bringing cost/value equity to the project through value engineering and life-cycle costing is often questioned by owners. It is important to show the owner that they are. To this end, it is

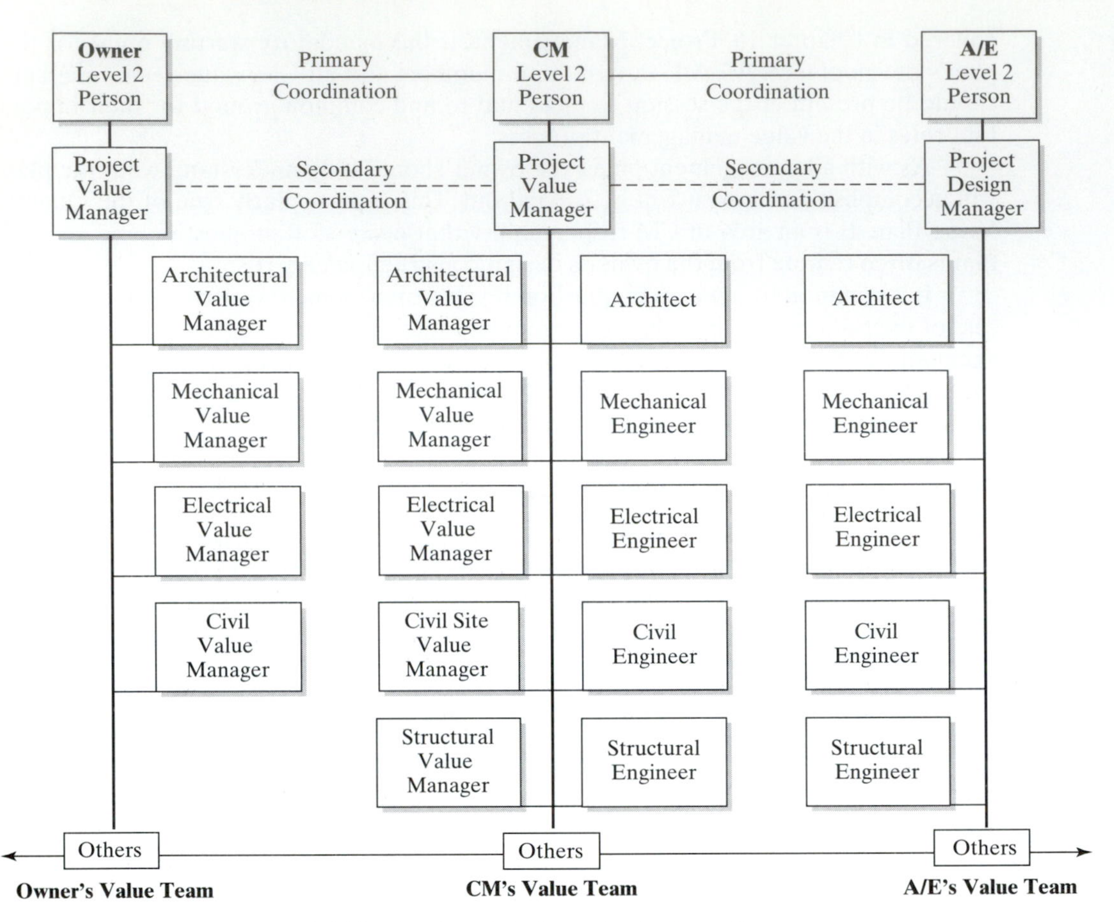

FIGURE 17.2 Typical Design and Value Team Structures.

suggested that some or all of cost/value studies be documented by the CM as they occur and filed in a design/value notebook, with copies of each study distributed to the owner and A/E. This documentation not only assures owners that the CM is providing cost/value studies, it also becomes an effective sales tool when looking for new business.

17.7 THE QUALITY CONFORMANCE PLAN

The goal of the Quality Conformance Plan is to extract specified materials and workmanship from the contractors functioning on the project. The plan must recognize the contractual responsibility of the contractors to provide specified quality and must not assign quality conformance responsibility in whole or part to parties other than the contractors themselves.

The Conformance Plan should include the usual preconstruction reviews and appropriate on-site and off-site testing. But it should also include a process for determining contractor conformance from the perspective of the drawings and the CSI

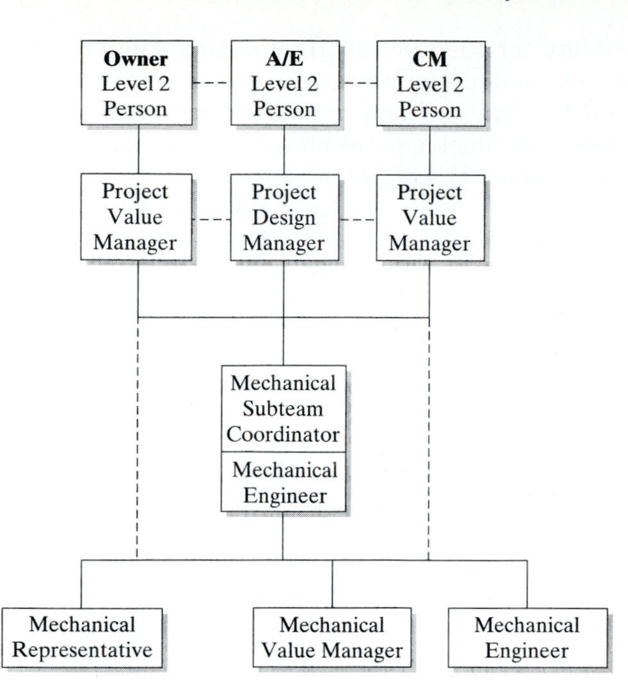

FIGURE 17.3 A typical Design and Value Subteam Structure.

Technical Specification Divisions 2 through 16 as developed by the A/E during the design phase.

17.7.1 Preconstruction Reviews

Preconstruction reviews include shop drawings, product data, and samples. As part of the Quality Conformance Plan, a procedure should be developed by the A/E and CM to expeditiously move these items through the review process. The A/E's contractual responsibility for technical review of shop drawings, product data, and samples provides ample reason for the procedure to be finalized to the A/E's satisfaction.

A collateral benefit of the CM contracting structure is the potential to accelerate the submittal review process. In the GC contracting structure, the GC is the focal point for submittal transmission. The GC receives them from subcontractors and suppliers, reviews them, and forwards them to the A/E who in turn forwards them to the appropriate design team member for technical conformance review. It is a five-step, linear procedure that starts with a supplier and progressively moves submittals to the ultimate reviewer. The return of reviewed submittals followed the same path in reverse. Even when using every available transmittal conveniences (FAX machines and same-day mail), the process consumes time in transit and often delays delivery of material/equipment.

Contractually, the A/E's review responsibility does not include checking the compatibility of interfacing components; each component is reviewed on its own merits.

Responsibility for checking interface compatibility is left to the GC as the party responsible for ultimate project conformance.

A GC's failure to completely or conscientiously check submittals often causes construction coordination problems, with accompanying delays. It is common for GCs to discount submittal obligations when pressured by other operating problems. GCs often have the attitude that someone in the submittal chain will catch what the GC may miss. Few general contractors perform preconstruction submittal responsibilities energetically or enthusiastically.

The specific contract responsibility and the thrust of the CM's purpose on the project can turn submittal reviews into the effective process it was meant to be.

17.7.2 CM Preconstruction Reviews

Figures 17.4a and 17.4b on pages 263 and 264 describe one way preconstruction submittals can be handled expeditiously and effectively. There are others that work as well. This system reduces traditional submittal turnaround time by having contractors send submittals directly to the A/E at the same time they are sent to the CM. This action permits reviews by the A/E and the CM to occur simultaneously and eliminates dependency on timely successive party action as well as one transmittal interval.

The time between the date a vendor's initial submittal goes to a contractor and the date the vendor receives a satisfactory review is essential to planned construction progress. Portions of the work that require reviewed submittals cannot be started unless a satisfactorily reviewed submittal is in the contractor's hands.

The fabrication or manufacture of many items will not commence unless the supplier has an "approved" submittal in hand. It is common for suppliers to designate delivery time as so many "days after receipt of an approved shop drawing." Consequently, the submittal review process could become a major factor in the project schedule.

A few other features of the submittal system in Figure 17.4 should be observed. The first is that two standard submittal forms should be developed by the project team. One is for use by contractors when they initiate a submittal; the other is for the A/E when returning submittals to the CM and contractors.

Computer software is readily available that facilitates customized, inexpensive forms. Because there will be numerous submittals from many contractors involved in several projects, it helps if the contractors, A/E, and CM have a simple visual means of recognizing which project the submittal belongs to and the status of the submittal when it arrives.

Color coding by project, as well as color coding multiple-copy forms, is a simple way to efficiently handle the volume of paper involved. It is an example of good management that simply smoothes the process. The use of bar codes can also keep an important paper trail conveniently in order.

The system in Figure 17.4 also provides a convenient approach to expediting the submittal process. As transmittal forms are received by the CM, they can be logged into a computerized expediting file and tracked while they proceed through the review process. The CM should be able to determine at any time, where each submittal is and, if it is bogged down, take appropriate action to expedite it.

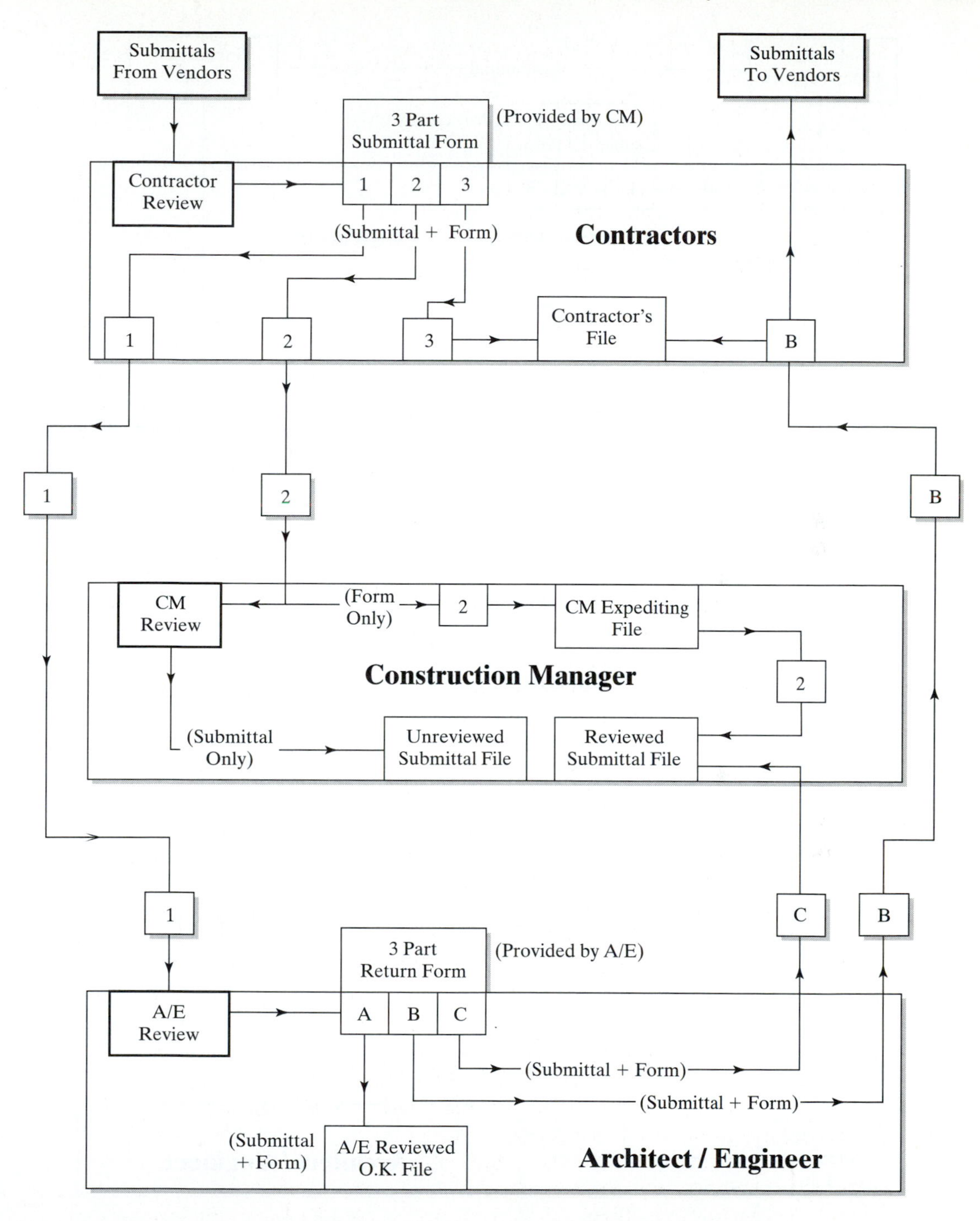

FIGURE 17.4a Preconstruction Submittal Review Procedure (for "Reviewed, O.K." or "Reviewed, O.K. as noted" submittals).

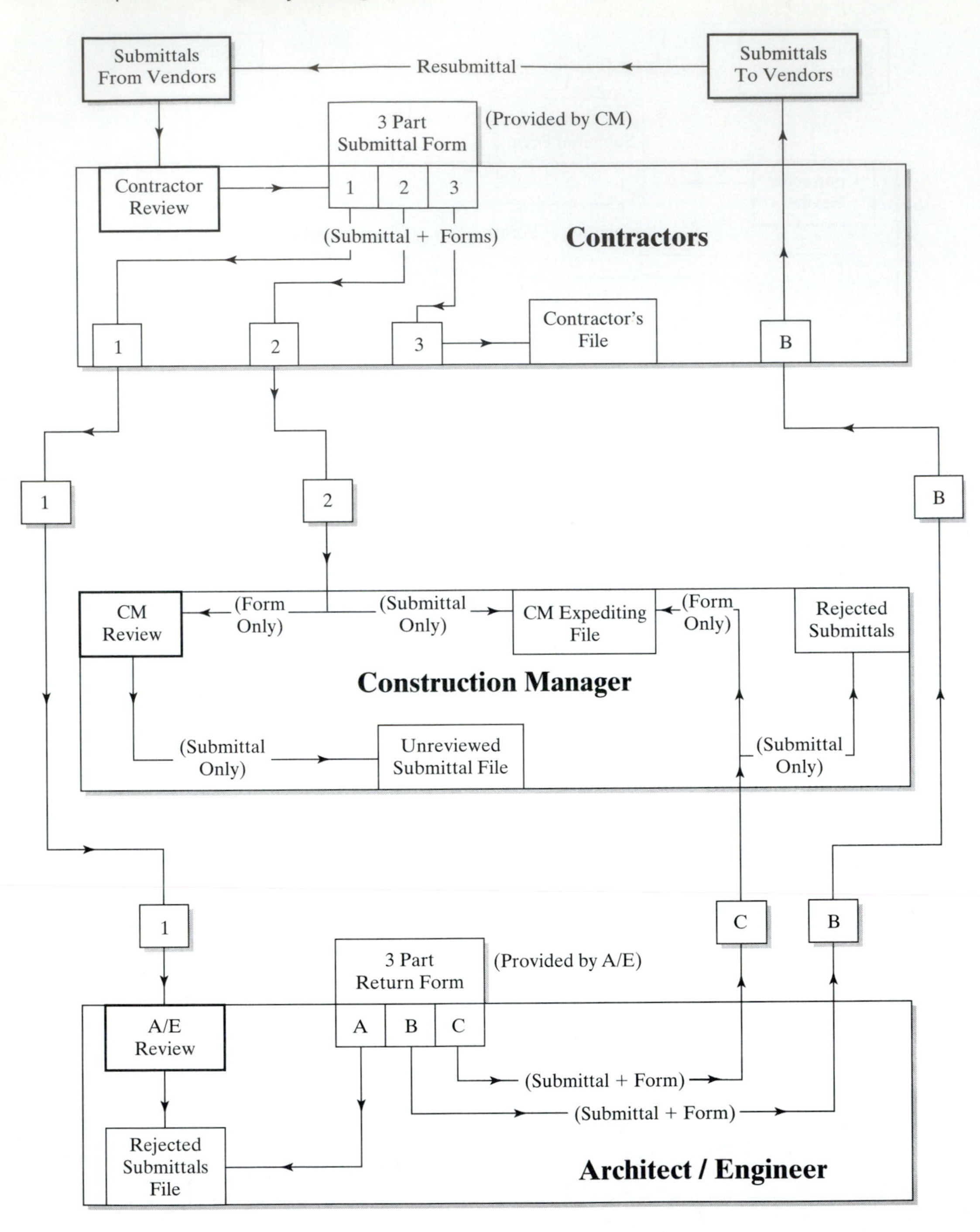

FIGURE 17.4b Preconstruction Submittal Review Procedure (for "Reviewed, N.G." submittals).

To accomplish this, the submittal forms provided to contractors should be consecutively numbered and distributed to contractors in blocks of numbers. This eliminates the option of recipients assigning their own logging numbers. The submittal number printed on the initiating form can be used by the A/E and entered on the return form issued by the A/E. Communication and tracking will be simpler with a single numbering system.

Another feature of the system shown is the positive separation of unreviewed and successfully reviewed submittals in the possession of the CM. If the transmittal forms provide sufficient information, the CM can file the forms coming from contractors separately from the submittal itself. This will prevent unprocessed information from mixing with processed information, especially when CM submittal review occurs both in the home office and the field.

Lastly, if the CM or A/E review uncovers a problem with a contractor's submittal, a telephone call might be sufficient to relate the specifics of the problem and perhaps arrive at an expeditious solution. It must be remembered that the A/E and CM are representatives of the owner and have a common goal in solving project-related problems. While a shortcut such as this would be risky in the GC system, it is a logical alternative in the CM system.

17.7.3 Quality Conformance Responsibility

There are two separate responsibilities when dealing with quality conformance: *providing* quality conformance and *determining* if quality conformance has been achieved.

The party that signs the contract for construction with the owner is responsible for providing construction that conforms to the drawings and specifications covered by the scope of work stated in the contract.

Determining if conformance has been attained by a contractor should remain the responsibility of the party that developed the drawings and wrote the technical specifications from which the contractor constructed and by which conformance must be measured.

Departing from these two well-founded principles when developing the Quality Conformance Plan is neither prudent nor appropriate. However, under pressure from disoriented owners or advice from those not thoroughly knowledgeable in the CM system, the departure sometimes happens.

17.7.4 Inspection by Convenience

The fact that the CM's Level 3 Person is in residence on the construction site tends to influence owners to assign day-to-day inspection responsibility to the CM. This assignment for shear convenience cannot be supported by logic. The day-to-day responsibilities of the site manager leave no time for more than cursory construction inspection opportunities.

This is not to say that inspection is not important. The on-site manager (as well as other team members on site for whatever reason) has a responsibility to point out nonconformance to contractors when observed. However, inspection's perceived importance converts to virtual unimportance when it is assigned as a convenience.

If the owner deems it necessary to have full-coverage, day-to-day construction inspection by the CM, additional personnel trained specifically as inspectors could be provided by the CM (and paid for by the owner). A better solution, however, would be for the owner to add an inspection firm to the project team. The inspection firm would be responsible for both visual and physical quality assurance and involved in the Quality Conformance Plan.

The standard CM approach to construction inspection calls for no more than a cursory involvement by the field manager but requires that the Quality Conformance Plan designate those specific construction activities where surveillance by the A/E, CM, or both is required. Deciding which activities should be listed is the prerogative of the A/E. Once the activities are decided upon, notice should be served on contractors through the contract documents.

17.7.5 Quality Conformance on ACM Projects

A closer look at the use of multiple contracts on a CM project should lessen the owner's concerns for quality conformance. Involving multiple contractors beneficially alters the manner in which contractor conformance is reviewed when compared with the GC system.

This benefit is derived from the fact that there will be as many substantial completion date inspections as there are prime contractors on the project. On the typical GC project, there is one substantial completion date for the prime contractor—the date on which the project can be beneficially occupied by the owner in whole or in part.

A contractor's application for date of substantial completion triggers several administrative events that grant, deny, or change the date requested. An inspection is conducted by the A/E and owner to determine if the contractor's work is acceptable, and a list of uncompleted work or work that requires correction is made (punch list). The dollar value is retained by the owner to secure final completion.

Obviously, when multiple prime contractors are involved, the accepted, literal interpretation of that definition no longer applies. Each contract has a unique date of substantial completion, many of which have nothing to do with physical occupancy by the owner. However, each application for a date of substantial completion must be followed up by an inspection of the contractor's work by the A/E and a decision on contract conformance. If there are thirty prime contractors on the project, there will be thirty substantial completion inspections (conceivably at thirty different times).

There is no comparison between the value to the owner of this multiple-quality conformance procedure and the single-quality conformance procedure stipulated for the typical GC project. One obvious advantage is that there is less chance that nonconforming work performed early in the project has been covered up and is not visible when the A/E inspection takes place toward the end of a GC project.

17.7.6 Substantial Completion Requirements

The conditions that determine contractor substantial completion should be defined in the contract documents and will vary from contractor to contractor, depending on the type of construction performed and the ensuing interfaces with other contractors. The procedures which lead to determining the conditions of contractor substantial completion should be a part of the Quality Assurance Plan.

One only has to take inventory of the variety of construction services performed by the many trade contractors on a project to determine the completion requirements for each. For example, the substantial completion requirement for a pile driving contractor actually coincides with final completion. If each pile was driven according to drawings and specifications, there is nothing left for the contractor to do except clean up and demobilize. Consequently, there is no need to establish extensive substantial completion criteria for a pile driving contractor or other contractors with obvious completion definition.

On the other hand, the contractor who performs masonry work has considerable control over quality conformance, which must be a consideration for substantial completion. The masonry contractor also has interface responsibilities with other contractors, which also must be validated before substantial completion is granted. Consequently, the masonry contractor, and all contractors with similar individual control over their work, should have very explicit substantial completion requirements in their contracts with the owner.

17.7.7 Contractor Qualification

The Quality Conformance Plan should include procedures that assure as much as possible that the contractors engaged for the project are competent and capable of conforming to the drawings and specifications.

Although legal restraints often suppress consequential prequalification of contractors on public sector projects, there is still value in alerting potential bidders to the fact that quality conformance will be a major performance requirement on the project. There are no laws against obtaining attested information from contractors with reference to their past performances. In fact, such information may be of significant value in the event that performance problems develop with a contractor who provided questionable information.

Where permitted, contractor qualification should be prequalification, not post-qualification. It is much less "traumatic" to deny bidding privileges to a contractor than to withhold an award from a contractor who submits the low bid. Comprehensive prequalification questionnaires should be developed as part of the Quality Conformance Plan. A procedure that will assure fair and equitable comparisons and noncontroversial selection of contractors who file them should be documented.

17.7.8 CM Qualification Efforts

Aside from contractor qualification by form, the CM must screen potential bidders in every other possible way to evaluate their availability and specific performance possibilities. The CM should research the contracting field in an objective manner and substantiate questionable references with additional effort.

There are a number of information sources available in every area: local trade contractor organizations, cooperative general contractors who have worked with various trade contractors in the area and beyond, and owners who recently built or have extensively built in the recent past.

A CM that is knowledgeable in the construction industry should have no more difficulty investigating and accurately assessing the contracting marketplace in areas remote to his home base than in his own backyard. In fact, the chance of a CM producing a more complete and beneficial objective assessment of contractors away from his home base is much better, due to the amount of original effort that will be expended and the absence of preconceived opinions of local contractors.

17.7.9 Contractor Supervision

A final inclusion in the quality conformance plan should be a stipulation to include the names of acceptable on-site supervisors of key contractors in their owner-contractor agreements. Instituted to remove self-serving supervisory changes from contractors, this practice has become quite common in the GC system.

One of the signal attributes of the CM system is that owners can select their project managers on the basis of qualification and demonstrated ability. To extend this attribute deeper into the system, it is logical for owners to have the option of controlling the key personnel that contractors assign to the project. This is not necessary for all contractors, only those who are pivotal regarding timely or quality performance. To accomplish this, the CM must extend contractor research and qualification to include the pros and cons of their supervisory personnel. The availability of these persons to supervise the construction should be factored into the contractor selection process, and contract provisions should be inserted to require compliance. For practical purposes, more than one preferred supervisor should be named for each contractor selected to function under this provision.

17.8 THE CONTRACTOR'S QUALITY PLAN

To enhance contractor quality conformance responsibility, each contractor should be required to submit a Contractor's Quality Plan—a detailed description of how the contractor plans to accomplish contract responsibilities and meet the specified quality conformance criteria.

A Contractor's Quality Plan should not echo the requirements of the drawings and specifications. It should explain how the contractor intends to cooperate with interfacing contractors and maintain construction progress, and it should detail the construction means, methods, and techniques to be employed.

Contractor Quality Plans are not new; they have been required by some owners for many years (mostly on engineering projects) and have proved their value time and again. The major deterrent to their more frequent use is that the GC system and its single prime contractor format turns creating a quality plan into a major burden for the GC on an architectural project.

On architectural projects, the majority of the construction will be performed by subcontractors. This means that each subcontractor must develop a plan and submit it to the GC. The GC must then compile all the plans into a single plan that covers the entire project. This becomes time-consuming for the GC at the start of a project when efforts are concentrated on getting construction underway. As a consequence, and in spite of their value, Contractor Quality Plans have not become a standard on GC projects.

In the CM contracting structure, where GC subcontractors are now prime contractors, creating Contractor Quality Plans is no longer a problem. Each contractor creates a plan covering her/his own work and submits it to the CM within a reasonable time before interfacing of the initial contractors begins. Contractors who do not interface with the initial performing contractors need not submit plans until later in the project.

Each plan is a fraction of a total GC plan, so none are significantly burdensome. However, the CM must see that the contract documents permit ample time to review the plans and coordinate requirements between interfacing contractors.

17.8.1 Preconstruction Meeting

Every contractor's Quality Plan should mandate that the contractor's supervisor meet with the CM's Level 3 Manager a week or so prior to bringing a crew on-site to construct. The purpose of the meeting is to review the terms of the contract documents, especially the drawings and technical specifications, and go over the contractor's previously submitted and reviewed quality plan. This preconstruction meeting between an individual contractor and the CM is of utmost importance and should not be waived under any circumstances.

One of the major problems with trade contractor performance is blatant disregard for a project's specific requirements. The constant repetition of specialized work, day in and day out, from project to project, installs a mindset that every project is the same, when in fact they may be quite different.

Most trade contractors need a wake-up call on CM projects, especially if it is their first involvement with CM or with the specific construction manager. The preconstruction meeting provides this opportunity.

17.8.2 Construction Means, Methods, and Techniques

It is common practice to place as much responsibility as possible for construction means, methods, and techniques in the hands of the contractors. Contractors are most familiar with what works and what does not and have the ability to come up with economical solutions to construction operations.

The CM system supports this. However, unchecked freedom of choice in this area can lead to problems. A contractor's construction operations intentions should be known before he/she moves onto the site and begins work.

As an example, it will help the CM's contractor coordination efforts to know the sequence and timing of pours the concrete contractor intends to make, the equipment to use, and the source of concrete supply. Coordination planning may change if a conveyor, concrete pump, or crane and bucket system are to be used. If a conveyor is the contractor's choice, additional information concerning how segregation is to be controlled will be good to know before the first pour is scheduled.

The contractor's Quality Plan should include a detailed explanation of how the work is to be done with little omitted on the pretense of common practice. This part of the plan serves three purposes. First, it greatly assists the CM in planning the coordination of contractor operations. Second, it provides an opportunity for contractors to pointedly review their proposed construction means, methods, and techniques in the context of the immediate project's requirements. And third, it instills confidence in the contractors that the project is to proceed in a controlled and orderly fashion—a way that will allow them to extract their expected profit.

17.8.3 Interface Actions

On all projects, efficient interfacing of trade contractor work is a major challenge to general contractors (or construction managers, as the case may be). The fact that trade contractors are prime contractors on CM projects provides opportunities to reduce the frequency of interfaces and directly enlist trade contractors as cooperators rather than competitors.

The contractor's Quality Plan should explain how the contractor plans to interface with any preceding, concurrent, and following contractors. The plan should state how the contractor plans to verify the condition and dimensions of preceding construction before beginning work, and how the progress, condition and dimensions of the work are to be documented as it proceeds.

As an example, in an attempt to reduce an interface, assume a masonry contractor has been assigned to set hollow metal door frames in masonry walls, a task normally assigned to a carpentry contractor. The masonry contractor's work-scope states this, and the technical specifications prescribe the dimensional tolerance criteria that must be met to successfully hang the doors in the future.

The masonry contractor must include in his Quality Plan how he intends to assure the contractor who will hang the doors that the frames have been installed according to specifications. This could be accomplished via dimension documentation, periodically verified by the door contractor while the masonry contractor is active on the project. If this is the solution, the door contractor must include in his quality plan that these verifications will be periodically made.

It would, however, be simpler if the CM verified the dimensions, and there is no reason why this cannot unofficially be done. However, ultimate responsibility for dimensional interfacing must remain with the contractors.

17.8.4 Contractor Quality Plan Submittal

For the group of contractors who must be on-site at the start of construction, the Contractor Quality Plans should be submitted to the CM as soon as possible after award. The plans for the remaining contractors should follow the first group on a schedule that permits CM processing prior to a contractor's move on-site. Submittal lead time should convenience the CM's need to review, coordinate, and comment on the plans and include time for the contractors to respond to the CM's comments.

The CM's responsibility to the contractor in the review process should be the same as the A/E's responsibility to the contractor in the shop drawing, product data and sample review process. The owner-contractor contracts should include similar language to eliminate CM liability for "approval." The CM does not accept responsibility for a contractor's success; the CM is using the information to manage the project.

The plans are important and should be listed as a contract requirement to assure results. It is also prudent to require contractors to have their plans completely processed as a condition for their initial progress payment or even the signing of a contract. In the case of the initial group of contractors, a processed Quality Plan could be a condition for award.

Contractor Quality Plans should reside in the CM's file at the project site to assist the CM's Level 3 Manager in construction planning and scheduling and for one-on-one use with contractors that require motivation.

17.8.5 Warranties and Guarantees

Although the warranties and guarantees provided by contractors should not be considered quality management tools, they do play a role in the overall quality management plan. They provide static risk protection for the owner against latent problems with workmanship, material, and equipment.

Two types of warranties/guarantees are generally required by construction contracts:

1. Warranties/guarantees provided by the contractor and usually backed by a surety bond that assures the owner that quality conformance failures (that reveal themselves during a specified period of time after substantial completion is certified or final payment is made to the contractor) will be corrected by the contractor without cost to the owner.
2. Warranties/guarantees provided by the manufacturers of A/E selected materials and equipment that assure the owner that if their products furnished under the contract do not perform to their warranted standard or capacity, corrections will be made without cost to the owner, providing the failure occurs within a specified period of time.

17.8.6 Team Involvement in Warranties/Guarantees

With the A/E's lead, the project team should see that all available workmanship, material and equipment warranties/guarantees are listed in the specifications. Procedures to accomplish this should be included in the Quality Design Plan. The CM should include the collection of required warranty/guarantee documentation from contractors (including their final disposition) in the quality management plan.

Warranties and guarantees are documents that remain dormant unless the occasion to activate them arises. They have explicit start dates and durations which limit the time period during which claims can be made against them. Their wording is precise regarding what is covered and what is not. While they provide certain protection to the owner, they provide as much if not more protection to their issuers.

To extract the most protection from warranties/guarantees, the technical specifications in naming the date or conditions that start their coverage should be specific.

Although most manufacturer's warranties/guarantees consist of unbending "boiler-plate" provisions, the start date or the conditions that start the coverage can sometimes be specified by the A/E. This should be done if at all possible.

It must be remembered that on multiple-contract CM projects, the traditional date of substantial completion is never certified. There is a date of substantial completion for each contractor. Consequently, a date of substantial completion may not be an appropriate date on which to start a warranty or guarantee.

For example, the site electrical utility contractor's date of substantial completion could occur before the project is 20% complete. If a transformer furnished by the contractor is covered by a one year warranty at that time, it is possible that the warranty period could expire before the project is complete and the transformer put into service.

In this situation, the warranty period should be specified to start on the date the transformer is put into service rather than the date of substantial completion for the site electrical utility contractor.

17.8.7 Warranty/Guarantee Follow-Up

As the final step in carrying out the quality management plan, the CM should place copies of all warranties and guarantees in a tickler file. Chapter 14, Material/Equipment Management, covers warranty/guarantee follow-up as part of M/E management.

17.9 CONTRACT DOCUMENT QUALITY

To this point, this chapter has covered the procedures for designing quality into the project and for extracting quality from the performing contractors. An additional area of quality should be considered: the quality of the contract documents published by the team that contractors will use to estimate the cost of the project for bidding purposes and to eventually construct the project.

The contract documents consist of the drawings and the project manual. The project manual consists of the contract form, general conditions, supplementary and special conditions, and the technical specifications.

In order to receive bids for the construction work, bidding documents must be distributed to contractors. Bidding documents consist of the contract documents, proposal form, instructions to bidders, and other information that clarifies bidding procedures.

17.9.1 The 4C Requirements

To properly serve the owner, bidding documents must meet the 4C requirements of suitable documentation; they must be:

- Clear—to ensure that all contractors interpret them the same way;
- Concise—so they can be properly interpreted in a reasonable length of time;
- Complete—so that the total requirement is presented;
- Correct—to ensure against errors and ambiguities.

To achieve this, a major review of the bidding documents is necessary. The review should start in the contract document phase of design, coincidental with the submittal of the first component of the bidding documents to the team for review. It should continue until all components have been submitted and reviewed and the bidding documents assembled. The must-date for review completion is in time for the issuing of a pre-bid addenda to bidders, if one is required.

The contract documents are the product of the project team. Each team member contributes to their development, and the checks and balances of team interaction help to assure that 4C requirements are met in their publication. However, the explicit expertise of each team member must exclusively prevail when it comes to the contract document's completeness and correctness.

17.9.2 4C Responsibilities

Figure 1.1, Responsibility Distribution; ACM, GC and D–B Systems (Chapter 1, Fundamentals of the Root Form of CM), places sole responsibility for design when using the ACM system on the A/E. Design is exclusively delineated on the drawings and prescribed in the technical specifications. Consequently, the CM and the owner cannot assume any responsibility for the completeness/correctness of the drawings or technical specifications when assessing the contract documents' quality. Both can and are responsible for the clarity of the drawings and technical specifications.

The CM has exclusive responsibility for construction coordination, and all the bidding document components that deal with construction coordination must be developed by the CM. Chapter 20, Multiple Bidding and Contracting, expounds on these components. For example, top priority for the CM is determinating and describing work-scopes. The CM is the only team member with expertise in this area, and the CM is responsible for the correctness/completeness of the work-scope descriptions and their interfaces. The A/E and owner are responsible for the clarity and conciseness of these bidding document components but not for their correctness or completeness.

The owner is responsible for the correctness and completeness of the bidding documents in total. Although the components will be developed by the A/E and CM, the owner must certify their legality and their conformance to the owner's contracting policy. To guarantee this, the owner's attorney must review the documents and make changes as necessary. The A/E and CM can only advise on the clarity and conciseness of the wording, not on its adequacy. There will be times when the A/E or CM might have to familiarize the attorney with terms and procedures used on ACM projects, but the attorney has the final say.

17.9.3 CM Input to Document Review

The CM's responsibility for bidding document review should be carried out in a manner that precludes the need for large organized internal meetings. Drawings and technical specifications reviews can be better achieved by individuals working alone; there is no need for collaboration unless questions arise on a possible correction. This more informal arrangement allows the reviewers flexibility to devote time to responsibilities that may be more pressing.

However, bidding document review cannot be accomplished without planning and assignment of responsibilities. Its importance and time requirements are too sizeable to do otherwise.

Bidding document reviewing is one CM activity that can productively utilize most all operations, resource, and administration staff. The review essentially covers the clarity and conciseness of the drawings and technical specifications (the bidding document components developed by the A/E). A review of the CM organization covered in Chapter 8 indicates that most persons employed in operations and resource, and some in administration, should be capable of this type of review.

It expedites the review process if the drawings and specifications are assigned by discipline to members of the resource group. Mechanical VE and VM persons should review the mechanical drawings and specifications, electrical VE and VM persons, the electrical sections, etc. It is not necessary that those assigned to review the bidding documents for clarity and conciseness to have been involved in the project they are reviewing. This review is strictly based on the documents as presented.

It is best if operations/administration persons who have been involved in the project are assigned to review the general conditions, supplementary and special conditions, proposal forms, and instructions to bidders. However, to get an objective opinion, it is best if persons who have not been involved in the project review those parts of the project manual developed by CM personnel (such as work-scope descriptions and supplementary and special conditions that apply to multiple contracting and bidding).

17.9.4 Communicating 4C Comments

One method of effectively communicating bidding document review information to the team is to use a standard form. On this form, each item that could benefit from a revision would be capsulized. The form also identifies the source of the information, if further comment is needed. Figure 17.5 on page 275 is a typical Review Comment Form.

Each reviewer would log his/her comments as each page of each assigned document component is scrutinized. Submittal dates would be scheduled during the document review period, on which currently completed forms would be forwarded to the team's document review coordinator (see Chapter 8, Figure 8.5). The submittal process would continue until the review deadline established in the program schedule.

The Document Coordinator (either a CM or A/E team member) would review the comments when received and forward them to team members with the authority to make changes, if necessary. If additional information on a comment is required, the reviewer can be contacted. A notation on the review form would indicate the final disposition of every comment. The Document Coordinator would maintain the review comment file.

17.9.5 Typical Bidding Document Review Comments

Figure 17.6 on page 276 shows examples of the comments that could be made during a bidding document review.

Bidding Document Review Comments

A/E Project No.______ CM Project No._____ Project Name:________________ Owner:______________

Date___/___/___ Reviewer: A/E CM Reviewer's Initials:______________ Phone:____/____-______

No.	Discipline	Document	Location	Review Comments	Dispo-sition

Number: Reviewers' sequence number
Discipline: Arch; Str; Mech; Elec; Civ; Other
Document: Dwgs; Tspec; GCspec; SSspec; PropF; InvB; Other
Location: Dwg#; Page; Para; SubPara
Disposition: Used; Not Used

FIGURE 17.5 A typical Bidding Document Review Comment Form.

05	Arch	Dwgs	A008	East Wing Plan: Total dimension (162′ – 0′) does not agree with subdimensions (24′ – 6″ + 24′ – 6″ + 48′ – 0″ + 48′ – 0″ + 16′ – 6″)	
06	Arch	TSpec	04200/3	Below grade foundation blocks specified as lightweight instead of standard.	
07	Arch	Dwgs	A001	Architectural abbreviations are difficult to read. Suggest a larger font be used.	

14	Civ	Dwgs	S002	Catchbasin #8 in Detail 1-04 is not located on the parking lot layout plan. (Should be in NE corner.)	
15	Civ	TSpecs	02050/9	Demolition requirements are not included for the guardhouse at East Gate.	

06	Elec	Dwgs	E008	Indicate new breaker size in existing switchboard "DSW" that feeds new panel "MBA."	
07	Elec	Dwgs	E009	Indicate A.I.C. rating of buss of existing 100A switch-board "DSW" for correct sizing of new breakers.	

01	LndS	Dwgs	L001	Add note adjacent to irrigation controller in electrical room to refer to Dwg E007 for electrical connection.	
02	LndS	Dwgs	L001	"Note 1" regers to an alternate bid item not in TSpec. TSpec 02480, para 2.07 Sod indicates type shall match existing. Delete note 1.	

FIGURE 17.6 Examples of typical Bidding Document Review comments.

CHAPTER 18

Resource Management

Construction projects are unique in that each is physically different and typically located on a different site. The only repetitive features are the types of construction resources, and while the types are the same, the required resources differ on every project.

The real challenge of construction is not its physical accomplishment but the planning and mobilization of the resources that are required, and these resources are key to project success. However, success cannot be claimed until resources are properly converted to actions during construction.

A resource is something that can be called upon when needed; a person, a firm, an organization, a piece of equipment, a tool, a schedule, a pool of funds, and so on. In specific matching terms; a consultant, a contractor, a labor union, a crane, a hammer, a Short Term CAP, a contingency, etc. are all resources. Chapter 19, Risk Management, states that the alternate title for a construction manager could be "Risk Manager." This chapter suggests yet another title: "Resource Manager."

18.1 THE RESOURCE MANAGEMENT AREA OF KNOWLEDGE

The resource management area of knowledge encompasses the selection, organization, direction and use of all project resources, both human and physical. The CM contracting structure places all consulting, design, management, contracting, construction and construction services in a cooperative or team environment, focusing team coordination activities on the construction manager. Additionally, the CM's own multifaceted resources must be maintained in the flow of the project. These ubiquitous obligations make resource knowledge and resource management essential parts of successful CM performance.

The CM must have the capability to proficiently understand, organize, and motivate the project's resources in order to extract the best results. The CM must possess good judgment and excellent communication skills, and exhibit leadership qualities in one-on-one and team situations.

The resource management area of knowledge includes but is not limited to an understanding of human resource disciplines, physical resource capabilities, organizational structures, human nature, conflict management, motivational factors, productivity factors, and overall human relationships.

18.2 HUMAN RESOURCES

For the purposes of this chapter, available resources can be conveniently categorized as human, physical, and financial. On ACM projects, the A/E and CM provide human resources, the owner provides human and financial resources, and contractors provide services that include both human and physical resources. Beyond the project team, construction support suppliers provide physical resources, sureties and insurance companies provide financial resources, and testing laboratories provide human resources.

It is important that the project team has a clear understanding of the contribution resources make to the project during design and construction. Contrary to the concept of quality management (where it is accepted that varying levels of performance by contractors do not exist, only conformance), the performance of resources, especially human resources, varies considerably. The project team must recognize the capabilities and inclinations of their resources when making decisions or assignments.

18.2.1 Management Level Leadership

The Management or Level 2 CM Person must have the acumen to quickly and accurately assess the abilities and demeanor of the owner's and A/E's team representatives. This is the best way to extract their maximum and timely performance during the course of the project.

The CM's team member quality assessment should begin at the brainstorming session, develop during the organizational meeting, and expand during subsequent meetings. Constant contact between team members will confirm or change first impressions as the project moves to the design phase.

Authority and assigned responsibilities are not adequate in themselves to provide management success in private industry. Few people produce to their capacity simply because they are ordered to do so. When managers succeed in extracting maximum performance, it is because of their ability to lead and the respect they earn from those they manage.

Understanding the problems and idiosyncracies as well as the abilities of those being managed makes managing easier and more productive. To this end, the CM should purposefully learn as much as possible about the team representatives of the A/E and the owner. Knowing what to ask for, and how and when to ask for it, has much to do with the quality and timeliness of a response.

It was pointed out earlier that the team's prime leadership should not be assigned to any one team member. The leadership role should be pro tem, shifting from one member to another and coming to rest on the member whose expertise is pertinent to the current circumstance. However, the CM's managerial leadership is germane at all times. Therefore, it is the CM's responsibility to constantly and consistently motivate team members to act cooperatively on the tasks at hand.

For example, during the design phase when the A/E is the primary team leader, the CM must motivate both the owner and A/E to perform according to the constraints of the program schedule and make decisions that comply with the budget. This must be accomplished in a timely, positive, and unobtrusive manner in order to maintain the team's rapport.

A similar challenge exists within the CM organization when the management level CM representative refers project matters to operations, resource, administration, and support persons within the CM firm. Although it can be assumed that the CM representative's leadership has been established by virtue of promotion, it still requires appropriate motivational tactics to expeditiously extract maximum performance from in-house personnel.

18.2.2 Administrative Level Leadership

Management of contractors presents a different challenge. The project team is composed of members with different abilities but with a common objective—to complete the project as designed and funded. The project team members are agents of the owner and are pledged to consistently act in the owner's best interests.

The construction team is composed of members with different abilities but with individual objectives: to complete their work-scope and extract a profit while doing so. Construction team members are independent contractors pledged to conform to the contract documents but when doing so, act in their own best interests.

The field CM can cite various terms of the contract documents to compel contractors to perform. However, ordering contractors to comply will not always produce success. Leadership earned as the result of demonstrated construction knowledge and an obvious sensitivity for trade contractor success is the determinant. This criteria should not be taken to the point where contractor performance and the contract requirements are compromised, but to the point where contractors understand that the field CM has concern for their best interests as well as the owner's.

Partnering, the formal team-forming process recommended for GC projects, is an inherent part of ACM projects and is most effective in field operations. The CM construction team is an intuitive partnership composed of the project team and led by the field CM and the contractors. By working cooperatively, all team members will derive the maximum benefit from the project.

For the contractor, the path to profit is efficient operations; doing things right the first time, according to the Short Term Construction Activity Plan, and in cooperation with the other contractors. A field CM must let trade contractors know that their efforts will not go unnoticed and that their individual success is closely tied to the success of other contractors and ultimately to the success of the project.

Chapter 11, Contract Management, and Chapter 22, Multiple Bidding and Contracting, expand on contractor motivation.

18.2.3 The Project Team

To extract the best from a project's available resources, persons assigned to the project team at all three levels should be professionally and socially compatible. This is a large order, considering the temporary conditions under which the parties are brought together. It is understood that perfect matchups are improbable, but every effort should be made to approximate these ideal conditions when making personnel assignments.

Project team members must function in concert. Their productivity relies on mutual respect for each other's ability, work ethic, and personality. They must be knowledgeable in their area of expertise, resourceful, have good judgment and excellent communication skills, and appropriate leadership attributes. Each project team member must also have a desire to cooperate.

The CM's hiring practices should take these attributes into consideration. A/E hiring practices, though not always oriented to the CM contracting system, usually

Level	Function	CM Title	A/E Title	Owner Title
1	Executive	Executive	Executive	Executive
2	Management	Project Manager	Project A/E	Project Representative
3	Administrative	Field CM	Field A/E	Field Representative

FIGURE 18.1 Descriptive three-tier project team nomenclature.

attract the same caliber person. Owners do not always have construction projects in mind when employing people—they have a business to operate and hire accordingly—but persons of the required caliber to serve on the project team can usually be found in their organization.

A descriptive nomenclature for the three-tier CM management structure is shown in Figure 18.1. Briefly, the management level manages the project and the Level 3 managers; the administrative level coordinates and administers construction; and the executive level deals with issues and problems that for whatever reason cannot be handled at the management level.

It is impossible to rate the three levels in order of importance because each contributes in its own way. However, the CM Level 2 Person is the most ubiquitous; the one involved in the project for the longest period of time; the one that interfaces with the widest variety of resources, makes the most decisions, and must possess the broadest range of knowledge and experience, and serves at the management level. Of all three, this position is unique to construction contracting; there is no prototype in the GC contracting system.

The Field CM concentrates on construction activities and contractor interfacing. Although the stage and the characters have been preset, the Field CM must put it all together and make it work. There is no doubt that the success of the project is primarily in the hands of the Field CM once construction begins.

The duties of the Field CM are listed in Chapter 8, The CM Organization. Although some appear to be similar to those of a superintendent in the GC system, the approach to their execution differs significantly, as the field CM's dedicated allegiance is to the owner instead of to a general contractor.

One of the early lessons construction managers learned was that the best GC field superintendents could not always be successfully converted to effective Field CMs. The superintendents that could make the switch understood the difference between the "carrot" and the "stick" and instinctively knew when and how to use one or the other.

The CM has little to say about the owner's and A/E's personnel assignments to the project team. However, the CM should do all possible to make them understand the importance of team member compatibility to make them realize that in-house position and technical competence are important but one or the other may have to be sacrificed for the sake of the project's success.

When a mismatch occurs, the CM has no choice but to somehow tactfully mold discordant owner and A/E representatives into a functioning team, and coercion will not work. Team mismatches challenge the CM's management ability to the fullest. They can only be solved by assigning the very best CM representatives to the three levels of the project team. This is a prime CM responsibility.

18.2.4 Executive Level Leadership

Leadership at the executive level requires greater involvement in in-house management than in team management. The executive level is usually not immersed in day-to-day project activity, unless the 2nd or management level persons encounter unsolvable problems that must be addressed. However, the decisions of the executive level team persons shape the future of the project by installing project policies and making human resource assignments.

Day-to-day involvement of owner, CM, and A/E executive level persons is usually of short duration but very important, especially in the early stages of the project. The contractual arrangements between team members, assignment of responsibilities, and adoption of project policies must be formulated between them before the Level 2 and 3 team members can become actively involved. The CM's executive level person must initiate the team concept and get the project team started quickly and headed in the right direction.

Executive level persons select the management and administrative level staff for the project. These selections should not be made without very careful consideration of the required technical abilities and necessary compatibility. Each executive must be intimately familiar with their firm's human resources and be able to properly assess the human resource needs of the project.

18.2.5 Designated Human Resources

Chapter 24, Acquiring CM Services, points out that it is not unusual for the selection of the CM's management and administrative level persons to precede signing the contract with the owner (and at the insistence of the owner, have the names of one or both persons inserted in the owner-CM contract). If naming is an owner requirement, at least one alternate should be designated in each category in the event that medical disability prevents performance. It is also advisable to include a provision that permits replacing a person for other reasons with the owner's concurrence.

If the owner and A/E are sufficiently convinced of the need for compatibility, it is entirely possible that the CM executive level person can persuade them to have the names of their representatives inserted in the contracts as well.

Preselecting and designating team members is desirable but not without drawbacks. It is common for CM firms to respond to owners' proposal requests without predetermining if they have the human resources readily available to handle the project (if awarded). Because it is extremely difficult for a CM firm to control its work load, clients must be sought after at every opportunity. When a project arises, it cannot be ignored, especially when a firm is in an expansion mode.

18.2.6 Human Resource File

As owners become more knowledgeable about CM and its critical dependency on competent and compatible Level 2 and 3 Persons, they will become more insistent on meeting the persons assigned to their project and having them named in the contract. It is not difficult to visualize an owner's potential reaction if the CM is asked to introduce their proposed management or administrative persons and the owner is told that the CM has yet to figure that out. Hopefully the CM has other projects which are coming to an end, making management and administrative persons tentatively available. However, the coincidence is rare, especially when administrative persons are firmly obligated to one project throughout the entire construction phase.

To avoid situations such as this, CM firms should create and maintain a dual file, one consisting of a current listing of administrative level applicants who have been interviewed and are under serious consideration for employment. The second file should be a preferred promotion list of in-house persons who are ready for advancement based on periodic performance evaluations. If it becomes necessary to identify the Level 3 manager and none are available, it is possible to insert one of the top candidates in either file, making the award of the contract a condition of employment or promotion.

Owners are inclined to accept an organized, preplanned approach to CM human resource needs. It demonstrates credibility by showing good management instincts and indicates a high degree of concern for quality in the selection process. Experience has revealed that it is better for the CM to be candid with owners about staffing shortages than to try to conceal a problem.

Of course, it will be necessary for the CM's executive level person to provide a viable and convincing plan to the owner and A/E that shows how the newcomer will be supported by the CM team members. This should not be a problem considering the explicit chain of responsibility and authority in the project team structure.

If the newcomer is the management level person, the executive level person will have to be more closely involved in the day-to-day activities. If the newcomer is the administrative level person, the management level person will have to be more closely involved in day-to-day site activities. If both are newcomers, it will be difficult but not impossible to provide a satisfactory plan.

18.2.7 Employment Agreements

Another consideration when naming specific persons in an owner-CM contract is an assurance that the person will remain in the employ of the CM for the duration of the contract. While employee departure may seem a remote possibility, employees have been known to accept other opportunities when seemingly obligated to finish the project they have been assigned to, and owners have been known to void CM contracts as a consequence.

To prevent the possibility of a default (in the event the named person decides to leave the CM's employ before the term of the owner-CM contract expires), a formal employment agreement should be entered into at least for the duration of the project. This binding commitment will preclude the named person's premature departure and the undesirable effect of an owner-CM contract default. However, it could affect the person's performance if there is a strong desire to leave.

18.2.8 In-House Resources

It has been pointed out that a CM is a firm, not an individual. The CM firm is a pool of interrelated human resources that perform in concert to meet the demands of a construction project. The CM's management level team member is the director, the one who coordinates the actions of the firm's human resources. As described in Chapter 8, The CM Organization, the operational persons depend heavily on the resource persons and to a lesser degree on administration and support persons to accomplish many of the project requirements.

The resource persons are schedulers, estimators, value managers, value engineers, and planners—persons who do the tasks that operational persons could do themselves but have no time to accomplish. An ancillary benefit of this arrangement is that resource persons become specialists in the resources they provide, making them more capable in these areas than the operational persons. Comparatively, operations persons are generalists in all resource areas; resource persons are specialists in at least one resource area.

18.2.9 Scheduling In-House Resources

With several projects running simultaneously, each in different phases, the timely utilization of resources should not be difficult. However, if two or three of the projects are in synchronization, timely utilization cannot be achieved without formal scheduling.

One resource person is usually designated as the chief or head person for the resource responsibilities. CM operations persons should only have access to resources through the chief resource person. This will provide control of resource commitments and ensure quality input.

In small organizations, this level of control is difficult to maintain because of the close relationship of all personnel; however, it is wise to establish control from the beginning in preparation for inevitable growth and expansion.

To provide better control of the in-house resource area work load, it is suggested that a composite schedule be developed that combines the program schedules of all projects into one master resource schedule. This is easily done, assuming a program schedule exists for each project currently under contract. The master resource or in-house resource schedule should be updated each time a project's program schedule changes.

Operations persons should consult the in-house resource schedule when faced with program schedule changes on their projects. The chief resource person should consult with operations persons to keep the resource demands of each project on their program schedules. It is obvious that one project's program schedule change could seriously interfere with the work load of the resource people and disrupt the scheduled progress of other projects. The in-house resource schedule will permit the chief resource person to load-level resource activities, prepare for high-impact time periods, and use low-impact time periods to good advantage. It is doubtful that an owner would excuse a schedule change that results from a CM's inability to manage its own work load.

18.2.10 Human Resource Education

Chapter 9, The CM Body of Knowledge, pointed out that there is no single source of CM knowledge; it must be gleaned from many sources and galvanized with experience.

The CM body of knowledge can be divided into two categories: technical knowledge and management knowledge. Management knowledge is primarily vested in a

CM firm's operations personnel and technical knowledge in a firm's resource personnel. Additionally, some administrative persons contribute management knowledge and technical knowledge to operations personnel.

Operations persons are technical generalists, and resource persons are technical specialists. Resource persons require a much greater depth of knowledge in the technical areas of CM services than the operations persons. The list of questions in Appendix A, A Suggested Technical Knowledge Base for CM Operations Personnel, illustrates the depth of technical knowledge that individual operations persons, as generalists, should have (which is far short of the knowledge the resource area personnel should have as specialists).

A fully-staffed CM resource department provides an ideal platform for transferring technical information to operational generalists and for exchanging management information with administrative and management personnel. Unlike design, where each discipline (structural, mechanical, electrical, etc.) is a separate licensed specialty, CM has no containment limits to dissuade education in all technical areas. On the contrary, the more knowledge an operations person acquires in the technical areas, the more valuable that person is to the CM firm and her/his project team.

18.2.11 In-House Education Program

Currently, a fully-staffed CM firm is the only complete source of CM knowledge. Although academia theoretically produces sound resource persons, it does not produce theoretically sound operations persons. To fill this critical void and inject the experience that galvanizes knowledge, every CM firm should have an on-going internal information and technology transfer program. The program will provide a viable path to upgrading and expansion, if not survival, in the CM marketplace.

The program should be developed and administered by an officer of the CM firm to give it prominence and credibility. The content of the program should coincide with the firm's current human resource needs and accommodate new employees and those seeking promotion. Instructors can be anyone in the firm whose expertise is considered superior and who has the talent to put lesson plans together and make informative, interesting presentations.

Most sessions feature resource persons transferring generalist-level knowledge to operations persons (i.e., computer-assisted scheduling instruction for field CMs and life-cycle costing examples for project managers). Improved methods for estimating the cost of field changes and extra work will make operations persons more effective. Additional knowledge of bonding and insurance will improve management confidence. Instruction in electrical and mechanical systems will, of course, make Level 2 and 3 Persons more credible when dealing with the A/E and with electrical and mechanical contractors. Other sessions, perhaps critiques of recently completed projects, where unique experiences can be shared and compared, will also prove productive.

There will be more sessions on the compiled list than time to present them. Attendance should be mandatory for those selected; the program should be a high priority. Some CM firms include one and two day retreats away from the office in their presentation schedule. This is an effective approach to learning (albeit expensive and disrupting for the field CMs). However, a retreat adds a social component to the learning sessions and produces positive ancillary results that cannot be gained from in-office sessions.

Whenever and wherever sessions are held, each should be well organized and presented in a business-like manner. The demeanor of the sessions contributes to their success. Continuing education should be fostered by the CM firm and become second nature to all employees.

Another learning opportunity that should be included is seminars presented by credible institutions and associations. Picking the right ones to attend is difficult without investigation; many seminars are presented to make money rather than to transfer knowledge. References from past attenders can assist in selection. Those who attend should pass on what was learned to others during in-house sessions.

Another significant benefit of a functioning information and technology transfer program is improved confidence when dealing with owners and A/Es as team peers. To effectively provide checks and balances in the team decision-making process, CM persons must be factually conversant on topics exclusive to the owner and A/E. This is particularly helpful when discussing design options with the A/E and project requirements with the owner.

Additionally, CM services must be perceived as above the competitive image of contracting services and establish a peer status that is accepted by A/E firms and owners without restriction. A broad base of knowledge and skills in the areas of management, business, finance, design, contracting, and construction will maintain the CM firm's image as an organization of informed experts and extract professional respect.

18.2.12 Association/Society Memberships

To stay current with the industry and in all CM service areas, selected CM personnel should be actively involved in related associations and societies. Membership in most construction industry groups is expensive but beneficial, providing membership includes active participation on committees and contributes to the firm's networking effort. At least one association or society exists for every human resource discipline required to provide CM services.

18.3 PHYSICAL RESOURCES

A project's physical resources consist essentially of general condition or construction support items and the construction site itself. Multiple prime contracts and the elimination of the general contractor leaves the manner of providing physical resources to the owner's discretion. The CM budgets, arranges for, and manages the project's physical resources according to the project team's decisions.

18.3.1 Construction Support Items

Physical resource management begins by determining the necessary resources and how long they must be available. This analysis produces the construction support budget that is tracked throughout construction as a budget management function. Physical resources are either provided by designated contractors as part of their work-scopes or provided directly by the owner and managed by the CM.

The construction support items on a CM project are the same as items on a GC project, but there is a difference in how they are paid for. A GC includes the cost of construction support items in the bid submitted to the owner and is paid for them if used or not. On multiple-contract projects, construction support items are budgeted and the owner only pays for those used or consumed.

Items such as sanitary facilities, temporary power, temporary water, field testing, signing, and site security are arranged for by the CM and paid for on a direct billing basis by the owner.

Items such as temporary lighting, wiring for temporary power, piping for temporary water, temporary roads and parking, security fencing, temporary stairs, specific safety items, work layout, daily and final clean-up should be assigned as part of the defined work-scopes of selected contractors and listed as Schedule-of-Values items.

Items such as temporary warehousing, temporary shelter, hoisting, temporary elevators, refuse removal, and first aid facilities can be assigned to a work-scope division contractor, bid as a separate contract, or retained as the owner's direct responsibility and managed by the CM.

If the project requires considerable crane service, it could be:

1. Provided and crewed by a crane contractor, available for the use of all contractors at no charge under pre-published provisions.
2. Provided and crewed by a crane contractor; available for the use of all contractors at a pre-published hourly rate paid directly to the crane contractor.
3. Rented and crewed by the owner; available for use by all contractors at no charge under pre-published conditions.
4. Rented and crewed by the owner; available for use by all contractors at an hourly rate paid to the owner.
5. Supplied and crewed as part of a multiple contractor's work-scope and available for the use of all contractors at no charge under pre-published conditions.
6. Supplied and crewed as part of a multiple contractor's work-scope and available for the use of all contractors at a pre-published hourly rate paid directly to the multiple contractor.
7. Provided and crewed by each multiple contractor according to each contractor's hoisting needs.

Of the possibilities listed, items 3 and 4 are the least favored on a risk-management basis because the liability for the crane, when idle or in use, would belong to the owner.

Item 6 could be the least expensive to the owner; item 7, the most expensive to the owner. Items 1, 2, 3 and 4 would be tracked as construction support expense. Items 5, 6 and 7 would be tracked as direct expense, unless the hoisting portion of the contractor's price proposal is listed separately from the construction portion of the work-scope.

Whichever approach is selected, the CM has the responsibility to assess the costs and liabilities, see that all of the physical resources are arranged for in the contract documents, and ensure that they are provided on-site in a timely manner.

18.3.2 Site-Use Plan

The construction site is a physical resource that requires proper management during contractor occupancy. The CM is obligated to see that this is accomplished in the owner's best interest.

To achieve this, a definitive site-use plan that anticipates the sequence and duration of contractor occupancy according to the detailed construction schedule should be developed. The plan should also consider the work the contractors will be doing and the office and storage space requirements during their time on-site.

Site-use plans have been in use for many years. However, the inclination is to have a site plan in mind but not on paper (except on very large or on site-starved projects). The complexity of a site-use plan relates to the location, condition, influences, and size of the site, and the limitations put on its use by the owner. When using the CM system, every project should have a site-use plan that is written down and distributed to contractors as part of the bidding documents.

On GC projects, the owner temporarily relinquishes the physical site to the GC, and the GC manages site use within the terms of the contract as the GC sees fit. Arrangements for subcontractors is the GC's responsibility, not the owner's, and if a site-use problem arises with a subcontractor, the GC is obliged to handle it.

Site-use plan development on a CM project is more critical than on a GC project. Each trade contractor is entitled to adequate access to and convenient use of the site. If adequate access and convenient use are not provided, the trade contractors can look to the owner for betterment or satisfaction. With the owner's best interests in mind, and the possibility that the CM could be held accountable if contractor site occupancy caused a problem, the CM has every incentive to develop a pragmatic plan that will facilitate site use by contractors.

A CM site-use plan is developed exactly the same as a GC site-use plan, with the possible exception of the amount of detail included. The CM plan should indicate the spaces reserved and the facilities available for each trade contractor while on site. If the site is large enough, one plan might suffice. If the site cannot accommodate all contractors at one time, it will be necessary to draw three or four chronological plans, showing space vacated as starting contractors complete their work and new space assignments and facility arrangements as subsequent contractors arrive on site.

Site-use plans should show important features such as site-access roads, parking areas, office trailer sites, storage areas, warehouses, power and water supply points and capacities, site lighting, telephone jacks, first aid facilities, potable water supply points, staging and shake-out areas, CM, A/E and owner offices, hard hat areas, barriers, fences, fire lanes, and sanitary facilities.

As soon as possible after an award of contract, each contractor should review the space and facilities shown in the site-use plan. If deemed inadequate, changes to better accommodate starting contractor needs should be made. As with scheduling construction, only the initial site-use plan must be validated to start the project. When it is satisfactory, amended plans can be issued to contractors who will subsequently occupy the site. When issuing amended plans, care should be taken not to change major commitments made to contractors in the plan issued to them for bidding purposes.

18.4 CONTRACTING RESOURCES

Contractors are resources available to the project to provide its construction. As with any resource, it is important to know its capabilities and potential for performance before putting it to use. Chapter 22, Multiple Bidding and Contracting, refers to the advantages of prequalifying trade contractors rather than postqualifying them. Either way, the qualifying process provides insight as to what to expect from a contractor during a project.

From a resource perspective, contractors should be viewed as a combination of physical resources and human resources. A contracting firm's overall performance reflects its executive leadership and operating policies, but performance on each project reflects the leadership and attitude of the person selected to run the project.

The firm's overall performance establishes its reputation. The average performance of the firm's project leadership establishes its overall performance. Although it is possible that every superintendent and foreman handles projects the same, the chances are that a range of on-site performance exists between them.

Qualification of trade contractors should not only establish a firm's reputation as a contractor but should also provide insight as to the best performing project leaders. This information can usually be obtained from A/Es, GCs, and CMs who have worked with the trade contractor in the past. If these sources do not produce answers, it may be possible to interview the project leaders the trade contractor is considering for the project.

When discussing an award with a contractor, every effort should be made to extract a firm commitment to provide one of the contractor's better performing project leaders—one compatible with the Level 3 team representatives, especially the field CM. It is not going too far to request that the person selected be designated in the owner-contractor agreement.

When acquiring contracting resources, the CM should do all possible to establish a cooperative relationship that will respond to the static and dynamic construction needs from start to finish. It is a given that construction projects are seldom completed without problems and times when the common goal of completion is obscured. It is under these conditions that the CM's effort to create and maintain the best possible relationship with trade contractors pays dividends.

CHAPTER 19

Risk Management

Risk management addresses the innate and coincidental risks which are part of every project. Static and dynamic risks are constantly present and must be disposed of, one way or another, in the manner which best serves the owner's interests.

Unlike the general contracting and design–build system where most all risks are assigned to an independent contractor for a fee, the CM system permits the owner to contractually shed risks which can more economically be assigned and retain risks which can more economically be managed.

The Risk Management Plan prescribes the means by which both static and dynamic risks are identified, evaluated, and handled throughout the project-delivery process.

19.1 THE RISK MANAGEMENT AREA OF KNOWLEDGE

The risk management area of knowledge encompasses the dynamic and static risks that are part of every capital expansion program. Dynamic risks (risks directly tied to team decisions) and static risks (risks simply inherent to a construction environment) must be identified, evaluated and disposed of in a manner which will minimize economic loss to the owner in the event a risk with attached liability occurs.

The construction manager must be able to anticipate and analyze static and dynamic risks as well as identify, evaluate, and recommend manners of their disposal in the best interest of the owner. Disposal can be accomplished by elimination, assignment, or by accepting and managing them to minimize the consequences if they accrue.

The risk management area of knowledge includes surety bonding and insurance in the static risk area, and contracting and construction processes and procedures in the dynamic risk area. The CM must understand construction related insurance coverages, performance bonds, labor and material bonds, bid security, and other forms of available surety protection.

The CM must thoroughly understand the dynamic risks inherent to construction contracting, contracting procedures, construction planning, construction means, methods, and techniques, and be able to evaluate potential consequences and offer advice for minimizing. This area also requires excellent communication skills and high ethical standards.

19.2 PROJECT DELIVERY RISKS

When contemplating risk management, the word "risk" could cause undue alarm to those in the CM contracting structure (especially the project team members and certainly the owner). Risk is a word that emits doubt rather than confidence. However, risk is a part of every business undertaking. A construction project simply has more areas where risk must be dealt with, none of which are less manageable than those faced in other business ventures.

When considering risk management, it helps to realize that a construction manager is essentially a risk manager and obligated to guide an owner through the project without suffering inordinate mishaps. A more accurate title for a construction manager would be "Project Delivery Risk Manager," as that is precisely the orientation of a construction manager's expertise.

19.2.1 Dynamic and Static Risks

Dynamic risks are risks that challenge the risk-taker's response to a speculative situation. Once a risk is identified, the risk-taker must evaluate its gain/loss-potential and determine how or if to take advantage of the situation. Dynamic risks are speculative risks.

A nonconstruction example would be whether or not (and if yes, how much) to invest in additional manufacturing facilities in anticipation of additional product sales. If sufficient additional sales materialize, the risk-taker made a properly calculated dynamic-risk decision and will experience a gain. If additional sales fall short, the risk-taker made a wrong dynamic-risk decision and will experience a loss.

A construction-oriented example might be whether or not to qualify the contractors bidding a project in a jurisdiction where qualification is permitted and contracts must be awarded to the low bidder. A decision to qualify bidders ensures that a contract will be awarded to a qualified contractor; a decision not to qualify bidders could result in a contractor being awarded that cannot not perform at an expected level. Recognizing that a specific dynamic risk exists, an educated evaluation of its consequences, followed by an experienced-based decision or action, will mitigate if not eliminate adverse consequences and increase the possibility of overall gain.

Static risks are fortuitous chances for loss without the opportunity for gain. Once a risk is identified, the risk-taker must evaluate its loss-potential and determine how to best dispose of it by one or more of the following alternatives: elimination, avoidance, prevention, reduction, assignment, expensing, or by retention and management. Static risks are pure risks, and unlike dynamic risks, have definitive solutions.

A universal example of a static risk is loss by fire. Fires are unpredictable occurrences which result in financial loss in capital property and loss of use. The common approach to handling the risk of loss from fire is assignment and prevention.

Properly placed fire and loss-of-use insurance can almost put an owner in the same financial position after the fire as before. As well, a credible fire-loss prevention program coupled with fire-proof construction and fire-protection devices will either reduce the premiums or increase the limits for a loss. As a static risk, there is no chance of financial gain and there is also no chance of major financial loss.

Financial reimbursement does not mean that a static-risk loss such as a fire has no negative consequences. Although the replacement cost of the building and rental costs for alternative space are provided by the insurer, time is not recoverable in kind. There is no way to recover time from an insurer, only dollars.

19.2.2 The Static-Dynamic Risk Connection

It is impossible to separate dynamic risks from static risks, because determining static-risk action depends on a dynamic-risk decision. In the fire insurance example, the decisions to assign the risk of loss by fire to an insurance carrier, and to select coverages and coverage amounts, are dynamic decisions, based on the risk-taker's evaluation of the risk involved and the cost of the available coverage options offered by the insurer.

Another example of a static-dynamic risk connection would be a company that is self-insured. In this case, rather than assigning the risk to an insurer and paying premiums, the risk-taker invests would-be premium dollars into managed prevention and absorbs the cost of losses when they occur. The risk-taker's dynamic-risk decision not to resort to a static-risk solution is based on the risk-taker's considered opinion that the costs of managed prevention and ensuing losses would be less expensive than the cost of insurance premiums, minus any reimbursements from losses which might occur.

In effect, risk-management decisions are a special category of value management decisions. Based on the circumstances and available experience and expertise, the issue of how maximum value can be received from project dollars spent by the owner is an important one.

19.3 MAJOR RISK-MANAGEMENT COMPONENTS

The major components of risk management are the:

- Awareness of risk
- Identification of dynamic risks
- Identification of static risks
- Risk disposal decisions
- Risk Management Plan including contractual assignments, project management procedures, and risk management meetings.

19.3.1 Awareness of Risks

Risks connected to the project-delivery process exist throughout the project and are not limited to construction. To those familiar with the general contracting system, the risks of GC project delivery are well known; they have been established through use of the system over the years. Many of the inherent and unavoidable risks of the GC contracting structure have been addressed in the standard contract documents used in the GC system.

For example, when the GC system is used in the public sector, standard contract documents stipulate that an awarded contractor must provide a labor and material bond to the owner. This bond is a static-risk covering device through which the owner assigns payment responsibility to a third party (the surety) if the general contractor fails to pay subcontractors and suppliers for work, services, material and equipment that have been incorporated into the project. Awareness that this risk can occur on any project prompted a standard clause to be written.

As stated in Chapter 3, The Development of the CM System, CM was organized to incorporate the best practices of general and design–build contracting and eliminate the undesirable practices of each by substituting new practices. A main consideration was how project-delivery risks were handled. Consequently, new approaches to handling risk have been installed in the CM system—some static, some dynamic—and risk awareness has become a major activity of the CM project team.

CM's unique contracting structure allows owners to participate in more risk-taking decisions than either of the other two systems. Consequently, CM places greater emphasis on risk awareness and offers many new opportunities to accrue cost savings from risk-management decisions.

19.3.2 Early Identification of Risks

The primary risk categories in every construction project are cost and time overruns, quality deficiencies, and business interruptions that result from project-related disputes. These can occur at any time during the course of a project, and risk awareness should focus on them at all times during the project.

It is assumed here that the construction manager is providing services during the feasibility phase. This is often the case because the information that the owner needs to make a decision of whether and how much to build can best be provided when both an A/E and CM firm participate in the feasibility study.

The risks which develop during the feasibility phase of a project are usually the combined result of over-optimism and incomplete or flawed information. Eventual problems in two of the four primary risk categories—cost and time overruns—can be eliminated (or at least mitigated) if the consequences are addressed before design starts.

The risks created by over-optimism and incomplete or flawed information can be reduced or eliminated through team member expertise and the checks and balances of project team action. Synergism should produce a viable project budget complete with contingencies and a practical project schedule with realistic time contingencies.

These two attributes are the goal of feasibility studies on every project, regardless of the contracting system used. However, CM is the only system with a contracting structure that assures single goal dedication to the task at hand with a minimum potential for conflict of interest and provides expert checks and balances during the feasibility phase.

19.4 IDENTIFICATION OF DYNAMIC RISKS

Dynamic risks are those that challenge the risk-takers response to a speculative situation. The project-delivery process has numerous dynamic risk situations, all of which can have positive or negative results. Once identified and evaluated, a procedural decision to bring about their most advantageous disposal usually can be made.

The CM system provides decision checks and balances and construction and contracting expertise during design, beginning with the schematic phase. As a responsibility of all project team members, and especially the CM, risk awareness adds a protective dimension to the early decisions which will affect the project in later phases (especially those relating to designability, contractability, and constructability).

The following is a small sample of the many risks that can be identified early in the project and positively disposed of via risk-management procedures.

- In the early design phases, value engineering and life-cycle costing studies reduce the dynamic risk of overdesign.
- Informed estimating controls the dynamic risk of exceeding the construction budget.
- Emphasized and contractor-collaborated scheduling avoids the dynamic risk of misdirected effort.
- Document reviews avoid the risk of error and ambiguity and reduce the risk of Change Order generation during construction.
- Multiple bidding ensures against the risk of not getting the lowest competitive cost of construction.
- Prequalification reduces the risk of awarding a contract to an unqualified contractor.
- Proactive bidder recruiting reduces the risk of not having adequate bidding competition.
- Direct contractor progress payments reduces the risk of liens made by contractors.
- Individual work-scope definitions eliminate costly contractor work-scope overlaps.
- Multiple schedules of values reduce front-loading on lump sum contracts.
- Team synergism and checks and balances reduce the risk of making bad decisions.

19.5 IDENTIFICATION OF STATIC RISKS

Static risks are fortuitous chances for loss without opportunity for gain. Every project has two major owner-risk exposures: monetary loss resulting from death, injury, or property damage; and monetary loss resulting from a performance failure on the part of a contractor or constructor.

Both of these risks can be transferred to a third party in exchange for owner dollars. Death, injury, and property-loss risk can be assigned to an insurance company. Contract-performance failure risk can be assigned to a surety company. The owner's associated dynamic risks affect the decisions of whether to transfer these risks and in determining the insured and bonded amounts.

From the owner's perspective, liability and property-damage insurance is mandatory on every project. Bonding is required on public sector projects but optional on private sector projects. When bonding is optional, an owner can use qualification of contractors on a performance and financial basis to decide whether or not to bond. The decision to bond is a response to a dynamic risk, and the bond itself is a static-risk coverage device.

19.6 SURETY BONDS AND INSURANCE

Although surety bonds and insurance are both static risk transfer devices paid for by premiums, the similarities end there. They function differently: insurance companies presuppose that losses will occur, surety companies do not.

Insurance is a pool of money, sustained by premiums paid by an analogous group of insureds, that is called upon to cover specified losses when they occur. Premiums fluctuate in response to losses as a means of maintaining the level of the pool. Insurance is a competitive business where insurers are constantly seeking new insureds. Premium charges are competitive from insurer to insurer and adjusted according to the loss experience of the insureds.

On the other hand, surety bonds underwrite a financial obligation of one party to another much the same as when a note co-signer backs up a borrower of funds. If the borrower fails to repay the funds according to the note's stipulations, the co-signer is legally obligated to do so. When this occurs, the co-signer's recourse is to seek restitution from the borrower.

In the context of construction, a surety bond is a pledge from a third party (the Surety) to complete a contracted obligation to a second party (the Owner) made by a first party (the Contractor) who cannot, for whatever reason, complete the contracted obligation to the second party. The surety is paid a premium by the contractor for providing the bond to the owner, and the cost of the bond is passed on to the owner as part of the contractor's cost of the project.

19.7 SURETY BONDS

Unlike insurance, a surety's acceptance of a contractor as a client is based on the contractor's financial resources and performance record, not on ability to pay premiums. The contractor's capacity in both of these areas must be established and maintained to the satisfaction of the surety if a surety-contractor relationship is to exist. Generally, sureties do not seek out contractors as clients; contractors must seek out sureties.

The assets of a contractor are an indication of ability to repay the surety if the contractor defaults on a project and the surety becomes involved. Consequently, a contractor's financial condition determines the size of the projects which the contractor can bid. A contractor with considerable assets will be permitted by the surety to bid projects that contractors with lesser assets will not. By establishing bonding capacities, sureties determine which contractors can bid which projects.

The contractor is obligated to repay all costs expended by the surety to complete the contractor's obligations, even if it forces the contractor into bankruptcy. If the contractor defers to its surety, even though the contractor fully repays the surety, the surety will sever its relationship with the contractor. Once dropped by a surety, a contractor will have great difficulty finding another surety willing to provide bonds.

Because bonds are required on most public sector projects and many private sector projects, losing a bonding source excludes a contractor from a large share of the construction market place. It takes considerable time and resources to reestablish a bondable position.

19.7.1 Surety Bond Types

There are many types of surety bonds and three of them are signally important to construction project delivery: bid bonds, performance bonds, and labor and material bonds.

Bid bonds replace cashier's checks on projects where a bid security is required as part of a contractor's bid. The bid security backs up a bidder's pledge to accept a contract award if offered. If the contractor refuses an award, or for some reason cannot enter into the contract, an amount equal to the difference between the contractor's bid and the next highest bid is forfeited by the contractor to the owner (usually as liquidated damages).

As liquidated damages, the amount forfeited cannot exceed the difference between the bids or the face value of the bid security, whichever is less. If a bid bond is provided as security, the surety is pledged to pay the difference, if the contractor fails to do so. The extent of the surety's involvement is stated in the bonding document.

A performance bond backs up an awarded contractor's pledge to complete his contracted obligation to the exact requirements and terms of the contract documents. In the event it is determined that the contractor will not or cannot complete his obligations, the surety is pledged to accept the obligation in kind for the contracted amount.

The surety has several options. The contractor can be retained by the surety and subsidized to complete the project. The surety can replace the contractor with another contractor(s). The surety can pay the owner the face value of the performance bond. In either of the first two options, the outstanding amount due the contractor at the point of default is paid to the surety when earned, according to the terms of the contract. The nature and extent of the surety's involvement and its specific options are stated in the performance bond.

A labor and material payment bond protects the owner from paying twice for the labor, materials, and services in project construction. In the event a party that does not have a contract with the owner but who has one with a party that has a contract with the owner is not paid by the party with the contract, the party that was not paid usually has a legal right to transfer the unpaid amount to the owner for direct payment.

Most states have mechanic's lien laws that allow unpaid parties to effectively become co-owners of an owner's property to the dollar value of the unpaid amount. To shed the co-owner's financial rights under the lien, the owner must pay the amount owed, regardless of whether or not it was previously paid by the owner to the party who owed the money to the claimant. The labor and material payment bond shifts the responsibility for payment to a surety, relieving the owner of the claim.

In some jurisdictions, liens against public property are not permitted. To provide the same financial claim opportunity to contractors and suppliers involved in public projects, parties who hold contracts with owners must provide a labor and material payment bond. The extent and terms of the surety's responsibility is stated in the bond provided to the owner by the contractor.

19.7.2 Surety Bonds in the CM System

The CM system provided a productive opportunity to reevaluate the traditional bonding concepts on construction projects. As explained in Chapter 22, Multiple Bidding and Contracting, multiple contracting is a major factor in providing this opportunity.

CM multiple contracting significantly reduces the size of construction contracts held by the owner, because each contract represents a portion of the total contract amount. A $10 million contract held by a general contractor is usually covered by a $10 million performance bond and a $10 million labor and material bond, although it is becoming common for the labor and material bonds to be reduced to 50% of the contract amount.

On a CM project with a multiple-contracting format, each contractor is only responsible for a portion of the work and is only bonded for that contracted amount. Consequently, the owner will have as many bonds as contracts, and the total of the bonded amounts will represent the total of the multiple contracts. It is common for the owner to hold thirty or forty bonded contracts when multiple contracting is used to its optimum.

The increased number and reduced size of the bonds when using multiple contracting does several things:

1. It covers the static risk of contractor failure on the project at the trade contracting level; the level where most construction contract failures originate.
2. It reduces the dynamic risk of awarding a contract to an unqualified contractor; the ability to be bonded can be viewed as an indicator of potential performance.
3. It reduces the size of a surety's potential liability on a single project (some sureties do not consider this a plus due to increased paperwork per bond issued).
4. In many instances it considerably lowers the cost of bonding to the owner; it is common for sureties to insist that general contractors bond their subcontractors as a condition for bonding the general contractor.

The first item mentioned above is a benefit to the owner. Contractor performance failure liability, while just as complete and secure as under a single bond, is spread among the multiple contractors. Nonperformance by a contractor can be quickly detected and dealt with directly in the owner's best interests. A general contractor who is having difficulty with a subcontractor will expectedly act in his own best interests, not those of the owner. If contractor replacement is required, it is much simpler to replace a trade contractor than a general contractor.

As mentioned, general contractors are often required to bond their subcontractors in order to be bonded for the total project by their surety; this is referred to as *double bonding*. It protects the general contractor and the general contractor's surety but provides no protection to the owner, in spite of the fact that the owner pays all costs of second-tier bonds.

Although the costs of bonds increase as the cost they cover decreases, CM multiple-contract bonding is less expensive for the owner than a GC single bond because of double bonding requirements.

An important point is that the sum of all the multiple bonds on a CM project will be less than the value of the single bond on a GC project. The overhead, profit, and general condition costs of the general contractor are part of the single-bond amount. On a CM multiple-contract project, there is no GC overhead or GC profit to bond, and the general condition costs are usually an owner-reimbursable expense projected as a budget line item. Surety bond economics are discussed further in Chapter 22, Multiple Bidding and Contracting.

19.7.3 Waiving Bond Requirements

When using the GC system, bonding of contracts is practically mandatory because of contract size. In the private sector, where contractor reputations are known to an owner, bonding is sometimes waived by the owner as a dynamic-risk decision. On public projects, bonding is usually mandatory on contracts greater than a statutory limit. Consequently, most public GC projects require total bonding.

CM multiple contracting on medium-size buildings may include contracts below the statutory limit. This permits an owner to exercise bonding prerogative on some of the contracts. The advantage of not bonding is in saving premiums. On a project where a sizeable number of contracts fall below the statutory limit, significant savings can accrue. A dynamic-risk decision must be made by the owner as to whether or not the savings are worth the exposure.

19.7.4 Obtaining Multiple-Contract Bonds

The advent of CM has created a new type of surety market by virtue of the size and number of the bonds required by the multiple contract concept. Most surety companies participated in the large, single-contract bond market. Few were interested in providing bonds on small contracts until the federal government began to assist small and minority contractors.

Additionally, losses in the early 1970s (when many contractors experienced the ills of overexpansion, material shortages, and labor problems) led to sureties becoming more conservative regarding who they bonded and for how much. Trade contractors were not considered a prime market for sureties, but the popularity of CM and government intervention in behalf of small and minority contractors convinced them to participate.

A collateral benefit of the CM system that is not appropriately appreciated is that multiple-contract bonding has forced trade contractors to move to a new level of business awareness. In order to be bonded, they have to demonstrate financial stability and performance ability; this improvement has proved beneficial for the construction industry and its users.

19.8 INSURANCE

Construction insurance is required in three areas: Owner Protection, CM and A/E Protection, and Contractor Protection. The CM system and multiple contracting do not change the traditional forms and coverages but deal with some of them differently.

Construction industry insurance is a highly specialized field and construction managers are usually not expected to have insurance experts on their staff. However, a broad knowledge of insurance is necessary to help the owner establish a static-risk protection program that provides effective coverage and is compatible with the CM contracting structure.

19.8.1 Types, Forms and Coverages

Types of Insurance

A. Property Damage
   1. **Forms** of Property Damage Insurance
      a. Standard Builder's Risk
         (1) **Coverage** of Standard Builder's Risk
            (a) Fire and Lightning
            (b) Extended Coverage
            (c) Vandalism and Malicious Mischief
            (d) Additional Endorsements
      b. All Risk Builder's Risk
         (1) **Coverage** of All Risk Builder's Risk
            (a) Standard Builder's Risk Coverage
            (b) Broad Form Coverage
            (c) Additional Coverage
      c. Boiler, Machinery and Power Plant
      d. Floater Policies
         (1) **Coverage** of Floater Policies
            (a) Contractor's Equipment
            (b) Transportation
            (c) Installation

B. Liability
   1. **Forms** of Liability Insurance
      a. Worker's Compensation
         (1) **Coverage** for Worker's Compensation
            (a) Injury, Disability, Death
            (b) Employer's Liability
      b. Comprehensive General/Public Liability
         (1) **Coverage** of Comprehensive GPL Insurance
            (a) Premises-Operations
            (b) Explosion, Collapse and Underground Damage
            (c) Personal Injury
            (d) Contractor's Protective Liability
            (e) Contractual Liability
            (f) Completed Operations
            (g) Umbrella Excess Liability
            (h) Following-Form Excess Liability
            (i) Automobile
                Medical Payments
                Physical Damage
      c. Professional Liability
         (1) **Coverage** of Professional Liability
            (a) Architectural/Engineering
            (b) Construction Management

FIGURE 19.1 Insurance types, forms and coverages.

19.8.2 Owner Protection

Whenever embarking on a building program, an owner is exposed to a variety of liabilities against which protection must be secured, regardless of the project-delivery system chosen. The CM contracting structure does not create exposures or risks which are not already present in the general contracting and design–build contracting structures, but it does create new contractual relationships (especially when multiple contracting is used) which must be accommodated.

It is the owner's responsibility to review, approve, and install an insurance program for the project that satisfies the owner's needs. It is assumed that an insurance agent or consultant will advise the owner, with assistance from the construction manager and A/E.

The types and forms of insurance usually arranged for, and supplied through contracted parties or directly by the owner, are:

- Liability Insurance
  - Errors and Omissions (Architect/Engineer; Construction Manager)
  - Comprehensive, General/Public Liability
  - Worker's Compensation
- Property Damage Insurance
  - All-Risk Builder's Risk

Errors and omissions. E&O (or professional liability coverage) is common in architectural and engineering practice. It protects the owner from liability when an A/E makes a mistake or forgets something related to design. It usually does not protect against A/E misjudgments in areas other than design, such as a imprudent contract administration decision or misdirection to a contractor in the field. For this reason—plus the opinion that A/E fees were not sufficient to compensate them for contract administration responsibility—A/E firms distanced themselves from situations where decision error was not covered. A/E firms, especially architectural firms, stayed away from construction site involvement as much as possible.

In fact, in the late 1960s, architects were considering the possibility of excluding contract administration responsibilities and terminating their services after the bidding phase. Concern for how shop drawing approvals and design changes were to be handled, and how the construction industry would respond to this break from tradition, slowed the decision-making process on this issue.

In the meantime, the CM system, which succinctly provided some of the services that A/E firms were trying to avoid, came into the picture. Now, if the A/Es could be paid a fee for taking the risks connected with contract administration and field direction, their concerns might be reversed. In fact, in the mid-1970s the American Institute of Architects published a set of standard contract documents for CM that allowed A/E firms to provide "contract administration" services for a fee as CMs.

The services rendered by a construction manager produced an additional professional liability policy: E&O insurance to cover the practice and performance of construction managers. This policy covered imprudent contract administration decisions and misdirection of contractors in the field but not design decisions.

It should be understood that E&O insurance does not protect the insured against negligence, only errors and omissions. As well, E&O insurance is not a hedge against mediocre CM or A/E performance. Most E&O policies have deductibles to get them into the cost-practical premium range. E&O policies always carry exclusions, and care should be exercised to evaluate the possible consequences of each.

An example exclusion in an E&O policy is as follows:

> "The Insuring Agreements, and all other provisions of this insurance, shall not apply to claims for or arising out of the advising or requiring of, or failure to maintain any form of insurance, suretyship or bond, either with respect to the insured or any other person."

Comprehensive General/Public Liability Coverage. Coverages under this form of insurance are provided by both the owner and the contractors in accordance with the contract documents. Double coverages are common in construction because of the number of parties simultaneously involved at the site.

When an accident occurs, it is often difficult to assign ultimate responsibility to any one party. Fault could be passed from one party to another, consuming considerable time and effort. In addition, numbers alone make it impossible to ensure that all the contractually required coverages are in force at any one time, or if the claim on a policy will exceed the insured amount. For this reason, it is usual that contractors indemnify the owner from liability by contract and that each insurance policy in force on the project have a waiver of subrogation clause to prevent "finger-pointing" reactions from one insurer to another. The subrogation waiver eliminates the right of the prime insurer to claim compensation from another insurer whose insured might have been involved in the occurance. The prime insurer pays the claim.

To eliminate liability for an uninsured loss or for an inadequate insured amount, an owner should carry coverages in amounts that will provide protection if contractually-prescribed insurance coverage does not. This insurance is in addition to requiring all contractors to name the owner as an additionally-insured party on their liability policies. Owner's protective liability coverage provides this protection.

Owner's Protective Liability. Eliminating a general contractor on a CM multiple-contract project removes the insurance layer between the GC and his subcontractors. To retain the credibility of the insurance network, the owner directly purchases coverage the GC provided.

Similar to the coverage referred to as "Contractors Protective Liability" (under the Comprehensive General/Public Liability form of insurance on page 298), Owner's Protective Liability coverage protects the owner against liability resulting from the actions or inactions of the contractor or contractors working on the owner's project.

The liability could be contributory, resulting from a contractor not carrying sufficient coverage to withstand a loss or working under an expired policy. It could also occur if the owner creates an attractive nuisance which results in injury to an on-site trespasser.

Owner's protective liability coverage is often ignored when using the GC system. This occurs on the theory that if high enough limits are required of the GC on his protective liability policy, and an owner indemnification agreement is incorporated in the contract, the owner will have adequate protection.

However, higher-than-normal limits and indemnification are additional cost factors to coverage which create an uneconomical cost/coverage value relationship. Neither protect the owner against inherent exposures, nor do they protect him/her in the event that contractor coverage lapses or is cancelled. By not carrying owner's protective liability on GC projects, the owner faces exposure that she/he could well do without.

Builder's Risk. During construction operations, the structure is vulnerable to certain perils such as wind, rain, fire, and other acts of God. Under the GC system, the general contractor is usually required to provide and pay for a Builder's Risk policy which protects the owner against financial loss from these perils. The general contractor obtains the policy and includes its cost in his bid or proposal to the owner. The premium indirectly paid for by the owner is based on the size and accident exposure of the project.

When using multiple contracts, the owner replaces the general contractor in many ways. In this case, the owner provides the builder's risk policy. The owner benefits by the opportunity to obtain the coverage competitively and minimize premium costs. The owner may save even more premium costs by providing the coverage through endorsement of existing property-protection policies.

19.8.3 Risk Packaging

The owner's purchasing power in the insurance industry often provides packaging possibilities with premium discounts not available to contractors. Carriers can base premium rates on special fillings and not on the manual listed rates applicable to contractors. The packaging of insurance requirements produces an effective competitive premium situation.

On large projects there may be an opportunity for the owner to arrange a "wrap-up" insurance policy; one that covers the owner, A/E, CM, and all contractors on the site. A wrap-up policy can reduce overall insurance costs and eliminate the question of whether contractors are currently and properly insured.

19.9 CONTRACTOR REQUIREMENTS

The following substitution for an insurance article, typically found in standard general conditions of the contract forms, has been previously used on multiple-contract CM projects. While this article should not be used without scrutiny and modifications (especially in the areas of insured limits), it stands as an example of the typical coverage on a medium-size ACM multiple-contract project.

INSURANCE

All reference to Insurance in the General Conditions shall be deleted. The insurance requirements shall be as follows:

1 CONTRACTOR'S LIABILITY INSURANCE

1.1 Each Contractor shall purchase and maintain such insurance as will protect it from claims which may arise out of or result from the Contractor's operations under the contract, whether such operations be by itself or by any subcontractor or by anyone directly or indirectly employed by any of them, or by anyone for whose acts any of them be liable. All insurance companies writing coverage for this project, shall be approved or licensed to operated within the State in which this project is located. Written certification of this requirement shall be presented to the Owner, Architect and Construction Manager.

1.2 The insurance required by Sub-paragraph 1.1 shall be as follows:

1 Worker's Compensation and Employers' Liability Insurance:

Worker's Compensation and Occupational Disease Insurance at statutory limits as provided by the State in which this contract is performed and Employer's Liability Insurance at a limit of not less than One Hundred Thousand Dollars ($100,000.00) for all damages arising from each accident or occupational disease.

2 Comprehensive General Liability Insurance covering:

A. Operations—Premises Liability: including, but not limited to, bodily injury, including death at any time resulting therefrom, to any person or property damage resulting from execution of the work provided for in this contract or due to or arising in any manner from any act or any omission or negligence of the Contractor and any subcontractor, their respective employers or agents.

B. Elevator Liability: including, but not limited to, bodily injury, including death at any time resulting therefrom, to any person or property damage resulting from operation or use of any elevator or hoist, if either or both are operated or used in connection with execution of this contract.

C. Contractor's Protective Liability: including, but not limited to bodily injury, including death at any time resulting therefrom, to any person or property damage arising from acts or omissions of any subcontractor, their employees or agents.

D. Products—Completed Operations Liability: including, but not limited to, bodily injury, including death at any time resulting therefrom to any person or property damage because of goods, products, materials, or equipment used or installed under this contract or because of completed operations, which may become evident after acceptance of the building, including damage to the building or its contents.

E. Contractual Liability: Each and every policy for liability insurance, carried by each Contractor and subcontractor as required herein shall specifically include contractual liability (hold harmless clause) coverage.

F. Special Requirements: The insurance required under 1.2.2 shall specifically include the following special hazards:
   1. Property damage caused by conditions otherwise subject to exclusions "X, C, U", explosion, collapse, or underground damage, as defined by the National Bureau of Casualty Underwriters.
   2. Property damage liability coverage shall be broad form coverage.
   3. "Occurrence" bodily injury coverage in lieu of "Caused by accident."
   4. "Occurrence" property damage coverage in lieu of "Caused by accident."
   5. EXCEPTION: Contracts that do not require excavation or underground work are not required to have the above "Special Hazards" insurance coverage under (1) above.

G. Limits of Liability: The insurance under 1.2.2 shall be written in the following limits of liability as a minimum:*

   Bodily Injury:

   $500,000 each occurrence

   $500,000 aggregate products

   Property Damage:

   $500,000 each occurrence

   $100,000 aggregate operations

   $100,000 aggregate protective

   $100,000 aggregate products

   $100,000 aggregate contractual

3 Comprehensive Automobile Liability Insurance covering:

A. All owned, hired, or non-owned vehicles including the loading or unloading thereof.

B. Special Requirements: The insurance required under 1.2.2 shall specifically include the following special hazards:
   1. "Occurrence" bodily injury in lieu of "Caused by accident."
   2. "Occurrence" property damage in lieu of "Caused by accident."

C. Limits of Liability: The insurance under 1.2.3 shall be written in the following limits of liability as a minimum:*

   Automobile Bodily Injury

   $500,000 each person

   $500,000 each occurrence

   Automobile Property Damage

   $1,000,000 each occurrence

4 Umbrella Clause: If an umbrella clause is written to implement the above prime coverages, the umbrella clause shall specifically state that the policy

---

*Note: Limits stated here are an example only. The owner must establish limits on a project basis.

is written on an "occurrence" basis. Umbrella Liability: $1,000,000 each occurrence and $1,000,000 aggregate.

1.3 No Contractor shall commence work under this contract until it has been approved by the Architect and Construction Manager, nor shall any Contractor allow any subcontractor to commence work until the same insurance has been obtained by the subcontractor. Unless exceptions are noted or specified, each and every Contractor and subcontractor shall maintain all insurance required under 1.2.1, 1.2.2 and 1.2.3 of this section for not less than one (1) year after completion of the contract.

1.4 Each Contractor shall file with the Owner, Architect and Construction Manager a Certificate of Insurance (available from the Construction Manager). Any certificate submitted and found to be incomplete will be returned as unsatisfactory. Certificate of Insurance shall contain a clause that coverage afforded by the policies listed will not be canceled or materially altered, except after 45 days advance written notice to the Owner, Architect and Construction Manager, mailed to the addresses indicated herein.

1.5 If requested by the Owner, Contractor shall furnish the Owner with true copies of each policy required of him or his subcontractors.

1.6 Each Contractor shall secure the following endorsements to each of the above policies:

"It is understood and agreed that the insurance company will give not less than forty-five (45) days advance written notice of any cancellation or material change under any of these policies to the Owner."

"In the event that such notice is not given to the Owner at least forty-five (45) days prior to cancellation or material change, the policy will continue in full force and effect for the benefit of the Owner as if such change or cancellation had not occurred."

2 OWNERS LIABILITY INSURANCE

2.1 The Owner shall provide Owner's Protective Liability Insurance as will protect the owner against claims which may arise from operations under this contract.

3 PROPERTY INSURANCE

3.1 BUILDER'S RISK COMPLETED VALUE INSURANCE:

The Owner shall effect and maintain builder's risk insurance for completed value coverage, and shall include the interest of the contractors and their subcontractors. This insurance is to be upon all the structures on which the work of all the contracts is to be done to one hundred percent (100%) of the insurable value thereof, including items of labor and materials connected therewith whether in or adjacent to the structures insured, materials in place or to be used as part of the permanent construction including surplus materials, shanties, protective fences, bridges, temporary structures, miscellaneous materials and supplies incident to the work, and such scaffolding, staging, towers, and equipment as are not owned or rented by the contractor, the cost of which is included in the cost of the work.

EXCLUSIONS: This insurance does not cover any tools owned by mechanics, any tools, equipment, scaffolding, staging, towers, and forms owned or rented by the contractor, the capital value of which is not included in the cost of the work, or any cook shanties, bunkhouses or other structures erected for housing the workmen. The loss, if any, is to be made adjustable with and payable to the Owner as trustee for the insured and contractors and subcontractors as their interest may appear, except in such cases as may require payment of all or a proportion of said insurance to be made to a mortgagee as its interests may appear.

19.10 THE RISK MANAGEMENT PLAN

The risk management plan should prescribe procedures that address the static and dynamic risks inherent to the project. The goal of the plan is to minimize the owner's exposure to risk from the start of design to occupancy and through the warranty period.

The plan should emphasize risk awareness. It should include procedures that will identify static and dynamic risks, evaluate their potential loss value, and prescribe ways to effectively dispose of them in ways that serve the owner's best interests.

The conventional means of risk disposal are to:

- eliminate the risk, by taking an alternate course of action
- shed the risk, by letting someone else bear the burden
- assign the risk to others, by agreement or contract
- retain the risk and minimize it through micromanagement.

The list of risks will be substantial because it should be as comprehensive as possible. It should be started at the brainstorming session and continued throughout the project. The CM should be the custodian of the list.

Every team member should be risk, quality, and safety-conscious and contribute to the list as risks are identified. The best approach is to always keep a "what if" attitude when planning action or pondering decisions. Individual team members should not evaluate risks to determine their suitability for the list. All risks should be forwarded to the Level 2 CM team member for evaluation.

The obvious risks inherent to a construction project can be identified by team members from their experience. Experienced CM firms can likely contribute a starter list accumulated from past CM projects. The team can add risks to the list as they are uncovered.

Risk management should be a standing item on every team meeting agenda, because risk-management decisions should be static whenever possible. Static decisions can only be made if the risks are identified early enough to facilitate team action.

The major risk-management solution is insurance; surety bonding is a close second. Loss due to accidents and non-performing contractors has the highest potential of all single risks. However, both these risks are static risks commonly dealt with on every project, CM or otherwise.

There are many other risks, as suggested earlier in this chapter and as eluded to in other chapters. Some of them will require extensive procedures and others will not. The goal should be to review all identifiable risks by priority and establish procedures to lessen the potential of each one.

Risk management is a part of contracting and construction that must be micromanaged. The team concept of CM makes that task easier to perform.

CHAPTER 20

Schedule Management

Scheduling has always been used as a management tool in the construction industry. As in any other industry, efficiently dovetailing operations and timely receipt of materials and equipment is mandatory if a project is expected to proceed smoothly.

However, unlike most other industries, construction has variables such as wind, rain, snow, and temperature that adversely affect scheduling. Most manufacturing sites eliminate these variables by physical enclosure, but a construction site is not conducive to similar controlling measures.

While the art and practice of scheduling has reached very high levels of sophistication, the variables convert construction scheduling into a less-than-perfect management tool. In spite of its recognized inadequacies, construction scheduling is considered a mandatory tool for managing the project-delivery process. There is no doubt of its value when properly used.

The CM contracting structure provides an ideal environment for the use of scheduling within its capabilities. The scheduling tool is widely used throughout CM practice and procedures.

20.1 THE SCHEDULE MANAGEMENT AREA OF KNOWLEDGE

The schedule management area of knowledge encompasses all aspects of scheduling throughout the project. Scheduling is the management tool that best represents the controlled operations philosophy of the CM contracting system. It combines the element of time with the project's resources from the start of design to owner occupancy. Scheduling eliminates or mitigates potential time-resource crises by predicting start and finish dates for intermediate project milestones. The use of scheduling is a means to an end, not an end in itself. It is a form of communication that should be presented in the simplest form with just enough detail to convey its message.

The CM must be able to apply scheduling as a major management tool, proficient in its use and applications, and sensitive to its ability to plan and predict. The CM must have the capability to design the Scheduling Management Plan for the project, select appropriate scheduling formats and techniques for specific project applications, and extract information from team members and contractors for scheduling purposes.

The schedule management area of knowledge includes a thorough understanding of scheduling techniques, from bar charts to precedence diagramming, including their use in planning, predicting, analyzing and tracking project activities and events. The CM must be computer literate in the area of scheduling, understand the fundamentals of modeling, be proficient in matching scheduling applications to requirements, and possess excellent communications skills.

20.2 SCHEDULING BACKGROUND

The introduction of Critical Path Method (CPM) scheduling in the 1950s as an alternate to traditional bar charts was greeted enthusiastically by the construction industry. Although CPM could not eliminate the variables that hampered accurate construction schedules predictions, it provided a sophisticated diagramming technique that clearly defined the activities that moved the project from start to finish and identified those events that controlled the duration of the project.

CPM was first used on construction projects thirty-five years before computers were universally available. Except where large computers were available, construction schedules were produced manually.

A large, complex project required a schedule that took considerable planning, diagramming, and time calculations. When the project proceeded according to the diagram, users had the tremendous advantage of accurately assessing progress and predicting the start and end dates of many activities as well as final completion.

However, if an activity sequence or duration changed due to one of the variables inherent to construction's exposure, the diagram became useless and had to be completely redone. The time, effort, and money required to manually update schedules each time a sequence or duration changed was discouraging, and this prompted many CPM advocates to revert to the more simplistic, manipulative bar graph schedules.

With all its promise, CPM scheduling remained practically dormant in the construction industry until economical computers became available in the 1980s. Computerization provided the means of conveniently and quickly updating schedules. As a result, CPM regained its popularity many times over. In fact, the self-serving interests of the computerized software producers has taken scheduling well beyond its practical use as an effective management tool for construction.

20.3 THE PRACTICAL USE OF SCHEDULING

CM scheduling efforts should always be viewed as a means to an end, not an end in itself. The "bells and whistles" of contemporary scheduling software make it very difficult to keep this tenet in view. Scheduling software users, caught up in the extensive capabilities of the programs, tend to use the available capabilities whether they contribute to effective project management or not. The result places restrictions on what must be a flexible management format that can be readily adapted to changing project conditions.

It should be remembered that scheduling is a basic management tool that should not be used beyond its practical capacity. Scheduling is a means of communication and should be used at its most *understandable* level, not its highest level of sophistication. Scheduling should be used to accomplish, not to impress. The CM should keep this in mind when carrying out scheduling responsibilities to the project team and contractors.

20.4 THE REASONS FOR SCHEDULING

There are five fundamental reasons for scheduling:

1. To **ensure planning** takes place before an activity is undertaken
2. To **give direction** to those responsible to execute an activity

3. To **consider alternatives** before an activity begins
4. To **compare** actual performance with anticipated performance
5. To **provide a record** of time-related performance

A sixth use of scheduling, as a sales tool, is important to a CM. Many requests for CM proposals ask for a schedule as part of the CM's submittal. When this occurs, the "bells and whistles" which are not necessarily valuable when using scheduling as a management tool can be advantageous in impressing owners.

20.5 ADDITIONAL ASPECTS OF SCHEDULING

Scheduling deals with all time-related aspects of a project, not just construction activities. The time-value of money (merging of cost information and time scheduling) produces time-oriented financial information that can prove vital to project financing and other cost-related project activities. Additional aspects are discussed in Chapter 15, Value Management.

20.6 THE FIVE BASIC SCHEDULES

There are five basic schedules that should be developed and used by a CM on every project. They provide a sound interrelated core for the successful execution of schedule management. Figure 20.1 diagrams the interaction of the five schedules.

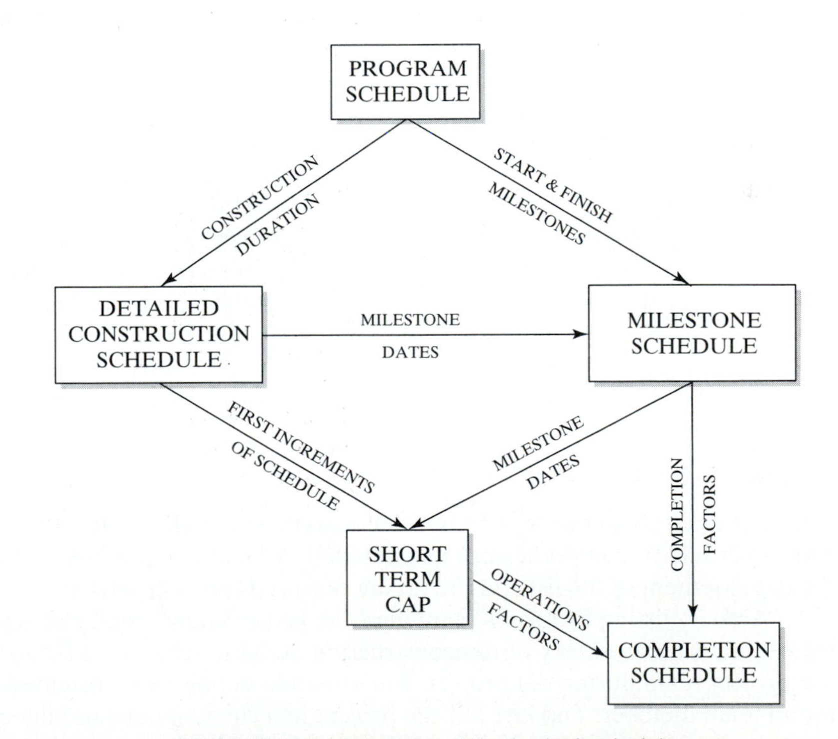

FIGURE 20.1 The interaction of the five basic schedules.

They are:

1. **Program Schedule:** Drives the project from start to finish.
2. **Detailed Construction Schedule:** Determines the time needed for construction.
3. **Milestone Schedule:** Drives the project during construction.
4. **Short Term Contractor's Activity Plan:** Plans contractor activities to meet milestones.
5. **Completion Schedule:** Coordinates construction completion and occupancy.

There are other schedules and schedule applications that can be helpful on a long or short term basis. However, these five basic schedules provide the CM with all the information required for successful schedule management.

As Figure 20.1 shows, the five schedules are dependent on one another for information. The initial schedule is the program schedule, developed as soon as the team is formed. The final schedule is the completion schedule, developed about two-thirds through construction.

20.7 THE PROGRAM SCHEDULE

Recognizing that planning activities and design activities need time-control as much as construction activities established the need for a specific schedule that could chart the preconstruction phase as well as the construction phase of a project.

The Program Schedule fills this need. It is used to define and track the actions of team members, and the construction phase, throughout the course of the project. The construction phase is included as an activity with a single duration for its accomplishment. A subsequent schedule (Detailed Construction Schedule) will breakdown the program schedule's construction activity in detail.

The CM system recognizes that time lost during any phase of project delivery is unrecoverable without additional cost to someone. Time lost during planning or design is no different than time lost during construction, but the consequences of a missed time-line obligation during design can cause an inordinate project delay.

Planning and design activities such as meetings to extract decisions or dates established for the decisions themselves are generally short term. Being unprepared to make a decision can cause an extended time delay (i.e., a regulating agency that meets monthly to grant approvals that will allow design to proceed could produce a 22 working day project delay, if the data required for approval arrives one day late).

20.7.1 Program Schedule Activities

Considerable thought should go into the activities included in the Program Schedule. The list should be comprehensive and detailed. All team members should contribute to the development of the list and the timing required for each activity.

Activity listing mandates planning (one of the main benefits of scheduling). The list becomes an inventory of nonconstruction activities that must be accomplished by the project team during the project. The schedule defines time-oriented requirements for all team members and lays out the project in a clear, understandable perspective.

Many Program Schedule activities are owner decision dates dealing with material/ equipment selection or approvals of design phases. Others are action dates such as confirming financing, mailing documents, or assigning project personnel. Still others are days set aside for meetings and team work-sessions, including value and quality management meetings and budget reviews. Regardless of an activity's duration, if it is a vital part of the sequential progress of the project, it belongs in the Program Schedule.

Program schedule activities are those necessary to keep the project running smoothly on both ends of the principal construction activity. These typically include: Construction Loan Approval, Bid Package Reviews, Agency Approvals, Material/ Equipment Selections, Building Permits, Bid Package Preparation, Contract Document Reviews, Bid Document Mailings, Team Meetings, etc.

20.7.2 Program Schedule Format

Computerized, precedence-diagrammed scheduling should be used for the program schedule. These techniques permit convenient and rapid updating which is especially beneficial when the schedule is young. There will be many activity additions and sequence and duration changes as the schedule develops. Many activities will be concurrent; however the schedule itself will be linear due to its decision-based character. Program Schedule printouts should be listings, not diagrams.

Program Schedule activities are not influenced by uncontrollable variables; they are mostly project team actions and interactions that are subject to the control of the individuals involved. For these reasons, plus the fact that the Program Schedule is formulated as a team, the Program Schedule has the best potential for staying on track of all schedules used on a CM project.

In spite of its high potential to stay on track, it should not be assumed that a Program Schedule will automatically do so. Like everyone else, team members encounter problems when trying to meet multiple commitments. Additionally, when team members plan, they tend to be optimistic about their own abilities. They frequently set timelines for themselves that are only attainable if outside influences are excepted. This tends to create an ideal schedule that could become problematic in execution. However, to counter this condition, team members can be expected to make an extra effort to keep their commitments.

20.7.3 Program Schedule Print-Outs

The computerized program schedule should permit the CM to obtain several print-out sortings that aid in identifying team activities. Besides the common list sort by work-item or activity number, which provides activity correlation with the precedent diagram, early-start list sorts for the team or individual team members should be available. These list sorts clearly identify the day-to-day activities of team members on letter-size paper without the need of a cumbersome diagram.

An early-start sort which lists the activities chronologically can be the guide for team members' daily actions. Every team member can determine her/his scheduled input to the project for any day by looking at the early-start sort. He/she can also see what other team members should be doing by reviewing their respective early-start print-outs.

Because of their use as a planner and motivator, and the dynamic character of the preconstruction phase of the project, early-start print-outs should be sent to each team member weekly. If an action date is missed, the CM should be given immediate notice. An alternate course of action can be instituted to bring the program activities back on schedule, and an immediate print-out showing the time-line consequences should be sent to team members.

Figure 20.2 is an example of a program schedule. The schedule was developed by drawing a precedence diagram and printing it as a list-type schedule. The sort in this case was "early-start by team responsibility." The activities are listed in date-sequence producing a critical path. In use, the schedule is a day-to-day calendar for team members (a list of what should be done today (March 28) and each day thereafter to keep the project on schedule). The (*) and (M) designate key dates.

Other sorts by work item, activity sequence, and team member responsibility can be printed if desired. One that shortens the sort in Figure 20.2 and serves the same purpose is the "early-start by team member responsibility." This sort would only include the activities for one team member. Each team member would get her/his own personalized day-to-day activity schedule.

By viewing this schedule print-out, the architect/engineer (AE), owner (ON), and construction manager (CM) can determine where and when their individual efforts should be applied and when team (TM) input is required. Responsibility assignments should agree with the responsibility chart developed in Chapter 16, Project Management.

It should be noted that the date of the schedule (March 28) coincides with the start date of the activity 00015 on this page of the six-page schedule. The preconstruction phase is on schedule as of March 28. The 000 start-float after each activity indicates that the required-start date and late-start date are the same. There is no room provided for a later start than the required one. This is especially critical on one-day duration activities.

20.8 DETAILED CONSTRUCTION SCHEDULE

One activity that should be on every program schedule is "Develop the Detailed Construction Schedule." The Detailed Schedule is a construction schedule for the total project or a phase thereof. Its prime purpose is to determine a practical duration for the completion of all construction activities; secondary but important uses are explained below.

Of all of the schedules used in the CM process, the Detailed Construction Schedule appears most traditional. The concept is to build the project on paper, step by step, with measured consideration given to the sequence and timing of activities.

20.8.1 Producing the Detailed Construction Schedule

With its flexibility and clear graphic presentation, precedent diagramming is an appropriate tool for this purpose. Bubble sheets (sheets that are printed with a blank bubble grid) can be used for planning and manual drafts of the schedule during development.

March 28, 1997 **Sequence Early-Start, Team Responsibility** **Pg 02/06**

Work Item Number	Duration Orig.	Duration Remain	Respons-ibility	Work Item Description	Required Start	Required Finish	Late Start	Late Finish	Start Float
00051	18	18	AE	Take Soil Boorings/Prepare Soils Report	28Mar97	20Apr97	28Mar97	20Apr97	000
*01000	180	180	X	Construct Parking Ramp Facility	04Apr97	16Dec97	04Apr97	16Dec97	M 000
00021	24	24	ON	Prelim Select Major Med/Non-Med Equip	05Apr97	06May97	05Apr97	06May97	000
00053	1	1	TM	Half Point Review: Schematics	05Apr97	05Apr97	05Apr97	05Apr97	000
00057	12	12	CM	Budget Estimate: Schematics Design Phase	05Apr97	20Apr97	05Apr97	20Apr97	000
00059	12	12	TM	Value Engineering: Schematics Design	05Apr97	20Apr97	05Apr97	20Apr97	000
00061	1	1	TM	Team Workshop: Schematics Design	21Apr97	21Apr97	21Apr97	21Apr97	000
00063	9	9	CM	Complete Detail Budget Estimate	22Apr97	04May97	22Apr97	04May97	000
00065	9	9	AE/CM	Prepare Schematic Presentation Materials	22Apr97	04May97	22Apr97	04May97	000
00067	9	9	AE/CM	Owner Reviews Schematic Design	22Apr97	04May97	22Apr97	04May97	000
00069	9	9	AE	Final Revisions: Schematics	22Apr97	04May97	22Apr97	04May97	000
*00071	2	2	AE	Present Schematic Design/Owner Approval	05May97	06May97	05May97	06May97	M 000
00073	51	51	AE	Design Development Documents Preparation	09May97	20Jul97	09May97	20Jul97	000
00081	51	51	ON	Final Select Major Med/Non-Med Equip	09May97	20Jul97	09May97	20Jul97	000
00101	24	24	AE	Bid Pkg #1: SITE/FDTNS Contract Documents	09May97	10Jun97	09May97	10Jun97	000
00103	1	1	TM	Bid Pkg #1 Documents: Half Point Review	25May97	25May97	25May97	25May97	000
00115	12	12	CM	Bid Pkg #1: Construction Estimate	25May97	10Jun97	25May97	10Jun97	000
00117	12	12	TM	Bid Pkg #1: Value Mgmt/Scope Verification	25May97	10Jun97	25May97	10Jun97	000
00201	20	20	AE	Bid Pkg #2: STR STL/LG LEAD ITEMS Contr Docs	06Jun97	01Jul97	06Jun97	01Jul97	000
00123	5	5	CM	Bid Pkg #1: Construction Estimate	13Jun97	17Jun97	13Jun97	17Jun97	000
00127	5	5	CM	Bid Pkg #1: Write Work-Scope Descriptions	13Jun97	17Jun97	13Jun97	17Jun97	000
00133	5	5	CM	Bid Pkg #1: Milestone Schedule	13Jun97	17Jun97	13Jun97	17Jun97	000
00075	1	1	TM	Half Point Review: Design Development	14Jun97	14Jun97	14Jun97	14Jun97	000
00077	26	26	CM	Budget Estimate: Design Development Phase	14Jum97	20Jul97	14Jun97	20Jul97	000
00079	26	26	TM	Value Engineering: Design Development	14Jun97	20Jul97	14Jun97	20Jul97	000
00137	1	1	TM	Bid Pkg #1: Team Workshop	20Jun97	20Jun97	20Jun97	20Jun97	000
00135	2	2	AE/CM	Bid Pkg #1: Final Revision, Site/Fdtn Docs	21Jun97	22Jun97	21Jun97	22Jun97	000
00203	1	1	TM	Bid Pkg #2 Documents: Half Point Review	21Jun97	21Jun97	21Jun97	21Jun97	000
00217	9	9	TM	Bid Pkg #2: Value Mgmt/Scope Verification	21Jun97	01Jul97	21Jun97	01Jul97	000
00219	9	9	CM	Bid Pkg #2: Construction Estimate	21Jun97	01Jul97	21Jun97	01Jul97	000
00141	3	3	AE	Bid Pkg #1: Print Drawings/Specifications	22Jun97	24Jun97	22Jun97	24Jun97	000
00143	3	3	CM	Bid Pkg #1: Print Proposal Forms	22Jun97	24Jun97	22Jun97	24Jun97	000
00145	1	1	AE	Mail Bid Pkg #1 Drawings/Specifications	24Jun97	24Jun97	24Jun97	24Jun97	000
00149	1	1	CM	Mail Bid Pkg #1 Proposal Forms	24Jun97	24Jun97	24Jun97	24Jun97	000
00153	10	10	X	Bid Period: Bid Pkg #1, Site/Fdtn Contracts	27Jun97	11Jul97	27Jun97	11Jul97	000
00155	1	1	TM	Pre-Bid Meeting, Bid Pkg #1	05Jul97	05Jul97	05Jul97	05Jul97	000
00157	3	3	AE/CM	Bid Pkg #1: Prepare/Issue Pre-Bid Addenda	05July97	07Jul97	05Jul97	07Jul97	000
00223	4	4	CM	Bid Pkg #2: Construction Estimate	05Jul97	08Jul97	05Jul97	08Jul97	000
00227	4	4	CM	Bid Pkg#2: Write Work-Scope Descriptions	05Jul97	08Jul97	05Jul97	08Jul97	000
00233	4	4	CM	Bid Pkg #2: Milestone Schedule	05Jul97	08Jul97	05Jul97	08Jul97	000
00237	1	1	TM	Bid Pkg #2: Team Workshop	11Jul97	11Jul97	11Jul97	11Jul97	000
*00161	1	1	TM	Bid Date: Bid Pkg #1, Site/Fdtn Contracts	12Jul97	12Jul97	12Jul97	12Jul97	M 000
00163	2	2	TM	Review Bids: Bid Pkg #1, Site/Fdtn Contracts	12Jul97	13Jul97	12Jul97	13Jul97	000
00239	2	2	AE/CM	Bid Pkg #2: Final Revision, Str Stl/LL Items	12Jul97	13Jul97	12Jul97	13Jul97	000
00160	1	1	TM	Early Award, Site Contractor (Bid Pkg #1)	13Jul97	13Jul97	13Jul97	13Jul97	000
00241	3	3	AE	Bid Pkg #2: Print Drawings/Specifications	13Jul97	15Jul97	13Jul97	15Jul97	000
00243	3	3	CM	Bid Pkg #2: Print Proposal Forms	13Jul97	15Jul97	13Jul97	15Jul97	000
00165	2	2	TM	Post Bid Interviews/Award Rec, Bid Pkg #1	14Jul97	15Jul97	14Jul97	15Jul97	000
00100	2	2	CM	Site Contractor Mobilization	14Jul97	15Jul97	14Jul97	15Jul97	000
00245	1	1	AE	Mail Bid Pkg #2 Drawings/Specifications	15Jul97	15Jul97	15Jul97	15Jul97	000
00169	1	1	ON	Contract Award, Foundations (Bid Pkg #1)	16Jul97	16Jul97	16Jul97	16Jul97	000

FIGURE 20.2 A typical Early-Start, Team Responsibility Sort, Program Schedule.

The scheduler determines the construction activities and puts them into sequence by entering activity descriptions in selected blank bubbles and connecting the bubbles with lines by precedence.

The scheduler can begin with a skeletal network, using broad-titled activities, covering the time allotted for construction in the program schedule. He/she can then fill in detail by converting the broad-titled activities to subnetworks. If time is available and the construction of the project is fairly conventional (or if the scheduler prefers), he/she can put the detail into the network as scheduling proceeds.

The choice of procedures is optional to the scheduler. However, projects with time constraints and those that have unique construction requirements or require an early scheduling effort can best be accomplished using the skeletal/subnetwork method. (See Figure 20.5 on page 321.)

20.8.2 Network Diagramming

When originally adopted by the construction industry, network scheduling was mainly used to schedule construction projects from beginning to end. It produced a "roadmap" of sequential activities with at least one sequence (a critical path) that, if followed by general contractors and their subcontractors, would achieve construction completion on time.

This schedule, developed before the start of construction, assumed contractor progress, anticipated inclement weather, and estimated the consequences of the problems inherent to construction operations. The schedule was often contractually imposed upon the contractors without their input and used in a dictatorial manner to "drive" the contractors from construction start to finish.

Prior to computerization, the initial construction schedule was often the last schedule, because manually updating a noncomputerized schedule tended to stall construction progress while updating was being accomplished. Updating a schedule (especially in the early stages of a project) sometimes took as long as initially creating the schedule.

Computerization reduced update time from days and weeks to hours and minutes. However, the assumed sequence of construction activities, and the predicted effects of weather and unforseen construction problems, seldom proved accurate, and the effort required to stay with the original construction sequence often became a serious problem.

The advent of the CM system and involvement of a construction manager effected a departure from conventional detailed construction scheduling. The CM is an advocate of both the owner and A/E; the CM has the same project goals as both and is a specialist in construction and scheduling. More importantly, the CM accepts scheduling as one of several management tools. It is a tool that guides rather than directs construction progress to a timely completion and can cope with the realities of a construction project. The roadmap concept still prevails, but the path between target activities can vary, and the target check-points are farther apart.

This departure from convention has produced a new scheduling philosophy, one that constantly extracts input from contractors and consequently is not considered dictatorial. Experience has shown that this approach to scheduling has a much better potential to achieve its purpose.

20.8.3 Developing and Using the Detailed Schedule

When using CM, the purpose of the Detailed Construction Schedule should be threefold: to determine a practical and acceptable construction time, to establish approximate dates for the completion of certain groups of construction activities, and to provide contractors with direction at the outset of construction.

All the precautions and practices of good professional scheduling should be followed. Scheduling should be done by an experienced competent scheduler who has intimate knowledge of technical construction and the project. (These requirements should be inherent in a CM organization.)

The Detailed Schedule can be manually produced, computer-generated, or both. Either approach is acceptable for its intended use. The schedule need not be a finished document; it must simply be readable and accurate within the limits of the information available just prior to bidding the project. It should be produced as conveniently as possible but not at the sacrifice of credibility.

The schedule should not be distributed to team members. It should be filed by the CM as a source document for the future if needed. It should not be updated once the required information is extracted from it or shared with contractors. Figure 20.3 is a sample list of Detailed Construction Schedule activities.

In addition to forecasting the duration of the project, the Detailed Schedule is the source of time-performance information that should be provided to contractors at bid time. The Milestone Schedule provides this information in the form of a key date listing.

20.9 MILESTONE SCHEDULE

Milestones are dates that will be used to measure actual progress with scheduled progress during the course of the project. The first milestone date is project start, the last is project completion. It is assumed that if each intermediate milestone date is met according to schedule, the required end date will be met.

*Award Contracts
Shop Drawing Approvals
Fabricate Open Web Joists
Order Hardware
Order Hollow Metal
Deliver Masonry Specialties
*Construction Mobilization
Deliver Hardware
Deliver Hollow Metal Frames
Deliver Masonry Specialties
*Earthwork
Concrete Footings, West Side
*Bearing Masonry
*Site Utilities
*Concrete Floor Slabs
Set Owner Equipment
Erect Steel Joists
Erect Metal Deck
Roofing and Sheet Metal
Install Windows
*Building Enclosure
Install Interior Partitions
*Wall Rough-ins
*Ceiling Rough-ins
Ceiling Finishes
Wall Finishes
Floor Finishes
*Punch Lists

*Activities that are appropriate for use in a Milestone Schedule.

FIGURE 20.3 A sample of Detailed Construction Schedule activities.

The purpose of the milestone schedule is to provide bidding contractors with adequate schedule information from which they can roughly estimate the time(s) when their services will be needed on the project, and to provide a means by which the project team can measure project progress (and consider alternatives if required).

20.9.1 Bidding Information

Bidding contractors should have construction schedule information to assess the demands of their time involvement in the project. General contracting bidding documents usually designate an approximate start date and a maximum construction period in days, or a specific start date and a specific completion date. A Detailed Construction Schedule is seldom published as part of the bidding documents or provided by a general contractor to the subcontractors as bidding information. Subcontractors traditionally determine their time involvement from the general contractor's start and finish information in the bidding documents. Over time, subcontractors have acquired enough knowledge of the construction process to closely estimate their time involvement, especially on conventional projects.

Seldom does a subcontract issued by the general contractor designate a specific start date or end date for a specific subcontractor (other than those in the GC's contract with the owner). Subcontractors would prefer specific time-related contract provisions to protect them from cost overruns caused by the operations of the general contractor or another subcontractor. However, general contractors prefer contract language that simply requires subcontractors to keep pace with construction progress.

When computerized scheduling became the rule rather than the exception, CM practice became schedule oriented to the point where many construction managers relied mainly on a construction schedule to manage a project. The speed and flexibility of computerized scheduling provided a false sense of control when in fact computerization only streamlined the scheduling process, not the construction process. The realities of contractor progress, inclement weather, and the problems inherent to construction operations still had to be reckoned with.

20.9.2 Milestones and Multiple Bidding Information

Many CM projects use multiple prime contracts and phased construction. On these projects, it is probable that none of the multiple prime bidders will stay on the project for the total construction period and only some will be on the project at the same time.

Although the start and end date of the project could be used by multiple prime contractors to determine their time involvement (as it is when they work for general contractors as subcontractors), the presence of a construction manager (an expert scheduler) suggests that more and more specific information be provided.

However, for the same reasons general contractors do not provide specific schedule information to bidding subcontractors, the project team should *not* provide too much schedule information to trade contractors prior to bidding. Even though more schedule information is available due to the comprehensive preconstruction scheduling responsibilities of the CM, the credibility of that information is still subject to the realities of the construction environment.

Consequently, it is impossible to unerringly predict precise contract start and finish dates for the thirty or more interfacing contractor work-scopes that comprise the construction. Missed contractual start and finish dates for whatever reason other than the contractor's own fault are potential sources for claims by the contractors against the owner. For this reason, it is obviously in the owner's best interest to issue construction schedule information prior to bidding that has little or no chance of generating a delay claim against the owner.

Milestone Schedule information can be used for this purpose. A Milestone Schedule, in list or network form, can be issued to all bidders. However, it is much wiser to issue selected, pertinent, milestone events on a contractor work-scope by work-scope basis. If a Milestone Schedule is issued for bidding purposes, it should be an edited version that only includes prudently defined milestones—those that cannot be construed as defining a specific contractor's work-scope start or finish date.

The concept of providing a minimum amount of schedule information may be difficult for super-scheduler CMs to understand. However, regardless of the CM's perceived ability to produce schedules, as the owner's agent the CM has the primary responsibility to act in the owner's best interests. It would not be in the owner's best interest to needlessly expose the owner to delay claims by issuing finite information that is so vulnerable to change.

20.9.3 Potential Schedule Liability

If the Detailed Construction Schedule has little chance of being followed precisely, the Milestone Schedule could have the same problem. It is one thing to develop a schedule to guide performance and quite another to expect specific dates to be met during performance. Contractor performance is especially unpredictable due to the many variables involved in the construction process. Experience using Milestone Schedule information clearly indicates that it should be presented as a rough guide to bidding contractors.

Milestone events should not be closely identified with individual contract performance. A "winter close in" or a "building enclosure" milestone is acceptable; a "complete roofing" or "install windows" milestone is not. The intent is to provide soft, interpretable information while not providing hard dates that can be legally construed as the time bracket for a specific contractor's work-scope performance.

Bidding information accompanying the Milestone Schedule information should make it abundantly clear to contractors that the information is no more than a guide. The chances of meeting milestone dates are subject to the inherent variables present in every construction project. Additional exculpatory clauses should be inserted to further establish the intended purpose of the schedule information.

20.9.4 Contractor Construction Schedule Input

Two additional requirements should be included in each contractor's proposal. Both tend to reinforce the intended use of Milestone Schedule information as a guide to contractors rather than a dedicated prediction of work-scope start and finish dates.

1. Each contractor should be required to accept or amend the schedule information contained in the issued Milestone Schedule at the time of bidding. The response will provide a practical check on the assumptions made by the scheduler and provide some assurance that bidders will review schedule information when assembling their proposals.
2. If bidders do not agree or cannot generally comply with the Milestone Schedule information as issued, they are required to state so and provide their reasons in their proposals. They are also required to provide alternate schedule information if they cannot comply. On projects with tight schedules, satisfactory schedule commitments by bidders can be used as a consideration for contact award.

20.9.5 CM Scheduling Philosophy

The CM philosophy advocated here establishes scheduling as a cooperative endeavor involving the multiple contractors and the project team, especially after contracts have been awarded. Inserting additional milestones to those used for bidding purposes enhances the schedule once construction starts. Using the milestone concept to its ultimate, construction operations are scheduled on a short-term basis, from milestone to milestone to accommodate unforseen contractor problems and actual project conditions. The goal of short-term scheduling is to maintain progress by meeting as many milestone dates as possible. Figure 20.4 is a sample list of Milestone Construction Schedule activities.

20.10 SHORT TERM CAP

The second use of both the milestone schedule and the detailed construction schedule is the development of the initial Short Term Construction (or Contractor's) Activity Plan, or Short Term CAP. The initial segment of the CAP is based on the first month or

Award Contracts
Shop Drawing Submittals
*Fabricate Open Web Joists
*Order Hardware
*Order Hollow Metal
*Deliver Masonry Specialties
Construction Mobilization
*Deliver Hardware
*Deliver Hollow Metal Frames
*Deliver Masonry Specialties
Earthwork
*Concrete Footings, West Side
Bearing Masonry
Site Utilities
Concrete Floor Slabs
*Set Owner Equipment
*Erect Steel Joists
*Erect Metal Deck
*Roofing and Sheet Metal
*Install Windows
Building Enclosure
*Install Interior Partitions
Wall Rough-ins
Ceiling Rough-ins
*Ceiling Finishes
*Wall Finishes
*Floor Finishes
Punch Lists

*Activities that are NOT appropriate for use as bidding information.

FIGURE 20.4 A sample of Milestone Schedule activities.

two of these schedules, after they are amended by the team to reflect information provided by the successful bidders. It is then adjusted by information revealed at post-bid and preconstruction meetings.

Predicated on the reality that construction schedules face many variables, many of which are unpredictable, the Short Term CAP is the ultimate on-site construction scheduling tool. Spanning a duration of approximately two months, the CAP should be produced and maintained in the field on a daily/weekly basis and used as a means of compensating for the negative effects of as many of the variables as possible. The Short Term CAP is a dynamic schedule within which activities can vary in both duration and precedence but only between two sequential milestones.

Information exchanged between the contractors' foremen and the Field CM at any time, but especially at weekly progress meetings, provides the data necessary for updating. As the current week ends, the next week (approximately two months in the future) is added to the end of the CAP. Using this short-term progressive process, a current viable construction activity plan covering the next two month period is constantly available to the contractors on-site.

Based on weather forecasts and the current knowledge of the variables, it is reasonable to assume that the week immediately ahead can be better planned than the next week, and that the next week can be better planned than the one after that. Excepting surprise, prediction accuracy diminishes the farther plans are made in the future. Consequently, as the next week is planned, a higher degree of refinement can be expected for planning subsequent weeks.

20.10.1 Target Activities

Milestones dates are the constant targets for all on-site construction planning. If each successive milestone date is met, the final completion milestone date will be met. If a milestone date is missed, time must be recovered through expedient planning between future milestone dates. If time cannot be recovered, for whatever reason, the completion milestone date must be changed.

The CAP provides contractors with schedule flexibility. It is common for contractors to have unforeseen problems (a piece of equipment that breaks down or workers that fail to report for work). Trade contractors especially have unique operational problems such as pressure to finish another job or a shortage of necessary material. The cooperative nature of short-term planning can often alleviate the full effects of these problems on the construction schedule.

When a contractor's problem is passed on to the Field CM and the other contractors on-site, a plan might be collectively found to work around the problem without loosing any time to the schedule. As peer participants in the scheduling and construction process, contractors are more inclined to help one another by changing planned activities for that day to offset a delay. Each contractor is aware that the next problem could be theirs and they too might have to seek relief through the help of their peers.

The Field CM should be the willing intermediary whenever short-term changes are suggested by contractors so long as the short-term change will provide a long-term solution to a schedule problem. It is to the team's advantage to mitigate or eliminate construction delays through cooperation rather than intimidation.

20.10.2 The Short Term CAP Format

Selecting the best format for the CAP depends on the scheduling capabilities of the field CM and the contractors' on-site superintendents or foreman. To be completely successful, the CAP should be based and maintained in the field, not at some remote office location. However, if CM field personnel are not proficient in scheduling, help can be provided by a competent scheduler at another location.

Computerized Precedence Diagramming (CPD) should be the first choice. Even though suggested changes will be short term, CPD has the ability to accept and quickly analyze "what if" situations and provide both short-term and long-term effects. Most CPD programs can print out a bar chart if a diagram format is better understood by contractors. It can be safely assumed that the scheduling expertise of on-site contractors will vary from high to low. Formats should be used that reach them all without exception.

If field personnel are not sufficiently proficient in CPD, a computer connection between the field and a proficient person in the CM organization can provide a satisfactory solution. However, the advantage of the CAP is that planning is done in the field, by field level supervisors. They know their individual capabilities as well as the constraints of the actual on-site conditions and are in a position of authority to propose and pursue a short-term schedule redirection.

Contractors' on-site supervisors are obligated to their employers to keep crews productively occupied toward contract completion and maximum profit. They know the current status of labor, material, equipment, and site conditions and understand how to work around problems that arise.

The field CM, on the other hand, has the most intimate understanding of the project and is in an excellent position to effect mutually productive redirection efforts when called for. On-site contractors tend to cooperate with the field CM's efforts when they realize the solution to short-term schedule problems will optimize the productivity of their personnel.

Figure 20.5 is an example of how the need-to-know information philosophy expressed in Chapter 13, Information Management, is incorporated into schedule management. The Milestone Schedule provides sufficient information to the executive level. The Detailed Construction Schedule information is sufficient to keep management level participants up to date, and the Short Term CAP activities are sufficiently detailed to provide on-site direction to contractors.

20.11 COMPLETION SCHEDULE

The fifth and final schedule that should be used is the Completion Schedule. It provides a timely and orderly transition from construction to occupancy or commissioning. Similar in detail to the Detailed Construction Schedule, and in content to the Program Schedule, the dates of its development and initiation should be activities listed on both the Program Schedule and Milestone Schedule.

The Completion Schedule is the sequence of activities that must be accomplished before a constructed facility can be beneficially used or occupied by the owner. Occupancy activities are not considered important during construction start-up opera-

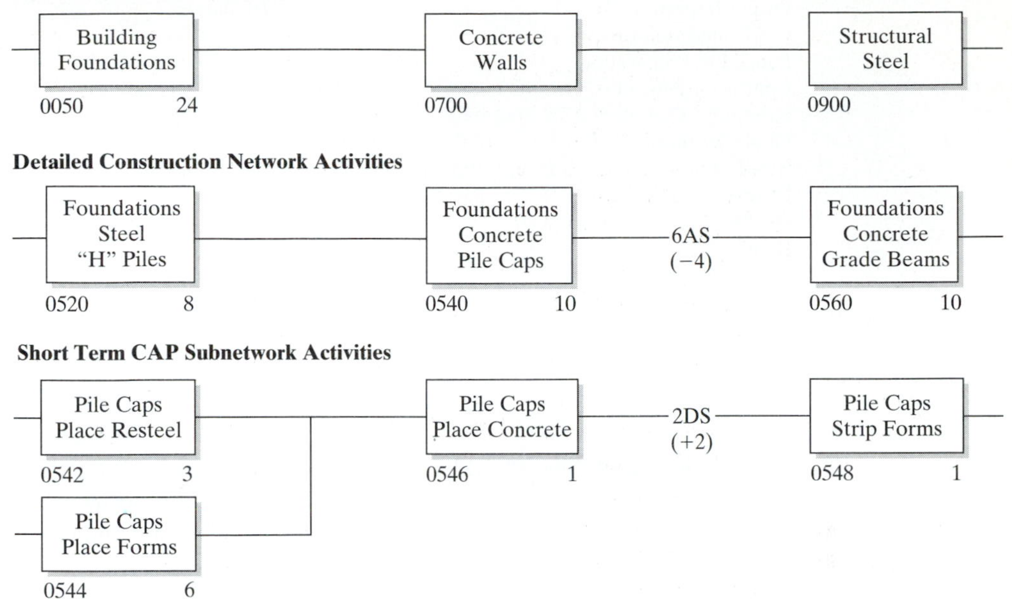

FIGURE 20.5 An example of schedule networks and sub-networks.

tions. However, sooner or later, these activities will become critical to the owner, and therefore they should be anticipated as part of schedule management. The Completion Schedule (or Occupancy Schedule) accomplishes this.

20.11.1 Completion Schedule Development

The first step in producing a Completion Schedule is to develop a complete list of owner, A/E, CM and other activities upon which occupancy is predicated. This list usually includes such things as inspections, permits, licenses, equipment/furniture delivery, computer and telephone installations, supplies and raw material stockpiles, trial operations, and any other unique owner start-up necessities.

Figure 20.6 is a short list of the many typical activities that should be included in a Completion Schedule.

The activities vary with the unique commissioning or occupancy requirements for a particular facility. Responsibility to arrange or execute some of these activities may have been assigned to team members in the responsibility matrix developed at the beginning of the project. However, at that time, team efforts were properly directed to getting the project started and constructed; some completion activities were probably overlooked.

One activity that should not have been overlooked was owner occupancy and its anticipated duration. Another item that should appear in the Program Schedule during the project is developing a Completion Schedule (positioned about two-thirds of the way through construction).

Fire Marshal Walk-Through	Signage Release
Owner Inspection	Signage Istallation
CM Completion List	Maintenance Manuals
Punch List Preparation	Approve Mechanical Systems
Contractor Punch List Completions	Approve Electrical Systems
Substantial Completion Certificates	Approve Plumbing Systems
Final Cleaning	Punch List Progress
New Furniture and Equipment Installation	Negligent Contractor List
Issue Seven-Day Letters	Systems Balancing
Health Department Certificate	Move Furniture and Equipment
Required Licenses	Occupancy Certificate
Local Angency Approvals	Test-Run Processes
Telephone System Installation	Spare Parts Inventory

Figure 20.6 Typical activities for the Completion Schedule.

The Completion Schedule should be developed backwards, from the required date of beneficial occupancy or use of the facility to the date of the initial activity in the occupancy sequence. This initial activity in the Completion Schedule will be dependent on one predetermined construction activity in the Milestone Schedule.

20.11.2 Initiating the Completion Schedule

The date toward the end of construction, when the Completion Schedule must be initiated to allow time for the completion of all Completion Schedule activities, is determined by overlaying the Completion Schedule on the Milestone Schedule. If the date of the predetermined construction milestone in the Construction Schedule coincides with the date of the initial activity in the Completion Schedule, the project is on schedule.

If coincidence does not occur, construction is either ahead or behind schedule. If behind schedule, the owner can accelerate construction so date coincidence occurs or change either the date or conditions of occupancy. If ahead of schedule, the owner can either occupy earlier than planned or slow down the occupancy process. In the latter case, slowing down construction should never be considered a viable option. Figure 20.7 graphically explains implementating and interfacing a Completion Schedule.

20.11.3 The Completion Schedule Format

The Completion Schedule should be in a computerized precedence diagram format to facilitate convenient development and provide timely updating and appropriate print-out sorting. Each team member should receive an original print-out and frequent updates, sorted by team and by early start. As stated, time lost during the performance of completion activities is no different than time lost during construction.

If problems exist, team meetings dealing specifically with completion activities should be held with increased frequency as the Completion Schedule initiation date approaches. The closer the project gets to the occupancy date, the more difficult it is to make completion activity adjustments without jeopardizing the occupancy date.

Interfacing the Completion Schedule with the Milestone Schedule and Short Term CAP is a simple and revealing process. The Short Term CAP is a forward pro-

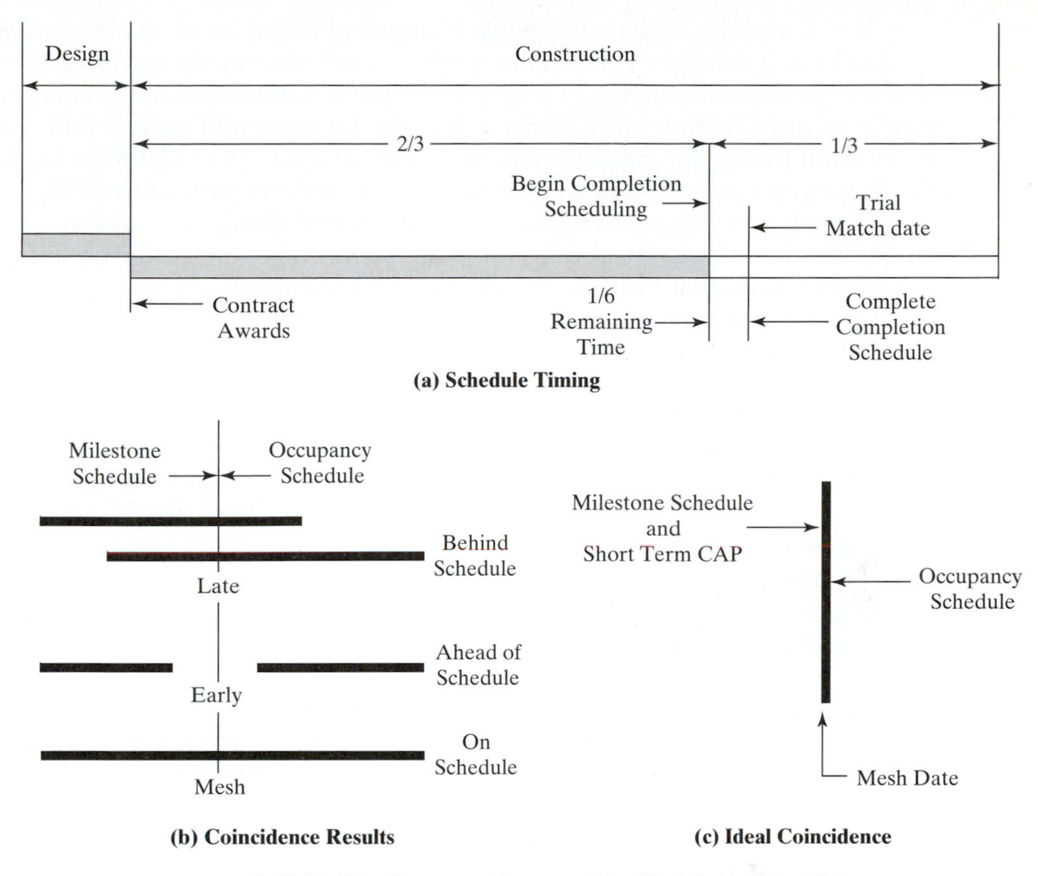

FIGURE 20.7 Graphic explanation of the Completion Schedule's use.

jected schedule of construction activities, headed for the future milestone called occupancy. The Completion Schedule is a backward-projected schedule of occupancy activities, starting from the occupancy milestone, headed for a designated mesh date approximately two-thirds of the way through construction. How close they coincide will determine the practical occupancy date, alert the team to required acceleration of construction or occupancy activities, or both.

20.12 SCHEDULE COMMUNICATION

Scheduling is a powerful management tool for the CM. Properly used, it is a major contributor to the success of a construction project.

Earlier in the chapter, overusing scheduling software by trying to use all the accessories software producers include (mainly to increase sales) was discussed. Scheduling software should be viewed by a construction manager as a hammer is viewed by a carpenter—a basic tool to be used for basic tasks.

It was also suggested that schedule information be passed on to others using the need-to-know philosophy covered in Chapter 13, Information Management. The idea is to only provide information to project participants that was within the limits of that party's assigned responsibility, to make it easier for parties to process and react to information from their defined perspective and management level. Producing schedules as networks, subnetworks and sub-subnetworks accommodates this need.

There is final suggestion regarding scheduling that should be considered. On the assumption that the quality of collective performance relies on a common interpretation of communication between parties, the CM should pay special attention to the schedule format or language used.

Schedules can be communicated in several different languages, each with its own level of sophistication: appropriately sorted lists, bar charts, with or without tales, and CPM Networks, precedence or arrow diagrammed.

Of these, lists represent the language familiar to people in general. Lists are commonly used for radio and TV, airline, bus and train schedules as well as many other services with preplanned timelines. List schedules should be used to communicate information to those involved in the project who cannot effortlessly understand bar charts or networks. Although it is difficult to develop schedules in the list format, it is very simple to networks to lists.

Bar charts represent the schedule language with the longest history in the construction industry. Simple bar charts are also widely used by the media to convey information to the general public. However, bar charts require enhancement to specifically convey schedule information on a construction project. If a simple bar chart can accurately convey schedule information, it should be used; if it cannot, a bar chart with tales should be used. Bar charts can be used to develop limited (perhaps short term) schedules and can be consistently interpreted by those who read them.

CPM network diagrams represent the schedule language that is unrivaled when used to develop and update schedules, but they are not recommended for communication between project participants, except as supplements to list schedules or bar charts, or for communicating with parties who will use the network diagram for schedule analysis.

The foregoing will cause consternation to more than a few advocates of network scheduling, especially those who favor displaying the diagram on the wall as a "roadmap" for completing the project. However, with the knowledge that sooner or later there will be an update, an understanding that once construction begins scheduling is used as a time problem solving tool, the realization that what has occurred to date is for record purposes, and that only persons with a high level of scheduling sophistication can interpret CPM diagram language, some traditional scheduler's opinions may change.

20.13 IN RETROSPECT

The five schedules—Program Schedule, Detailed Construction Schedule, Milestone Schedule, Short Term CAP, and Completion Schedule—are interdependent, yet each serves a single useful purpose. Experience has shown that the comprehensive scheduling effort suggested here upgrades CM schedule management success.

Separating scheduling components into their own entities tends to revitalize scheduling responsibilities by providing new beginnings and fresh approaches to problem solving. To contemplate a single schedule approach that would combine the functions of the five schedules into one is baffling after using the multi-schedule approach on even one project.

20.14 USE OF SCHEDULES BY THE CM FIRM

To this point, schedule management has been discussed from the perspective of CM services provided to owners (external scheduling in support of the project-delivery process). Schedule management should also be practiced within the CM organization as internal scheduling that can improve the construction manager's performance on current projects.

External scheduling assists in making team decisions in the other areas of knowledge that are schedule dependent: Budget, Decision, Material/Equipment, Project, Value and Resource Management areas. When the team is faced with time-related decisions, schedule information or schedule applications are expected to be provided by the construction manager.

Examples of time-related decisions in other management areas are cash-flow predictions, life-cycle cost studies, time criteria for expediting, least-cost analysis, and resource leveling. The CM's scheduling expertise should play an important part in executing these activities. It is expected that familiarity with these types of schedule applications is an integral part of the CM's capabilities.

20.14.1 The Internal Master Program Schedule

The CM organization covered in Chapter 8 identifies five departments: executive, administrative, operations, resource, and support. Of these, the resource department is composed of persons with varying expertise whose input is required on every project, at times dictated by each project's progress.

Allocating resource personnel could be a problem if the schedules of two or more backlogged projects require the input of the same resources simultaneously. To prevent this from occurring without some warning, resource-leveling techniques can be used within the CM organization to anticipate overloads on specific resource personnel and manage the resources in a productive fashion.

If each backlogged project is under the control and guidance its own Program Schedule, combining the Program Schedules of all projects into an Internal Master Program Schedule could identify when specific resource personnel input is required, provide time to have adequate personnel available, or provide time to slightly adjust one or more project Program Schedules to ease the problem.

The Internal Master Program Schedule would promote the same level of management to the CM's organization as the CM provides to its clients under service contracts.

CHAPTER 21

Safety Management

Safety in construction has always been of significant concern to the industry and the government. Construction sites have the infamous reputation of being one of the most accident-prone of all work places in the United States. Persistent efforts continue within the industry, and by federal and state authorities, to make construction sites safer. Although progress has been made, it remains apparent that zero lost-time accidents and deaths on a construction site are often the result of good fortune as well as conscientious practices.

From a practical standpoint, construction accidents can be reduced in number and severity but never fully eliminated. Construction is simply a dangerous occupation, one that employs people with normal human frailties. Once this bleak reality is acknowledged by the authorities, greater progress can be made to reduce the frequency and severity of accidents.

When an accident occurs, the responsibility of a construction manager is controversial due to the persistence of obscure interpretations of contract language and a biased opinions of a CM's inherent duties and physical capabilities on a construction site.

In the early years of CM's development, a construction manager was considered by most (especially the legal system) as a close kin of a general contractor. Although wrong, this was understandable, because for years CM was considered undefinable, not only by the courts but by most persons inside and outside the construction industry.

As a new and different contracting system, every dispute involving a CM broke new ground during litigation. Without a consensus authority to turn to, attorneys and courts provided their own definitions of CM and interpreted early obscure contracts accordingly. It is accurate to assume that many of the early cases involving CM would have been decided differently if current CM contract documents had been used.

Within our legal system, it is difficult for courts to change the bases for their decisions (i.e., cases are decided on the recorded findings of previous cases of a similar nature). Even though it is obvious now that there is both a legal and functional difference between a construction manager and a general contractor, current court rulings still defer to the past for many of their decisions.

On GC projects, the co-occupancy of the construction site by separate business entities and the labyrinth of contracts and agreements that tie them together exposes practically everyone to the possibility of citation when an accident occurs. This of course includes the possibility of becoming involved in any ensuing litigation initiated by the injured party. As subcontractors, trade contractors are held accountable to the GC and the GC is held accountable to the owner. Both are held accountable to safety authorities. In essence, the effectiveness of the safety program is the discretion of the GC's.

Even though the construction site co-occupancy numbers are essentially the same, the labyrinth of contracts and agreements is simplified on CM projects by eliminating one contracting tier. The Second Tier contractors on a GC project become Prime or First Tier contractors on a CM project, and the form of contract they hold with the owner is virtually the same as that which a GC would hold with an owner.

The fact that each trade contractor has an identical and direct contract with the owner simplifies assigning safety responsibilities to them. Coordinating responsibilities of the CM facilitates the integration of each contractor's effort in the safety management program as well.

21.1 THE SAFETY MANAGEMENT AREA OF KNOWLEDGE

The safety management area of knowledge encompasses safe practices at the construction site in accordance with the prevailing regulations in the area of the project. The CM has the responsibility to promote safe site conditions by example and urge contractors to have organized safety procedures in force. Although each contractor bears the responsibility for the safe practices of its own employees, the CM has the responsibility to coordinate safety requirements common to all contractors and to see that safety provisions are included in construction contracts.

The CM must be familiar with construction practices, safety programs, practices, procedures, and administration and safe environmental conditions at the construction site, as well as an understanding of the 1970 Occupational Safety and Health Act (OSHA) and the safety/health regulations in the area of the project. Excellent communications skills and high ethical standards are required.

21.2 CONSEQUENCES

The severest consequence of an accident obviously falls on the injured party, and in the case of death, on the injured party's relatives and friends. However, the CM, A/E, owner, and contractors can suffer consequences as well, albeit of much different severity. These consequences will be monetary, in the form of fines levied by safety compliance authorities and damages awarded by a court to the victim or the victim's survivors. Fines can be sizeable, but damage awards can be formidable.

21.3 SAFETY LAWS AND REGULATIONS

The Occupational Safety and Health Act (OSHA) was enacted by the U.S. Congress in 1970. The Act covered most all industries, including the construction industry. Its purpose was to establish working environs that decreased worker exposure to unhealthy and unsafe conditions. It also created, within the framework of the U.S. Department of Labor, the Occupational Safety and Health Administration whose responsibility was to establish safety and health standards and promulgate the rules and regulations to implement them.

OSHA legislation permits individual states to establish and enforce their own safety standards and regulations so long as they are as strict or stricter than the federal versions. Most states have done this and have taken over the enforcement of the regulations within the framework of state government. Enforcement includes inspections, investigations, record keeping, and the levying of fines on those who commit violations.

When targeting the employer's premises as the location for compliance, OSHA saw little distinction between construction sites and manufacturing sites, yet there was a great difference between the two. Manufacturing sites are essentially permanent locations, manned by a static, long-term work force. Construction sites are unique for each project and employ a changing, short-term work force. Control of health and safety conditions at a manufacturing site presents a stationary target; control of health and safety conditions at a construction site presents a rapidly moving target.

It took several years before OSHA regulations for construction were brought into practical focus. Contractors noted that abiding by OSHA's initial regulations increased the cost of construction by 30%. A cooperative effort between contractors and OSHA brought the regulations within reason, but there is still no comparison between the diligence required to maintain site-safety on construction sites and on manufacturing sites.

OSHA Construction Safety and Health Standards, 29 CFR, Part 1926, and OSHA Safety and Health Standards, 29 CFR, Part 1910, with subsequent amendments, contain the minimum requirements. Copies of these CFRs, individual state regulations, and any other prevailing rules and regulations should be part of every CM's library.

21.3.1 Legal Consequences

In our legal system, anyone can sue anyone with the slightest justification; by precedent, accidents routinely provide justification. It is reactively assumed by a victim's attorney that accidents are always caused in whole or in part by actions or inactions of another party or parties. To compensate for the pain and suffering caused by the accident to the plaintiff or plaintiff's family, those deemed responsible must pay monetary damages.

Plaintiffs' attorneys implicate as many defendants as possible to enhance the chances for collecting damages. When named as a defendant, whether at fault or not, the party has no choice but to hire an attorney and pay the sizeable legal expenses connected with a defense. Although the courts do not always assign responsibility to CMs, A/Es, and owners when accidents occur, their record to the contrary is enough to influence a victim's attorney to include them as culpable parties.

It is difficult for a CM to establish and maintain an effective safety-management program without being exposed to consequences when an accident occurs. The courts seem to take the position that an effective safety program is one that produces zero accidents, not one that keeps them to an absolute minimum under the circumstances. The courts go to great lengths to compensate an injured party, often discounting the pragmatic circumstances that caused the accident as well as the explicit responsibilities of the CM, A/E, and owner as detailed in their contracts.

Many times the injured person is the one clearly at fault; no one else could have prevented the accident when it occurred. However, this argument seldom stands up to the manifest sympathies of the court for workers who are injured or killed. Reasons can always be found to implicate others who in some remote way did or did not do something that contributed to the worker's unsafe act.

21.4 THE QUANDARY

On an ACM project, CM personnel have no contractual jurisdiction over contractor personnel, only over the contractors themselves. The Field CM cannot give orders to workers on the site—only the worker's supervisor has the authority to do so. When an unsafe condition is observed by a worker or a worker's supervisor, it is the responsibility of each to report the condition to the CM and to refrain from being exposed until the condition has been corrected.

Consequently, exercising safe practices is exclusively vested in the workers themselves and the supervisors who provide their direction. Only when worker safety is threatened by actions or inactions of another contractor or by conditions at the site can the worker and the worker's supervisor not be considered in control of their safety. The CM, A/E, and owner should only be culpable when a condition which is their responsibility contributes to an accident at the site.

The responsibilities of the CM, A/E, and owner for the site conditions should be categorically stated in their contracts and in the contracts entered into by contractors. The owner should be responsible for consequences relating to such things as latent site conditions and concurrent site occupancy; the A/E for consequences relating to design and specifications; the CM for consequences relating to safety management (in all cases, only to the point the conditions are considered controllable).

Currently and unfortunately, sound logic and contract provisions notwithstanding, the CM's, A/E's, and owner's responsibility for safety on the construction site is actually determined by the courts in the litigation following an accident. The inconsistencies of court decisions preclude the establishment of standard guidelines for CMs to explicitly follow. Briefly, the responsibility of owners, A/Es, and CMs for construction site safety is subject to the vagaries of the law.

21.5 APPROACHES TO SAFETY

The best approach to an accident-free project is as follows:

- establish safety as a state-of-mind of all persons and parties involved in the project, especially workers
- contractually identify where the responsibilities for safe practices abide, in clear, concise provisions
- have all parties diligently carry out the responsibilities assigned to them.

This approach is similar to the one suggested to produce designed quality in the constructed project. Like the pursuit of zero defects, the pursuit of zero accidents requires the constant awareness and involvement of all parties and individuals on the construction site. Everyone must be aware that an accident is an event waiting to happen. All involved must take a relative responsibility for prevention and carry out those responsibilities with assurance that others are carrying out theirs.

Responsibility for the measures that produce a safe construction environment should not only be conclusively established by contract—they should be just as clearly understood by all parties involved in the project before construction begins. If properly handled, when an accident occurs there should be no need for interpretation to determine culpability.

Site safety should be a topic on the agenda of preconstruction meetings held with contractors before they occupy the site. In a format similar to the contractor's quality plan covered in Chapter 17, Quality Management, a Safety Plan should be submitted in writing by each contractor and discussed with CM, A/E, and owner representatives to a point of mutual understanding.

The team's Safety Management Plan should incorporate a detailed safety responsibility matrix chart that expands the broad safety responsibilities assigned in the project team's responsibility chart (see Chapter 16, Project Management). This new chart should be discussed at the preconstruction meeting to eliminate misunderstandings relating to contractual and pragmatic safety responsibilities of the CM, A/E, owner, and contractors. After agreement, the matrix should be endorsed by each party and become a formal memo of understanding or, if possible, a part of each contract by amendment.

It is evident that the Safety Management Plan must be micromanaged by the team from its initiation during contract development to its conclusion when construction is completed.

21.5.1 Posturing the CM's Safety Involvement

The suggested posture of the CM with regard to safety is stated in the second paragraph of the Safety Management Area of Knowledge: "The CM has the responsibility to coordinate safety requirements common to all contractors and to see that safety provisions are included in construction contracts." Owner-CM contract provisions, CM procedures, and CM actions on site should conform to this scope of involvement.

It would be grossly insufficient to merely state in owner-contractor agreements that "contractors must comply with all prevailing health and safety regulations at the site of the project" without describing CM, A/E, and owner participation. Although a statement to that effect should be included, the contractor should also be made fully aware of how site safety is to be managed under conditions of co-occupancy (who is going to do what in the pursuit of zero accidents).

To appropriately serve the owner's, A/E's, and CM's best interests, the CM should formulate a detailed Safety Management Plan, reduce the plan to specific provisions, and incorporate the provisions in the supplementary and special conditions of the general conditions of the contract for construction; subject of course, to project team approval. All three team members must agree to carry out their responsibilities and to do no less and no more than stated except in the event of imminent danger.

21.5.2 Pragmatic Involvement

Project team members, especially the CM, should realize that doing *more* than required by the contract provisions can produce results just as punitive as doing *less* than required. Although this doesn't seem reasonable, the attitudes of the law are such that a party can be penalized for doing more than originally agreed to, regardless of the benefits to the other parties. The reasoning is that if a party undertakes an unassigned responsibility, others will assume that the responsibility has been assigned to that party and rely on it.

Standard CM contract provisions do not establish CM field persons as safety inspectors who are required to patrol the site, looking for unsafe conditions and safety violations. CM field persons have too many specific tasks to perform for this type of activity. If safety inspection is assigned to the CM through an amendment to the owner-CM contract, the CM will have to provide additional persons on site to perform that duty. This situation is identical to that of inspecting quality explained in Chapter 17, Quality Management.

If CM personnel on the site see an unsafe act or observe an unsafe condition, they have a moral commitment to intervene, even though there is no contractual authority to issue orders to workers to avoid the condition or to correct it (unless of course, it is clearly an emergency-producing imminent danger). Intervention by the CM simply demonstrates a positive, natural concern for the well-being of others.

However, where imminent danger is not an issue, repeated acts of similar intervention by the CM could imply assumption of responsibility to lookout for unsafe acts and conditions in the future. Although this is a responsibility for which the CM did not contractually subscribe, it is now assumed in the eyes of the courts. If an accident occurs, there is a very good chance that the CM's failure to properly perform the *assumed* responsibility could put the CM and owner in a culpable position.

21.5.3 Site Presence

The fact that the CM is required to have a full-time presence on site seems to automatically make the CM culpable when an accident occurs. As far as plaintiff's attorneys are concerned, anyone on the site with supervisory responsibility when an accident occurs must be implicated in some way. Contract provisions to the contrary, attorneys can stretch the wording to come up with interpretations that open the door to CM liability. Their interpretations often put the CM in a position of responsibility without authority.

In their contract administration role on GC projects, architects discovered long ago that the only way to lessen their exposure to accident liability was to be on site as little as possible. Standard contract documents support this by not committing the architect to a set number or frequency of site visits. The chances of evading culpability are improved if an accident happens when the architect is not on site. It should be noted however, that it is common for an architect to be named as a litigant even if none of their personnel were on site at the time of an accident.

Unfortunately, it is a rare occasion when an accident occurs on site when CM personnel are not present. When construction is in progress, site presence is a standard contract requirement and a major attribute of the CM system. Responsibility as a safety inspector is not a standard contract requirement. Construction managers have no choice but to carefully prepare for the safety realities of CM practice, both contractually and pragmatically.

21.6 MAJOR SAFETY MANAGEMENT COMPONENTS

The principal components of safety management are the Project Safety Management Plan and the Contractor's Safety Management Plans.

21.6.1 The Project Safety Management Plan

The project team members should accept the fact that it is impossible to completely isolate themselves from safety involvement on a CM project, just as it is impossible to do so on a GC or D–B project. From this perspective, the team with the lead of CM should:

1. use risk-management techniques to identify and evaluate potential exposure to accidents on the project
2. with concurrence, determine the most pragmatic approach to safety under the conditions which will prevail on the project
3. procure insurances with appropriate limits to mitigate the cost of claims if they occur
4. see that specific safety requirements and responsibilities are inserted in owner-contractor agreements
5. require every contractor to submit a safety plan explaining how it intends to comply with OSHA and State safety requirements
6. commit to a team-defined safety approach, doing no more or no less than the commitment requires
7. document the project's Safety Management Plan, incorporating items 1 through 6, and include it in the CM Project Manual.

21.6.2 Some Responsibility Rationale

Project team member safety responsibilities should strictly follow the essential elements of service assignments under the ACM contracting structure. The essential elements of service are covered in Chapter 1, The Fundamentals of the Root Form of CM. The CM's responsibilities are in construction coordination, project management, and contract administration; the A/E's, in design, project management and contract administration; and the owner's, in contracting. No one on the project team has responsibility for construction in the ACM format.

Confining safety responsibilities to these areas provides a broad guideline for specific team member roles for project safety. One major responsibility of all team members is to promote a site that is a safe workplace for all workers; another is to instill the state-of-mind safety concept that collectively influences safe practices on site.

Responsibility for worker safety is primarily assigned by law to employers. This is supported by Workers Compensation insurance where only the employer is obligated to pay the costs of an injured worker's medical treatment and rehabilitation. Team members are directly responsible for the safety of their own employees but not directly responsible for the safety of workers employed by others.

21.6.3 Developing the Safety Management Plan

Development of a Safety Management Plan should be started early in the design phase but need not be completed until the plan's contractor-related safety provisions must be incorporated into the contract documents. However, to keep pace with the development of the other related management plans (such as risk management and contract management), an early start is beneficial.

21.6.4 Evaluating Accident Exposure

Evaluating exposure is a risk-management procedure. Each project has its own exposure to safety and health problems. As soon as practical, team members should discuss the probable construction processes to be used and the conditions under which they will be executed. The goal is to identify worker exposure to the inherent dangers of the construction methods which will probably be used and the unique dangers of the project's environment. Although little design has been done when the first review takes place, sufficient information is known to identify and evaluate the probable exposures to risks. Future reviews will enhance earlier assessments.

Will the construction be spread out over a large area? Is it low-, medium-, or high-rise construction? What is the nature of the foundation excavation or trenching work? What type/size of equipment will be used? How will the site be accessed and used by contractors? How will construction areas be accessed by workers? How far away are emergency services located? Will hazardous materials be handled? Will explosives be used? How long will each high-exposure construction process last? Which construction processes control the project? Is the construction time adequate? Will shift work be required? What weather conditions will prevail? Will wood concrete forms be used? To what extent will inflammable materials be used? Is demolition required?

The responses to these questions will provide insight to potential exposures and assist in identifying the physical problems that could contribute to safe practices on site. They identify unique safety needs, if any, and surface problem areas that require special attention. There will be periods of time during construction as well as certain activities that are more prone to accidents than others and will require special attention. All of this information should be taken into consideration when developing the Safety Management Plan.

21.7 THE TEAM'S APPROACH TO SAFETY

Due to the potential for their own involvement in an accident, the A/E and owner should assist the CM in formulating the Safety Management Plan. Each team member is a stakeholder in site safety and should be sure that their interests are protected, at least to the limits in their contracts.

A critical element of site safety is specifying certain requirements as contractor responsibilities in the contract documents. To accomplish this, the provisions in the standard contract documents must be appropriately amended, with the cooperation of the owner and the A/E. As long as the amendments do not increase the liability exposure of the owner and A/E, the changes should be acceptable to them.

The CM's role in safety requires definition beyond that found in standard contract documents. The owner-CM agreement and the supplementary and special conditions to the general conditions should contain provisions that assign the CM responsibility to perform the following:

1. determine site-wide safety requirements and assign them to contractors in work-scope definitions
2. monitor contractor compliance with OSHA required paperwork such as reporting, posting, and record keeping
3. organize, attend, and record periodic contractor safety meetings at the site
4. require each contractor to submit a safety plan to the CM and discuss the plan with the CM before occupying the site
5. recommend amendments to the owner-contractor agreements that more specifically define contractor safety responsibilities.

21.7.1 Designating Responsibility

There are safety requirements that affect more than one contractor and must remain in place for extended periods of time (items such as temporary stairs, ladders, barricades, fire and life safety equipment and signs) and other requirements that could be more economically provided by one contractor for the benefit of all.

These items should be identified by the CM and either included as part of a specific contractor's work-scope or accepted as the CM's responsibility. The requirement should be to provide the item(s) according to OSHA standards and maintain the item(s) in good repair for a specified period of time. The cost of the requirement should be part of the contractor's contract price and listed in the contractor's schedule of values as a payment line item. Items deemed to be the CM's responsibility should be included in the owner-CM agreement by amendment and funded in the Construction Support Budget.

21.7.2 Monitoring Paperwork Compliance

OSHA requirements entail considerable paperwork which contractors must submit and maintain in a timely manner. There are also mandatory postings for individual contractors. Although these are not accident prevention measures, they are requirements which must be complied with under penalty of a fine.

The CM should coordinate site-wide postings to assure compliance and eliminate redundancy and be available to check contractor paperwork to improve proper and timely compliance by them.

21.7.3 Safety Meetings

Periodic on-site safety meetings propagate the state-of-mind philosophy that is the key to safety performance. In addition to the safety meetings which contractor supervisors hold with their own personnel, the contractors' supervisors functioning on the site at the time should be required to participate in a brief biweekly project safety meeting, chaired on a rotating basis.

The CM should help each supervisor organize the project safety meeting, prepare the agenda, and obtain visual aids such as video tapes and handouts. If necessary, the CM should assist the chairperson during the meeting. The agenda should include lost-time accident updates from each contractor and the status of safety items for which they are responsible. Suggestions to improve safety should be encouraged. Meetings should not exceed one hour, and the minutes of each meeting should be recorded and distributed by the CM.

Contractor safety meetings should be encouraged. The CM could require contractors to submit meeting schedules and spot check to assure they are being held according to schedule. Contractor safety meeting requirements should be prescribed in the supplementary and special provisions of owner-contractor contracts.

To emphasize the safety philosophy, contractor safety meetings and attendance at project safety meetings could be line items in each contractor's schedule of values. Contractors would be paid for completing meeting requirements as they are paid for completing construction work.

21.8 CONTRACTOR SAFETY PLANS

Contractor safety plans are similar to contractor quality plans covered in Chapter 17, Quality Management. Their purpose is to involve contractors in safety before they move on site, to give them a prescribed course to follow once they get there, and to identify those contractors who require assistance or motivation regarding OSHA safety compliance and the philosophy of project safety.

Safety plans provide an indication of how a contractor intends to comply with OSHA requirements. More importantly, they provide insight into each contractor's attitude toward safety. When inadequacies are found in a plan, corrections can be made by the contractor before becoming actively involved with other contractors on site. The time spent by contractors putting safety plans together and by CM plan reviews could be the most productive time of all, if an accident is prevented as a result.

Contractor plans should not iterate OSHA requirements, they should indicate how the contractor intends to meet OSHA requirements in concert with other contractors on the site. Each contractor's plan will differ according to the type of work they will perform and the conditions under which it will be performed.

Safety plans should acknowledge safety requirements provided by contractors for use by all contractors. A summary of these could be listed in CSI Division I, General Requirements, under an additional number and heading "01900, Safety," or existing "01040, Coordination." The safety plan could be specified under "01300, Submittals."

The major portion of the plan should describe how the contractor intends to execute the work, identify the unique hazards involved in the work the contractor will perform, and explain how these hazards will be avoided or mitigated in response to OSHA requirements and safe practices.

Writing a contractor's safety plan is time consuming. It would be a prohibitive requirement if an original plan was required for each CM project in which a contractor becomes involved. However, only the first plan will require an inordinate amount of effort. Once written, a safety plan will require only minor changes for future CM projects they become engaged in.

To provide format consistency, safety plans could follow the index for OSHA Construction Safety and Health Standards, 29 CFR, Part 1926, Sub-part C through W. An added section pertaining to recording and reporting is included at the bottom of the listing.

The headings of the subparts of 29 CFR Part 1926 are:

C General Safety and Health Provisions
D Occupational Health and Environmental Controls
E Personal Protective and Life Saving Equipment
F Fire Protection and Prevention
G Signs, Signals, and Barricades
H Materials Handling, Storage, Use, and Disposal
I Tools—Hand and Power
J Welding and Cutting
K Electrical
L Ladders and Scaffolding
M Floor and Wall Openings, and Stairways
N Cranes, Derricks, Hoists, Elevators, and Conveyors
O Motor Vehicles, Mechanized Equipment, and Marine Operations
P Excavations, Trenching, and Shoring
Q Concrete and Masonry Construction
R Steel Erection
S Underground Construction, Caissons, Cofferdams and Compressed Air
T Demolition
U Blasting and the Use of Explosives
V Power Transmission and Distribution
W Rollover Protective Structures; Overhead Protection, and Records, Notices and Reports.

The fact that contractors have functioned under OSHA requirements for decades should diminish fears that contractors will be overwhelmed by the safety plan requirement. Additionally, a contractor's operations will involve only some of the subparts listed, subparts they should already be familiar with from past construction operations. A safety plan simply requires contractors to document their current safety practices.

For example, the masonry work-scope contractor would address to one degree or another subparts:

Q Concrete and Masonry Construction
O Motor Vehicles, Mechanized Equipment
L Ladders and Scaffolding
I Tools—Hand and Power

G Signs, Signals, and Barricades
H Materials Handling, Storage, Use, and Disposal
E Personal Protective and Life Saving Equipment
C General Safety and Health Provisions
D Occupational Health and Environmental Controls, and Records, Notices and Reports;

of these, all contractors would probably address subparts:

O Motor Vehicles, Mechanized Equipment
I Tools—Hand and Power,
H Materials Handling, Storage, Use, and Disposal,
E Personal Protective and Life Saving Equipment,
C General Safety and Health Provisions,
D Occupational Health and Environmental Controls, and Records, Notices and Reports.

21.8.1 An Example of A Plan's Content

A plan should include how the requirements of subpart E, Personal Protective and Life Saving Equipment, will be met. Subpart E states equipment requirements and when it must be used but does not stipulate how the equipment should be made available to workers.

The contractor's plan should state whether personal safety gear such as head, hearing, eye, and respiratory protective devices will be supplied by the contractor or by the worker. If supplied by the contractor, will they be given to workers or loaned to workers? Will they be stored on site during non-working hours? If stored, where? What are the conditions of replacement? If supplied by workers, how will the contractor verify suitability and condition? Who is responsible for maintaining devices in working order? Who is responsible for the availability, condition, and use of worker's personal safety gear?

These may seem like unimportant considerations from the perspective of the total project, but the unavailability of regulation, properly maintained, personal protective devices could produce serious injury in the event of an accident.

An example of what might be covered under subpart N, Cranes, Derricks, Hoists, Elevators, and Conveyors, would be an update on the mechanical status of equipment moved onto the job. For a crane, for example, the following questions could arise: Is the crane rigged according to manufacturer's specifications? Has it been load rated? Are the ratings still valid? Is the rating based on manufacturers data or data obtained by testing? If tested, who conducted the tests? Will the crane be tested before being put into service on this project? Who will be responsible for meeting OSHA regulations when the crane is in use? For what purposes will the crane be used on the project? Any other information that establishes the utility of the crane will be pertinent.

21.8.2 Safety Plan Responsibilities

It is the contractor's duty to anticipate safety requirements through planning, not simply react to the needs of an immediate situation. To this end, the contractor should provide a plan that demonstrates planning to meet OSHA requirements, one that precludes surprises when construction begins.

A CM should accept the responsibility to see that contractors submit plans that categorically anticipate OSHA safety requirements and that contain enough information for the CM to predict a successful experience during construction. The CM should review each plan as part of preconstruction activities, and generally comment whether the plan is sufficiently detailed and has adequate coverage. Reviewed plans should be returned to contractors with questions and comments regarding completeness but not corrections or suggested corrections. The plan must remain the contractor's plan under all circumstances.

21.9 CONTRACT AMENDMENTS

After reviewing the risk-management study used to determine the team's approach to safety, several amendments to the owner-contractor agreements may be deemed appropriate. Care should be taken not to overdo amendment writing to a point where they cloud rather than clarify the intentions of the Safety Management Plan. Amendments should be narrow-scope definitions of already established broad-scope requirements, not new requirements in themselves.

One amendment that should be included is an explicit statement of who is responsible for the safety of contractor employees, worded to the effect:

> "The contractor shall not assign employees to work in areas and locations that are unsafe. The contractor's supervisor shall verify that the conditions under which their workers will be working fully comply with OSHA regulations.
>
> Supervisors who deem an area or location unsafe shall not permit their workers to begin or continue to work until the problem is eliminated and compliance is apparent. Supervisors shall immediately inform the CM representative on site, in writing, of any noncompliance observed.
>
> Workers who deem an area or location unsafe shall not start or continue to work in that area or location and shall immediately notify their supervisors of the unsafe conditions."

21.10 SAFETY MANAGEMENT PHILOSOPHY

Construction managers should do all possible to minimize the probability of being held liable for accidents. Exculpatory clauses, designed to isolate the CM from any responsibility whatsoever, have been tried repeatedly and do not work. The courts will not accept clauses of this nature because it appears to them that the CM firm's contracted involvement in the project suggests that at least some responsibility for safe practices are on the CM's shoulders regardless of how remote they might be.

Consequently, the CM should purposefully become involved in site safety in defined areas where the authority to exercise control and unilaterally make decisions is reserved exclusively for the CM. As part of the CM's synergistic team relationship with the owner and A/E, the CM should make sure that the responsibilities they assume have the same logical self-reliance.

Contractors should be assigned responsibilities using the same logic; what they can control, they should control. Their authority gives them the responsibility for construction means, methods, and techniques and the manner in which they utilize their workers and equipment. Safety regulations are oriented toward contractors. They should be required to accept their responsibilities for safety and conscientiously carry them out.

It is not simple to write the necessary provisions in the supplementary and special conditions that will properly and effectively assign safety responsibilities to contractors. It is much easier to write contract provisions that properly and effectively assign safety responsibilities to the CM, A/E, and owner. However, the time and effort spent on both will provide a significant benefit if an accident occurs.

The next step is to write the Safety Management Plan and the contractor safety plans. This will be difficult the first time it is instituted, for both the CM and the contractors. The CM, A/E, and owner should lead by example by submitting their own plans covering their on site personnel.

The final requirement is that everyone faithfully perform their responsibilities. This requires considerable effort, especially from the CM, and may tax the execution of the Safety Management Plan to its limits. However, the proper assignment of responsibility has no value without execution. The CM must lead the way by example as well as by direction.

COMMENT: Any chapter on construction safety would not be complete without suggesting that CM practitioners do all possible to bring reality into the monetary consequences of unsafe practices. The court system refuses to be realistic when allocating responsibility when an accident occurs. Either sympathy for the injured party or the fact that insurance funds are available to cover monetary awards, eclipses the reality that the injured party is often the one that caused the accident in whole or in part.

Worker "A" who sustains a head injury from a tool that was dropped by worker "B" from a scaffold above, should be held accountable if he/she was not wearing a hard hat and was located in a properly marked overhead danger zone. Worker "B" should be held accountable for carelessness while using the tool. Worker "A"'s employer should be held accountable if a hard hat was not available. Worker "B"'s employer should be held accountable if the scaffold caused the tool to fall, if the zone below the scaffold was not marked, or worker "B" was not instructed in the use of the tool.

It shouldn't be difficult for the courts to see that the parties responsible for the accident are workers "A" and "B" and, possibly, their respective employers. To assign even nominal responsibility to the owner, A/E or CM who have absolutely no authority over workers "A" and "B" can only be founded on the misplaced sympathy of the court, the unfair concept of "deep pockets," or the fact that insurance policies, not those responsible, would pay damages. Damage awards made by courts heavily favor plaintiffs such as worker "A."

Construction managers and architects who provide services on ACM projects, should be proactive in an effort to convince courts that the contractual assignments in the ACM contracting structure removes them from culpability in cases similar to the one stated above.

CHAPTER 22

Multiple Bidding and Contracting

Of the many contracting innovations of the CM system, multiple bidding and contracting appears to be the practice of most benefit to owners. Switching from the informal assembling of trade contractor proposals under the GC bidding format to a formal competitive bidding format under the CM system has produced significant cost savings and produced sizeable ancillary benefits.

The primary cost savings are in the efficiency of the multiple-bidding process and the basic competition it creates. Most of the ancillary benefits accrue from converting trade contractors to prime contractors—a move that sanctions improved project control and management.

Multiple bidding and contracting is not a simple format to install and manage. Awareness of its flexibility and limitations is essential. Implementing the format successfully requires construction industry experience and insight as to how it functions, especially at the project location. It is also a test of the CM's micromanagement skills.

Once experienced and used, the multiple bidding and contracting format is no more difficult to install than a single prime contract. However, experience never replaces familiarity with local construction contracting practices, and anything less than micromanagement of the format will not be enough to make it work.

22.1 WORK-SCOPES

When using the multiple bidding and contracting format, the foundation for success is determining and expressing the work-scopes or divisions of work. Without experienced input here, the format has little chance for success, regardless of how well it is managed during construction.

All the requisites listed under the contract management area of knowledge in Chapter 11 are necessary to successfully execute the multiple bidding and contracting format. Those in the industry who believe the format is too complicated or difficult to properly install and manage are perhaps not qualified under the area of knowledge or oblivious to its record of successful use by competent construction managers.

The concerns most often expressed about the format are: (1) it is not possible to define work-scopes accurately enough to eliminate overlaps and gaps between the work-scopes, and (2) it is not practical to manage multiple contractors without holding their contracts. Work-scopes can be defined accurately, and multiple contractors can be effectively managed. Literally thousands of successfully completed CM multiple contract projects bear witness to this statement.

22.1.1 Work-Scope Lists

Before work-scopes can be written, the number and content of the work-scopes must be determined. Obviously, number and content are related; the factors that determine number and content are extensions of the decision to use multiple contracts.

Consideration should be given to the project location, type, and size, the availability, capacities, and proficiency of available contractors, and the construction industry practices that prevail in the area of the project. Analysis of this data will precipitate a decision on the practical breakdown of work-scopes.

Work-scopes or bid divisions should be determined on a project basis. Experience has shown that uniform or standardized work-scopes are not practical because of the atypical character of each project. Standardization within the CSI Masterformat's 16 divisions is not possible because work-scopes commonly cross division lines as well as typical trade contractor performance lines.

A productive work-scope breakdown will minimize frequent and difficult contractor construction interfaces and provide each contractor with as much construction continuity as possible. From an economic perspective, it will create as few subcontract relationships as practical. For example, setting hollow metal door frames should be part of the masonry division work-scope even though the carpentry trade claims the work. The mason contractor can arrange to have the frames set by anyone it wants according to progress needs without dealing with an owner-contracted interface in the work-scope of another contractor. Work-scope assignments such as this are schedule-effective; they give contractors more control over their own progress. They additionally reduce finger pointing in the event interfacing contractors have different priorities.

22.1.2 Economic Work-Scopes

An example of an economic decision would be to establish work-scopes for *supplying* hollow metal frames and doors and for *supplying* the required hardware. The owner can purchase these items directly from a supplier and benefit from competitive bidding and the elimination of a contractor profit layer. The mason is assigned care, custody, control, and installation responsibility for the frames when they are delivered to the site. The care, custody, control, and installation of hollow metal doors and hardware when delivered to the site is assigned to the carpentry work-scope contractor. It is the A/E's and CM's responsibility to coordinate shop drawing review and delivery scheduling with the supplier and the two work-scope contractors.

On public projects, where owner material purchases may not be subject to sales tax, an economic benefit could accrue to the owner through a direct purchase work-scope for the material. This would be appropriate on items such as structural steel and large mechanical equipment or items that have long-lead delivery times and must be purchased before contracts for construction are let, to avoid a delay in construction.

Experience has shown that some construction managers have overstepped the limits of contracting propriety and the law by having the owner directly purchase as much material and equipment as possible—even to the point where all material and equipment was purchased on owner-issued purchase orders that simply copied the specific requisition requirements of contractors.

An astute CM should not let direct owner purchasing exceed the practical limit. If owner-instigated, the CM should advise the owner that overusing this option will produce a negative economic effect rather than a positive one.

Without the cushion of a mark-up on a reasonable amount of material and equipment, contractors will increase their mark-up on labor to replace the missing cushion. Labor-only work-scopes are only common in a few trade contracting areas (structural steel erection, carpet laying, casework assembly, and long-lead item installation, to list a few). To extend the list beyond the normal construction contracting practices in the area of the project could have an adverse effect on contracting economics.

Additionally, the sales tax agency could rule that the owner is taking advantage of the direct option to evade the payment of sales tax. One early CM adventure on a public school project, where virtually all the material and equipment installed by contractors was directly purchased on requisition information supplied by contractors, was brought to task by a sales tax agency. The authorities ruled this a pass-through operation undertaken by the owner solely to evade tax laws. Upon audit, the owner was required to pay all the accrued sales tax savings to the state. The legal costs and resulting construction disruption added to the owner's problems.

It should be noted that each state has its own laws governing sales tax on public capital expansion projects. In general, sales tax is required on material purchased by the contractor, the amount of which is passed on to the owner as part of the contract amount. Only material and equipment supplied without the involvement of on-site labor is usually free of sales tax. Nothing should be assumed.

22.1.3 Business Style Work-Scopes

The practical limit of work-scope definition should be controlled by the practices and the business styles of area contractors. This can be explained by examining the concrete and mechanical trades.

In some areas, there are contractors that do either flatwork or structural concrete but not both. In other areas, concrete contractors do both. If the trades are separated in the area of a particular project, structural concrete and flatwork concrete should be separate.

By separating the two work-scopes, the criteria of reducing potential subcontracting situations to a minimum is served and competition will be generated.

If a structural concrete contractor also does flatwork, or vice versa, that contractor can bid either or both work-scopes individually, or both work-scopes with a combined deduct if it is not interested in accepting award for only one. The combined bid, less the deduct, will be compared with the two lowest separate bids or any other combined bid to determine if one or two contractors will be awarded the structural and flatwork contract.

Another example involves mechanical trades. These trades include heating, ventilating, air-conditioning, plumbing, piping, fire protection, mechanical insulation, underground utilities, instrumentation, and balancing. If contractors are available for each trade separately, eight or nine bid divisions would be practical. This will not exclude a full-service mechanical contractor from bidding each work-scope and providing a combined deduct. In practice, many mechanical contractors operate as

Nos.	Work-Scope Descriptions	Work-Scope References
1	Excavation/Site Work	2, 3, 5, 7, 8, 17, 36
2	Site Utilities, Sanitary/Storm	1, 3, 5, 15
3	Site Utilities, Water	2, 15
4	Roads and Parking	1, 3, 5
5	Site Concrete	1, 2, 3, 4, 6, 7, 45
6	Electrical	5, 15, 20, 25, 46
7	Building Concrete	1, 19, 42, 44, 46
8	Building Masonry	15, 16, 23, 25, 33, 44
9	Structural Steel	15, 10, 11, 12, 13, 46
10	Metal Joists	9, 12
11	Metal Deck	12, 14
12	Steel Erection	7, 9, 10, 11, 13
13	Miscellaneous/Ornamental Metals	7, 8, 15, 20, 23, 24, 25
14	Roofing and Related Sheet Metal	6, 8, 11, 15, 19, 23, 24
15	Plumbing	2, 3, 27
16	Fire Protection	3, 6, 47
17	Elevator	6, 7, 8, 9, 13
18	Metal Panels	8, 9, 13, 20, 24
19	Carpentry and Millwork	14, 20, 32, 33, 34, 38, 39, 44, 46 (as noted)
20	Plaster/Wall Systems/Acoustical	6, 16, 23
21	Aluminum and Glass	41

FIGURE 22.1 A typical Partial Work-Scope or Division of Work List (with interfacing work-scope cross references).

mechanical-general contractors and subcontract to another contractor those mechanical trades not performed by internal employees.

Figure 22.1 is an example of a partial Work-Scope or Bid Division list for a typical medium-size building (46 separate work-scopes were used on this project). The work-scope references clarify contractor interfaces for bidders. The references are also very valuable when writing the descriptions to make sure all items are included somewhere in the listed divisions.

22.2 BID DIVISION OR WORK-SCOPE DESCRIPTIONS

Multiple bidding and contracting success is also vested in the accuracy and completeness of the work-scope descriptions. These narratives provide definition and uniform data on which bidding contractors can accurately calculate their dollar proposals and submit them competitively with confidence.

Compared to the way that general contractors receive offers from potential subcontractors, CM multiple bidding has a distinct and unique competitive edge. It is formal, fair, competitive, definitive, and responsive.

The four-sample bid divisions in Figure 22.2 demonstrate the use of a simple effective format and indicate the required content for complete definition. The "Excluded, Included, Also Included" format shown here was first introduced in 1972 and is in use today. It is a simple way to point out where and how traditional trade lines are crossed, where work-scopes begin and end, and who the interfacing contractors will be.

Bid Division 07, Building Concrete

Excluded:	Machine excavation (B.D. 1, Excavation and Site Work); moisture protection (B.D. 44, Waterproofing and Damp-proofing); sound isolation layer (B.D. 42, Floating Floor).
Included:	All concrete and concrete-related work physically associated with the building proper, including but not restricted to footings; retaining walls; walls; slabs on grade; structural, composite and supported slabs, equipment bases, entrance pads, platforms, etc. including hand excavation and back-fill, forming, form lining, re-steel, curing protection, finishing, etc., as shown and specified.
Also Included:	Art court concrete; foundation, insulation, expansion joint filler and sealer; maintenance pumping as required (Sec. 02200–3.3); required sand cushion under slabs. Layout, Housekeeping and Final Clean-up.

FIGURE 22.2a

Bid Division 08, Masonry

Excluded:	Preformed vapor barrier (B.D. 44, Waterproofing and Damp-proofing); furnishing of hollow metal door frames (B.D. 35, Hollow Metal Doors and Frames); furnishing of access panels (B.D. 23, Heating and Air Conditioning: B.D. 24, Ventilating and Sheet Metal; B.D. 25, Temperature Controls and others).
Included:	All masonry and masonry-related work required including but not restricted to interior and exterior walls, partitions, parapets, dividers, etc., including brick, block, mortar, reinforcing, accessories, insulation, flashing, cut-stone, grouting, etc. as specifically covered in Sec. 14100, 14215, 04220, 04420, 04610.
Also Included:	Installation of hollow metal door frames, access panels and other items embedded in masonry; foundation insulation. Layout, Housekeeping and Final Clean-up.

FIGURE 22.2b

Bid Division 12, Steel Erection

Excluded:	Furnishing of structural steel, joists and deck (B.D. 9, Structural Steel; B.D. 10, Metal Joists and B.D. 11, Metal Deck); setting of anchor bolts (B.D. 7, Building Concrete); installation of miscellaneous metals (B.D. 13, Miscellaneous and Ornamental Metal).
Included:	The erection of all structural steel, metal joists and metal deck as required, as shown and as specified in Sec. 05101, 05110, 05210 and 05300, complete in all respects including bolting, welding, cutting trimming as required.
Also Included:	Touch-up painting; Layout, Housekeeping and Final Clean-up.

FIGURE 22.2c

Bid Division 09, Structural Steel

Excluded:	Open web joists (B.D. 10, Metal Joists); decking (B.D. 11, Metal Deck); erection (B.D. 12, Steel Erection); anchor bolt installation (B.D. 5, Site Concrete); secondary supportive members (B.D. 13, Miscellaneous and Ornamental Metals); sales Tax.
Included:	The furnishing of all required structual steel framing members as required (not specifically excluded above) including anchor bolts, leveling plates, bearing plates, columns, support members, etc., as shown and as specified in Sec. 05101; completely fabricated, marked and delivered to the Erection Contractor (B.D. 08, Steel Erection) at the job site, including all connection assemblies, bolts, and accessories.
Also Included:	Shop painting, Erection drawings, Anchor bolt setting drawings.

FIGURE 22.2d

FIGURE 22.2 Typical Work-Scope Descriptions for Multiple Contracts.

22.2.1 Writing Work-Scopes

Writing work-scopes is not a difficult task. However, it does require the following: (1) comprehensive review of the completed drawings and specifications, (2) perceptive knowledge of technical construction, (3) ability to visualize the construction process, and (4) the capacity to put the requirements into writing. The work-scope writer(s) mentally builds the project and describes in words exactly what each contractor will be required to do.

Writing work-scopes should be the responsibility of the management or Level 2 CM Person. However, it is practical to have more than one person writing work-scopes for a project. The outcome is more consistent if writing assignments are doled out by groups of work-scopes that interface or within the framework of principal elements such as site development, architectural, mechanical, electrical, etc. The final step should be to have one person, preferably the management or Level 2 CM Person, edit and coordinate the completed work-scopes to ensure continuity and style.

Writing work-scopes for the first time can be intimidating, but once involved in the process, the writing flows more readily and completing the task is not as difficult as some anticipate. What is difficult is allocating enough *continuous* time to get the job done. It is not a task that is conducive to interruption. A block of time sufficient to complete the task should be included in the program schedule for this activity, and strict adherence to that schedule should be observed by the CM.

22.2.2 Ancillary Benefits

It was mentioned at the beginning of this chapter that multiple contracting provided ancillary benefits that were not factored in or anticipated when the CM system was developed but surfaced when the system was put into use. Writing work-scopes contributes one of these benefits; one that has signal value and contributes to the success of the CM system.

The purpose of the CM's drawing and technical specification review is to write work-scopes. However, the review process accomplishes much more. It presents the opportunity for the CM to find inconsistencies in and between the drawings and specifications that otherwise would not have been noticed until the documents were used by bidding contractors, or even later during construction.

Two other CM drawing and technical specification reviews also contribute: one by the scheduler when designing the detailed construction schedule (described in Chapter 20, Schedule Management), and another by the estimator when completing the work-scope budgets (covered in Chapter 10, Budget Management). The A/E's requisite pre-bid review notwithstanding, many inconsistencies will be exposed during the CM's reviews. The result is a briefer pre-bid addenda and fewer contract changes during construction.

22.3 BIDDING AND BIDDER LISTS

When the work-scope list has been finalized and descriptions outlined, the search for bidders should begin. The CM's goal should be to ensure that each work-scope will be bid by at least three competent contractors who have room in their workload, available staffing, and are interested, if not eager, to submit a proposal for the work-scope.

On private sector projects, bidders can be invited, located by advertisement, or both. On public sector projects, laws usually permit any contractor who so wishes to submit a bid. There is often a requirement to advertise the project according to stated rules. However, the requirement should not preclude the CM from searching for additional bidders.

The goal of the bidding process is to generate a competitive situation that is very favorable to the owner and to ensure that a cross-section of available trade contractors submit proposals in each work-scope division. To achieve this, contractor bids from beyond the immediate locale may have to be solicited by the CM.

22.3.1 Set Aside Work-Scopes

Multiple contracting provides a simple way to involve targeted contracting groups in the project. Regardless of why they are targeted, it is possible to reserve specific work-scopes for exclusive bidding by contractors in that group.

Deciding which work-scopes to set aside depends on the match-up of contractors in the target group and the requirements of the work-scope. If a certain goal must be reached (such as a percentage of project cost), work-scopes whose combined amount will meet that percentage can be reserved for bidding by the targeted group.

22.3.2 Bidder Qualification

Regardless of whether the project is public or private, each contractor on the list should be required to submit a statement of qualifications—one that will clearly establish ability to perform. In the private sector, submitting a completed qualification form can be a prerequisite to obtaining bidding documents. This is a simple way to ensure that all bidders are qualified.

In the public sector, it is not always possible to limit bidders through a qualifying process. However, in some public jurisdictions it is permissible, if the qualifying instrument is overt, equally fair to all, and can be evaluated mathematically without any influence of opinion. If it is possible to pre-qualify contractors, it should be done.

There really is no choice between pre-qualification and post-qualification. Prequalification is much preferred by contractors. Post-qualification forces contractors to go through the bidding process, spending time and money before they find out if they are qualified or not.

If deemed unqualified, some contractors will go to great lengths to change the ruling so they can be awarded the contract. The owner may be brought to court with the burden to prove the contractor unqualified. The contractor may file an injunction preventing an award until the matter is resolved. The disappointed contractor's efforts are fueled by the fact it is the low bidder and also intensified because of the time and effort spent to estimate and submit a proposal.

This is not to say that on a public project, a trade contractor who is disqualified before bidding will not go to great lengths to be permitted to bid. However, because of the smaller stake in the project, a trade contractor is less adamant than a general contractor would be and is less likely to pursue reinstatement as qualified. The team would have to face this problem as part of risk management (covered in Chapter 19).

A contractor qualification form should be as brief and to the point as possible. This is especially true if there is plenty of work in the area for contractors to bid on. There is no doubt that filling out a qualification form is intimidating to contractors and a deterrent to the bidding process. The CM must convince potential bidders that its purpose is to stabilize the competition and to make sure that competent contractors will not have to compete with incompetent contractors.

There will be times when the qualification requirement should be dropped to ensure competition. This is another risk-management decision, although it is certainly preferable to have unqualified competition than no competition at all. However, the team should not decide to sacrifice qualification without a thorough review of the situation.

One partially effective substitute for qualifying contractors is requiring labor/material and performance bonds. While it would not be accurate to say that the ability to furnish bonds relates directly to a contractor's potential performance, there is a remote relationship between a good performing contractor and his bondability.

When relying on bondability as a qualifier the credibility of the surety should be closely checked. There are dependable and undependable bonding sources. It is suggested that a list of acceptable sureties be included in the contract documents and that bidding contractors use one of these. Lists of prequalified sureties compiled by units of government are good sources for qualifying purposes. If a contractor has a working relationship with a surety not on the list the contractor should not be banned from bidding. However, efforts must now be made to review the qualifications of the surety as well as the contractor.

To obtain bonds, a contractor must at least have his financial "house in order" and a good record of completing contracts. Although sureties mainly dwell on financial strength and the ability to repay the surety for costs incurred as a result of contractor default, there is concern for the performance of contractors seeking bonds. In a contingency situation, a contractor who can provide labor/material and performance bonds from a reputable surety can generally be considered a good risk.

In the event a contractor who is assessed to be sub-par ends up on the project, one thing remains to be done. The CM should devote an inordinate amount of time tracking the contractor's performance to avoid adverse consequences. If trouble is anticipated, it can usually be prevented. A little counsel from the team can go a long way to upgrade a mediocre contractor's performance. If the contractor clearly lacks technical know-how, the CM can convert a mediocre performance to a good performance by providing leadership and technical guidance.

22.3.3 Pre-Bid Meetings

Prior to receiving proposals (about half-way through the bidding period), a pre-bid meeting should be held with all contractors planning to bid the work. Although the meeting need not be mandatory, prospective bidders should be urged to attend, especially those who have not previously been involved in a multiple-contract CM project.

The structure of the meeting should favor the contractors whose work-scopes interface. With the number of work-scopes to be discussed and the number of contractors who will attend, one continuous meeting dealing with all work-scopes would be exceptionally long and a frustrating waste of time for contractors.

The suggested solution is to hold two or three separate meetings which immediately follow each other during the course of a day. The first meeting could include the work-scopes dealing with bringing the building out of the ground (such as demolition, excavation, site utilities, roads, parking, drainage footings, foundations. ground slabs, etc.); the second with work-scopes such as masonry, structural concrete, structural steel, plumbing, HVAC, electrical, fire protection, etc.; and the third with the remaining work-scopes.

The groupings are readily available from the referenced work-scopes on the project work-scope list (Figure 22.1). Some contractors may opt to be present for two or more meetings, if they see a benefit. However, it would only demonstrate extremely poor management if painting contractors had to sit through pre-bid discussions on pile driving or excavation work-scopes.

The CM, A/E, and Owner should be represented by all three levels of their project persons to explain their special involvements and answer questions pertaining to the conduct of the project. The pre-bid meeting is the time for the CM firm to step forward and convince contractors that the project will be managed properly. Many of the contractors will be working with this CM for the first time, and others will be getting their first exposure to CM multiple bidding and contracting.

In areas where CM is not common, much of what is discussed at pre-bid meetings will deal with project procedures. Attendance will depend on whether or not the contractors bidding the project have had previous experience with the construction manager or the CM system itself. In the more technical work-scopes dealing with electrical and mechanical systems, technical questions will be raised which will require attendance by the team's engineers and their CM counterparts.

Pre-bid meetings are productive, especially when well planned, advertised, and managed. They are usually well attended, especially when the project is unusual or complex. If the project is ordinary, or if contractors are familiar with the CM firm, contractor attendance should not be relied upon as a measure of bidding interest in the project.

Regardless of contractor attendance, minutes of the meeting should be kept by the team's designate and sent to all contractors on the bid list the day after the meeting. The minutes of the meeting can serve as a pre-bid addenda if it is so designated. Whichever procedure is used should be specified in the bidding documents.

Contractor attendance at the pre-bid meeting should be recorded by the CM. The reasons why a contractor did not attend the meeting could be important and should be noted in the contractor's bid-list file. If a contractor has lost interest in bidding, the CM should know this as soon as possible.

After the pre-bid meeting, and the issuing of any pre-bid addenda, the next criteria for success is the degree of competition in each work-scope when bids are received. The CM should make every effort to keep contractors interested and primed to submit a proposal.

22.3.4 Sets of Bid Document

Unless the project is fast-tracked or phased, it is highly recommended that a complete set of documents be provided to every bidding contractor. Separating the documents to match work-scopes or groups of work-scopes exposes the credibility of the bidding process to unnecessary risk. It also involves considerable extra effort by the A/E and

CM, if the separations are to be complete without errors and omissions that could create problems when the documents are put into use.

Trading printing, distribution and administration costs for document sets (that provide uniform information to contractors and more competitive proposals) is purely a risk-management decision.

The number of work-scopes or bid divisions, added to the fact that all bidders submit proposals directly to the owner, require a large number of contract documents be printed and distributed. Assuming there are 50 work-scopes, and three or four bidders each, about 175 sets of documents will be required just to accommodate bidding contractors. Additionally, complete sets will be required for plan rooms, team members, and agencies where document filing is required. On a public project, the number could increase, if bidding interest is high. The number of documents is about seven times that needed to bid a traditional GC project.

On large and complex projects, with many sheets of drawings, unprecedented printing costs will accrue, and increased logistics of distribution, deposits, and returns must be worked out. These costs and requirements notwithstanding, the expense of document reproduction and handling will be returned through the inherent economies of the multiple-bidding process.

22.3.5 Printing and Distributing Bidding Documents

The responsibility chart covered in Chapter 16, Project Management, should clarify how printing and distribution will be handled. Long before the project reaches the stage where these activities are scheduled, the process will have been decided upon and responsibilities assigned.

The handling of bidding document deposits is worthy of comment. Document deposits should be the rule rather than the exception. Although documents will be distributed to trade contractors who are not used to having documents of their own, and plan room document access is available, it is important that each bidding contractor has a complete set of documents to clarify the work of interfacing contractors and to study the general and special conditions requirements of the project manual.

On a CM multiple-contract project, trade contractors are prime contractors, not subcontractors. On public sector projects, their proposal will be submitted once in open competition with other contractors bidding the same work-scope(s). They must conform to the Instructions to Bidding to be considered a viable low bidder.

Documents should be obtained only from the CM or A/E. Cash deposits should not be permitted. Documents should only be released upon receipt of a check drawn on the account of the bidding contractor in favor of the owner. All deposit checks should be kept on file by the document distributor and returned to the contractor when the documents are returned in reissuable condition. The deposit amount should not exceed the actual reproduction cost.

22.3.6 Submitting Proposals

Contrary to the traditional way in which trade contractors offer proposals to GC contractors, direct bidding by trade contractors demands the same formality as when GC contractors submit bids to owners. The number of bids being received is so greatly increased that special preparations are required if proposal submission is to be a

Proposal Form (Proposal Due:__________________)

Submit Proposal To: (Owner)

____

____

For: ____________________ (Project)

____________________ (A/E)

____________________ (CM)

Submitted by: (Contractor)

____

____

Phone: ____/____-______

Bid Division(s) Attached

_______, _______, _______,

_______, _______, _______,

_______, _______, _______.

Contract Documents

We have read and understand the Contract Documents, including the Instructions to Bidders, Advertisement, Special and Supplementary Conditions, Proposal Section, General Conditions, Technical Specifications, and the Bid Division(s) covering the work required in this proposal.

Addenda

We acknowledge receipt of Addenda ____, ____, ____, ____, ____, ____, ____, ____.

Construction Management

We understand the Construction Manager's function as described and referred to throughout the Contract Documents. We realize that each Bid Division Contractor is in fact a prime contractor and not a subcontractor working through a General Contractor.

Bid Division Responsibility

We recognize that the scope of the work within Bid Divisions represents a construction unit that is not necessarily restricted to single trade performance and our proposal includes the work of all trades required to fully and successfully complete all of the work required in the Bid Division(s) we have bid. We have included all work stated in the Bid Division(s) Proposals attached hereto.

Progress Schedule

We have reviewed the Milestone Schedule and hereby (ENDORSE) (AMEND) (*cross out one*) it with regard to the work of the Bid Division(s) we have bid. If you chose to Amend the schedule, please define your amendment here:

____

____

Combined Bidding Deduct

We have chosen to submit proposals for more than one Bid Division and offer a deduct of ____________________ Dollars ($________). If we are awarded the combined work covered in the following Bid Divisions ________ ________ ________ ________. We understand that the amount stated above will be deducted from the combined total of the Bid Divisions we have submitted proposal for.

Agreement

This proposal, if accepted by the Owner, will be used as the basis for a contract directly with the Owner. Therefore, the undersigned agree(s) to accept a contract for the work covered by the proposal, in accordance with the Contract Documents.

The Owner reserves the right to accept or reject any and all proposals with or without cause.

Name of Bidder ____

Address ____

By ____ (Signature)

____ (Type or Print)

Title ______________________________ Date ____/____/____

FIGURE 22.3 A typical Proposal Form for Agency–CM Multiple Contract Bidding.

(envelope flap)

Proposal Enclosed
Do Not Open

Project name: ______________________________

Bidder's name: ______________________________

Work-Scope Proposal(s) enclosed: #_____, #_____, #_____,

#_____, #_____, #_____, #_____, #_____, #_____,

#_____, #_____, #_____, #_____, #_____, #_____,

Combined deduct enclosed: (circle one) YES NO

FIGURE 22.4 The back of a typical proposal submittal envelope.

credible and orderly procedure. If the entire project is to be bid at one letting, 175 to 200 proposals may be received.

It is recommended that work-scope proposal forms (see Figure 22.3) be distributed to contractors by the CM, separate from distribution of the drawings and project manual. It is suggested that a special envelope (see Figure 22.4, page 351) be provided to contractors that will guarantee the identification of the work-scope(s) proposal(s) enclosed and that envelopes only be made available to contractors on the bidders list.

Proposal forms consist of two parts: a page similar to Figure 22.3 which covers all proposals, and an attached Bid Division page(s) that precisely iterates the Bid Division Description and includes spaces where contractors can enter their dollar proposal, signature, and any other information the team deems necessary on a Bid Division basis. All information on each bidder's proposal form that can be pre-entered by the CM should be pre-entered to avoid errors and misinterpretations. Proposals should be custom assembled by the CM for each contractor on the bid list and reflect only those Divisions the bidder has prearranged with the CM to bid.

22.3.7 Receiving/Opening Bids

The increased number of bidders on multiple-contract projects also requires changes in the traditional way of receiving proposals. Whether the project is phased or not, preparations must be made to accommodate a large number of proposals. On fast-track projects, the major bid package will include 20 to 30 work-scope divisions and 30 to 45 proposals. On non-phased projects, the number of proposals could be four times as great.

When receiving multiple bids, a single bid date and time is customary but not necessary, even on projects that are not fast-tracked. Bids could be received on different days and times, for whatever reason. The only requirement is that all bids for work-scopes designated to be opened at a specific time must be opened and read at that time. This is usually mandatory on public sector projects and is good CM practice on private sector projects.

There is an exception to this requirement.If the proposal envelope is marked that more than one work-scope proposal is enclosed, and if YES is circled on the combined bid deduct question, that envelope should not be opened until the last of the work-scopes listed on the envelope is scheduled to be opened. If there is more than one work-scope proposal enclosed, and NO is circled, the envelope should be opened when the first work-scope proposal is scheduled to be opened, but only that proposal should be read. The remaining work-scope proposals can be read when their turn comes up.

Figure 22.4 shows the back of a typical proposal envelope. To maintain control of the bidding process and keep track of the bidder list, all the information on both front and back should be filled in by the CM, *not* by the contractor. It is best if the color of the envelope is vivid (perhaps red or orange) to provide fast identification.

The front of the envelope should be used normally, showing the owner's name and address and the bidder's return address. The sealed envelope may be hand-delivered or sent. If the proposal is sent, it should be enclosed in another envelope supplied by the bidding contractor.

The opening of bids should be spaced and grouped to preclude the need for bidders to be present for the entire length of the proceedings. Assuming each bid requires approximately two minutes to open, read and record, and that each work-scope division will yield three to five proposals, about seven and a half work-scope divisions can be comfortably processed in an hour. The schedule for opening bids should take this into consideration and advise contractors accordingly. A single-phase project, with fifty divisions of work, could consume an entire day to open all bids. Unless a bidder is bidding the entire project, which is an available option, presence at all work-scope openings is a waste of time. If bidders know when their proposals will be read, it will save them a lot of time (experience has shown that they appreciate this).

Time does not permit reading all information; all that can be accomplished is reading the pertinent numbers into the record. At public openings, this procedure may be questioned and time will have to be allowed to read all of the information requested on the proposal form. This process will take longer and arrangements made to accommodate the requirements of each bidding situation.

One way to shorten the time required for opening and reading proposals is to process groups of work divisions simultaneously. All this requires is duplicate team staffing to cover two or three locations at the same time. It takes a while to grasp all of the possibilities connected with multiple bidding. Although it is a lengthy process, if micromanaged it can be executed without confusion.

22.3.8 Proposal Review

As bids are opened and read, designated team persons should be checking bids previously opened to determine if they are complete and that all required information has been included. This should be done simultaneously at a near location. If bidders want to be present while this is checking is being done, they should be accommodated. The apparent low bidders are really not identified until their proposals have been deemed in conformance. Traditional checking is all that is required.

22.3.9 Award Recommendations

After bids have been received and reviewed, the CM should recommend bidders for award to the A/E. The CM's input should include an assessment of qualifications, a determination of proposal completeness, and a cursory evaluation of the dollar proposal as compared with the CM's work-scope estimates. If qualification forms were submitted with proposals, these should be reviewed. In the private sector, if voluntary alternates have been submitted, they should be preliminarily reviewed.

The CM's review will determine if there is a need to recommend a post-bid conference with certain bidders to clarify any problems and with apparent low bidders that were not subjected to qualification procedures. Bid verification and evaluation are important parts of competitive multiple bidding. At this point in the CM process, the project team is searching for dutiful contractors for the construction team.

22.4 POST-BID MEETINGS

The technical aspects of the proposals being considered should be checked at post-bid meetings for completeness and accuracy. Each proposal should be compared with the work-scope description and the CM's construction estimate. The CM's estimates should not be considered targets for bidders to shoot for, but they are a big help when reviewing dollar proposals for errors and omissions.

This is a unique exercise that can only happen on a CM multiple-contract project. It confirms the integrity of the low qualified proposal in each work-scope division. If a significant variance occurs between the CM estimate and the contractor's estimate, it is a signal that one of the two contains an error in quantity, work-scope, or judgment. The possibility that an arithmetical mistake caused the difference exists as well.

On public sector projects, it is usually mandatory that bidders provide bid security—most commonly a surety bond (Chapter 19, Risk Management). On private projects, bonds are provided at the discretion of the owner. The bond offers protection from loss by the owner in the event a contractor refuses or for some other reason will not or cannot enter into the contract.

The ways in which bidding errors occur and are corrected vary widely. They could be a judgment error, an omission, or an arithmetical mistake. In the public sector, resolving bidding errors is sometimes dictated by rule or statute; in the private sector, it is not. In both sectors, the final resolution is usually in the hands of the owner and legal counsel.

For the benefit of the project, it is advisable in all instances that owners reject low bids that contain significant errors of *any* kind as long as the contractor can explain the error and agrees to the rejection. If the contractor supplied a bid security, it is suggested that the owner forego collection unless its substantial or required by law.

The "loss" sustained by the owner when rejecting a multiple-contract bid is minimal in terms of the total project budget. It should be kept in mind that one work-scope division represents only a portion of the total project. Above all, it should be remembered that the purpose of the bidding process is to assemble a group of capable contractors—contractors satisfied with their dollar proposals and enthused about the prospect of earning a just profit.

22.4.1 Award Paperwork

When post-bid interviews are concluded and award recommendations have been made by the CM to the A/E and the A/E to the owner, distribution and collection of documents can begin. The exact steps to accomplish this should be found in the Responsibility Chart.

Usually the CM assembles completed sets of contract documents, gathers required submittals from contractors, insurance certificates or copies of policies, Schedules of Values, labor/material and performance bonds, and other submittals that are required or determined necessary at the post-bid meetings. The CM reviews them with the A/E and forwards them to the contractors for signature.

Upon return to the CM, the documents are sent to the owner for signature and then returned to the CM to be deposited in the project information file for safe keeping and future reference. The project information file could be located in the office of the A/E or owner; whichever location is the most secure should be the choice. It is also advisable to have, at another location, a duplicate file containing all legal documents.

Sometimes a letter of intent or a notice to proceed letter is composed by the CM and sent to contractors under signature of the owner. These letters advise bidders of the status of their contracts, list any conditions for signing, and in the latter case, the conditions required before starting work at the site. Insurance certificates are mandatory if work is started on-site before a contract is signed.

22.5 ON-SITE COORDINATION

The CM has the responsibility to coordinate trade contractors on site. The major difference between the GC and CM contractor coordination is that the CM is *required* by law to first look after the owner's best interest and then his own. As an independent contractor, the GC is *allowed* by law to look after his own best interests before those of the owner, providing the GC stays within the terms of the contract for construction.

The CM must have the specific authority to coordinate the contractors in the field. This authority must be clearly stated in the contracts between the owner and each contractor and should be stressed at pre-bid and post-bid meetings. It is the CM's responsibility to see that adequate commanding language is included in the contracts. Exactly how this is to be done should be part of the CM's internal contract management plan. Chapter 11, Contract Management, provides insight to coordinating contractors during construction.

22.5.1 Direct Relationships

The same trade contractors that provide subcontractor services to general contractors on GC projects provide prime contractor services to the owner on a CM project. The contractual relationship between owner and each contractor is direct. This is both helpful and convenient to both parties during the course of the work.

If a problem arises with a contractor during the project, the full impact of a direct contract can be used to effect solution. In essence, the owner has the same provisions in the contract with each trade contractor on a CM project that the owner would have with the single general contractor on a GC project.

Warranties and guarantees provided by trade contractors are directly administered. Performance and labor/material bonds protect the owner if a trade contractor defaults. Each trade contractor's liability and property damage insurance protects the owner as an additionally insured. Monthly progress payments are made directly to contractors entitled to them. The replacement of a trade contractor for cause can usually be swiftly accomplished. Contractors who dutifully perform can be duly compensated; contractors who do not can be appropriately dealt with.

The activities of contractors are scheduled and coordinated by a competent full-time field person who has the same goal as the owner—the completion of the project within budget, on schedule, and in conformance with the contractor's contract for construction.

22.6 CONTRACT DOCUMENTS

The standard contract documents that have served the GC system for many years obviously cannot serve the CM system without considerable change. To understand what these changes should be and why they must be included is to understand CM multiple contracting.

The best approach to understanding is to compare CM multiple-contracting document provisions to the provisions in documents used in general contracting (the system best understood by construction industry participants, including owners). Although GC documents depend a great deal on precedent practices, CM documents cannot.

In the early days of CM, general condition documents were not available for CM projects. Additional and substitute information had to be provided in the supplementary and special conditions to the general conditions of GC contract documents to create CM contract documents. One practice used to prod bidding contractors to thoroughly read the new material was to give it the heading "Proposal Section," a title that was guaranteed to attract close attention.

The Proposal Section in Appendix D provides example provisions used to convert standard general condition documents for GC contracting into CM general condition documents prior to 1975. In that year, standard CM contract documents were issued by two of the major construction industry groups in the United States. Other industry groups have followed suit, providing easy access to the CM system in today's construction marketplace.

However, when using any set of standard contract documents, care should be taken to make sure they convey the exact requirements of the owner. Chapter 11, Contract Management, expands on this advice.

22.7 MULTIPLE CONTRACTING ECONOMICS

An original purpose of CM was to reduce overall project costs by improving the project-delivery process; by permitting greater owner involvement, more contracting options, and a higher level of management during all phases of the project. Reducing contract costs through the multiple bidding and contracting process surfaced as an ancillary benefit after the system was in use.

The cost comparison in Figure 22.6 on page 358, between a single contract GC approach and the multiple contract CM approach to the same project, demonstrates how favorable economics are extracted from a multiple bid CM project. The data in Figure 22.5 support the basis for the cost reduction claim.

To provide credibility to the comparison, conservative values have been used in both calculations. Contractor overhead *and* profit (Items III and IV) total $610,500 on

Project	Location	Bids Received CM	GC	% Diff.
1. Multi-Purpose Building	Cadillac, MI	265,704	306,000	15.2%
2. Harrison Elementary School	Port Huron, MI	420,026	477,400	13.7%
3. Forest Area Schools	Fife Lake, MI	1,622,463	1,854,000	14.3%
4. Kalkaska Elementary School	Kalkaska, MI	726,223	927,350	27.7%
5. Otsego City Building	Otsego, MI	177,443	224,202	26.5%
6. Munising Elementary School	Munising, MI	616,483	801,036	29.9%
7. Harbor Springs High School	Harbor Spring, MI	698,475	1,118,435	61.1%
8. Rowan County High School	Morehead, KY	3,603,464	4,352,000	20.8%
		$8,126,281	$10,061,323	
		Difference:	$ 1,935,042	
			$10,061,323	= 19.2%

FIGURE 22.5 CM Contract Costs vs. General Contract Costs (data recorded in the public record).

an $11,000,000 project; a new money return of $5^1/_2$%. The contingency (Item V) is not only a conservative 1%, it is used in both models and does not affect the comparison except in the value of the labor and material and performance bonds.

There are similar trade-offs in Supervision and Field Office Expense (Item IIA) and Contracting Overhead (Item III). These amounts totalling $568,250 are carried as CM costs of services and project overhead (Item VIIA). The difference of $91,750 (Item VIIB) is conveniently translated into CM profit and overhead simply to round out a conservative, assumed CM lump sum fee for the project of $660,000.

The project value differential (Item IX) falls in the range of the 5% cost effectiveness assumption used to calculate the CM contract costs in Items IB, C & D. A 5% reduction in contracted costs is conservative based on the public records in Figure 22.5.

Note that the bond premium (Item VIB) is based on the combined value of multiple contracts as adjusted for the higher premium rate charged for smaller contract values. The value of the multiple-contract bond is $9,439,430 (Item I CM column plus Item VI CM column) and the value of the single-contract bond $11,008,658 (Item VIII GC column).

Although the $111,930 premium for multiple-contract bonding is higher than the $65,658 for a single-contract bond, it is common for sureties to require general contractors to bond subcontractors. Called double bonding, this practice provides no more protection for the owner even though the owner pays the premiums. If the GC bonded all subcontractors, single-contract bonding costs would increase by $140,250. This would make CM multiple-contract bonding $93,978 less expensive than single-contract bonding and increase the project value differential to 6%.

The data in Figure 22.5 clearly show that multiple contracting with defined workscopes produces considerably more than 5% savings in bare contracting costs when competitively bid against general contractors. Even if the data are not accepted, the project value of the model still shows a savings of $221,728 less the contingency of $97,500 (Item VB) if it is expended. This is attributed to the elimination of profit on profit alone.

The information tabulated in Figure 22.5 was gathered during the late 1970s and early 1980s in Michigan and Kentucky. The projects listed are all public sector projects where pre-qualification was not permitted and any contractor who wanted to could bid. Bids were opened and read in public.

In each case, a general contractor decided to submit a combined proposal for several or all divisions of work. The bidding process specifically allowed this to take place. General contractors submitted separate bids for each division in which they were interested and tied them all together as a package by providing a combined deduct if awarded all in their package.

Some additional comments pertaining to Figure 22.6 are listed below.

Item	*Comment*
IA-CM	Does not apply (N.A.); this work commonly done by a GC is competitively bid in ACM.
IB-CM (l)	Includes IA-GC $1,500,000 (b) and IB-GC $4,450,000 (c). The total $5,950,000 is reduced 5% based on the results of multiple bidding demonstrated in Figure 22.5.
IC-CM (m)	Reduced 5% by separating the mechanical work into 5 to 7 separately bid work-scopes.
ID-CM (n)	No change from ID-GC (e) because electrical work-scope separations are limited.
IIA-CM (p)	CM field office expenses are part of the CM fee VIIA-CM (p, q).
IIB-CM	CM construction support costs are the equivalent of GC support costs.
IIIA-CM (q)	CM contract administration costs are part of the CM fee VIIA-CM (p, q).
IIIB-CM (r)	CM general overhead costs are part of the CM fee [VIIB-CM(v)].
IV-CM	Contracting profit is N.A. The CM works for the fee stated in VII-CM.
V-CM	The contigency would be approximately the same. The GC contingency is paid to the GC whether used or not. The CM contingency belongs to the owner and only costs incurred are paid.
VIA-CM	N.A. (There is no total project bond on an ACM multiple prime project.)
VIA-GC	Single contract bond fee. The bonded amount is $11,008,658, the total of I-GC(a) + II-GC(f) + III-GC(g) + IV-GC(h) + V-GC(i) + VI-GC(j).
VIB-CM	Multiple contract bond fee, The bonded amount is $9,439,430, the total of I-CM(k) + VIB-CM.
VIB-GC	Assumed N.A. It is common to include the bonding of GC subcontractors which would produce a sizeable entry here. However, this cost model assumes there will be no subcontractor bonding.
VII-CM (u)	The $660,000 fee for this project includes all CM costs. VIIA-CM(p, q) [$568,250] is a derived number determined by adding

Item	Project Value: ± $11,000,000 (P.V.)		Single GC Contract	Multiple CM Contracts
I	Cost of Construction		$ 9,750,000 (a)	$ 9,327,500 (k)
	A General Work Performed by GC		1,500,000 (b)	N.A.
	B General Work Subcontracted by GC		4,450,000 (c)	5,652,500 (l)
	C Mechanical Work Subcontracted by GC		2,500,000 (d)	2,375,000 (m)
	D Electrical Work Subcontracted by GC		1,300,000 (e)	1,300,000 (n)
II	Construction-Related Costs		485,000 (f)	265,000
	A Supervision and Field Office Expense		220,000 (p)	N.A. (p)
	B Construction Support Items		265,000	265,000
III	Contracting Overhead		348,250 (g)	N.A.
	A Contract Administration	(c,d,e) 0.5%	41,250 (q)	N.A. (q)
	B General Overhead Costs	(a,f) 3.0%	307,000 (r)	N.A. (r)
IV	Contracting Profit		262,250 (h)	N.A.
	A On Electrical and Mechanical	(d,e) 2.0%	76,000	N.A.
	B On GC Work with Own Forces	(b) 5.0%	75,000	N.A.
	C On GC Work Subcontracted	(c) 2.5%	111,250	N.A.
V	Construction Contingency		97,500 (i)	97,500 (s)
	A Contractor's Risk	(a) 1.0%	97,500	N.A.
	B Owner's Risk	(a) 1.0%	N.A.	97,500
VI	Bonds		65,658 (j)	111,930 (t)
	A Single L & M and Performance	(a,f,g,h,i) 0.6%	65,658	N.A.
	B Multiple L & M and Performance	(k) 1.7%	N.A.	111,930
VII	Construction Manager's Fee (est.)	(P.V.) 6%	N.A.	660,000 (u)
	A Cost of Services and Project Overhead		N.A.	568,250 (p,q)
	B CM's Profit and General O.H.	(CM cost, p,q,r) 16%	N.A.	91,750 (v)
VIII	Totals		$11,008,658	$10,461,930
IX	Differential (−5% Project Value)			(−) 546,728

FIGURE 22.6 Single GC Contract vs. Multiple CM Contract Cost Model.

IIA-CM(p) and IIIA-CM(q) together. VIIB-CM(v) [$91,750] is a derived number determined by subtracting $568,250 from the lump sum CM fee of $660,000. Therefore, amounts p, q and v are not typical sub-sums of a bonafide CM fee.

CHAPTER 23

Marketing and Sales

Proficiency in marketing and sales is essential to the success of any construction management firm. In simple terms, you cannot provide services until you have a client to perform them for.

Unlike contractors, who are extensively sought out by owners to bid projects, construction managers like architects and engineers must proactively seek clients. There are advertisements and invitations issued for CM services but only for a fraction of the projects that use CM services.

Every CM firm should have a well-conceived marketing or business development plan and an aggressive sales or new business acquisition program.

23.1 MARKETING AND SALES

Although often used interchangeably, marketing and sales are two different activities. Marketing is a prerequisite of sales. Marketing can be defined as determining what to sell, how much to sell, and who to sell to; it is a science. Sales is the series of acts that carry out marketing conclusions; it is an art.

Neither one is less important than the other, and although they closely interact, each requires its own requisite talents. A person who excels in one often does not excel in the other. In the case of CM, experience has shown that the selling of services presents the greatest challenge.

A well-conceived marketing plan can be developed in-house by committee as an annual or biannual activity, chaired by a person who understands the fundamentals and purpose of marketing. On the other hand, sales requires individuals with abilities not usually found in a construction-oriented organization. Selling CM services requires a talent as well as construction management skills and knowledge, a talent that ultimately determines the success of the CM firm's sales strategy.

Of course, the selling of CM service is not solely the responsibility of a talented salesperson. The marketing plan must correctly target the type, size, and location of potential projects, and the services being sold must have demonstrative success and obvious salability. Success is an attribute that can only be acquired through the past performance of the CM firm. Salability of services is dependent on the flexibility and clarity of the management format used by the CM firm. The salesperson must have a well developed and dependable service to sell.

23.1.1 CM Personnel Involvement

Although it may seem contrary to what has already been said, all persons in the CM firm should consider themselves salespersons while performing CM services. The image of the firm is projected by those who represent the firm. If clients are impressed with the CM personnel with whom they come in contact, their references will make it easier to sell CM services.

Additionally, all CM personnel should be aware that marketing and selling are essential to their firm's success. In performing their duties, they should be conscious of construction marketplace trends and alert to information on future projects. When personnel realize that their future is vested in the future of the CM firm, they quickly become part of new business activity. Their passive but important contributions will provide focus to the firm's marketing and sales efforts.

23.2 MARKETING PLANS

A marketing plan should be a candid, objective statement of a firm's practical goals and how they are to be attained. It is an honest representation of a firm's capabilities; it focuses only on the realities of the marketplace, and it is based on facts, not speculation. When properly developed and prudently followed, marketing plans provide coherent direction for future business development and organizational stability. When done superficially, they can mislead and adversely affect the economic future of the CM firm.

Developing CM marketing plans in-house by committee appears to be the currently popular approach, and there seems to be two reasons for this inclination: a misconception of the marketing plan's importance, and expense. The former is usually a result of inexperience; the latter, understandable but not necessarily prudent.

Consultants are available to help CM's develop a marketing plan, and their services constitute money well spent. However, exceptional efforts should be made to screen consultant candidates. CM is a unique practice within a unique industry. Only consultants who have compiled good records developing marketing plans for construction managers should even be considered.

If the plan is to be developed in-house, one person should organize and lead the effort. It is advisable for that person to learn as much as possible about marketing and marketing plans. Many books are available and seminars are conducted throughout the country. However, care should be taken to choose sources that deal with services, not products.

23.2.1 Marketing Plan Development

In essence, development requires collecting and interpreting information. Knowing the information needed, how and where to get it, and what to do with it after it is in hand obviously requires considerable marketing knowledge.

The logical initial step is to determine the legitimate strengths and weaknesses of the CM firm as a business entity and as an organization of diversely talented individuals.

If the plan is to be authentic, the strengths and weaknesses of the firm and its members must be objectively established. This step will determine the firm's collective capability and its capacity to provide credible CM services to clients. It will identify deficiencies and what might be done to eliminate them. As well, it will identify project types and sizes that match the firm's ability and the forms and variations of the CM system that could be provided on those projects.

Demographic information should be obtained to identify types, sizes, locations, and ownership of projects that used or are using the CM system, the forms and variations of CM used, and the reception CM received on those projects. The geographic area canvassed is an estimate at this time. It should be larger than one that is summarily deemed comfortable for the firm.

In addition to collecting demographic information on existing CM use, it is necessary to forecast the future volume of construction by type and size and the potential for the use of CM in the area. Aquiring this information is difficult without experience or advice; a marketing consultant's services will prove valuable.

23.2.2 Using The Information

The next-to-last step is to correlate the demographic and forecast information with the measured ability of the firm. This will provide insight to new business opportunities available to the CM firm, as it is currently staffed and oriented, and will indicate organizational and philosophical changes that would have to be made to compete for clients on a broader basis or wider geographic area.

CM firms should use the results to determine a rationalized direction for the future. Some will:

1. limit the size of their geographic market area
2. have no interest in certain forms and variations of CM
3. show a preference for public or private sector projects
4. confine themselves to certain project types and sizes
5. expand their capacity to capture a bigger market share
6. maintain their current work load level
7. curtail CM operations to concentrate on other opportunities.

23.2.3 The Written Plan

The final and most important steps are to formulate a marketing plan, reduce it to a course of action, and put it into operation as soon as possible. These steps should be the direct responsibility of top management because the plan may require staff changes and will require a relatively large budget to cover selling activities. This is especially true if the firm is instituting a formal CM marketing/sales program for the first time.

The Marketing Plan should state the firm's new business goals and strategy for achieving those goals for approximately the next five years. It should be reviewed and clarified each year to keep up with changing demographics, forecasts, and the firm's maturing capabilities. The plan should concisely define the market areas, project types, project sizes, ownership sectors, and the services to be offered by the CM firm. The plan should name the person in charge of all sales or new business activities and prescribe the general procedures that will be used for selling services, including the extent and format of reporting and record keeping. Job descriptions should be written or modified to assign responsibilities for sales activities to specific personnel.

The fact that selling is an art rather than a science requires a certain amount of flexibility in adopted procedures. However, reporting and record-keeping requirements should not be flexible. Selling of CM services to a client may require considerable time, sometimes years, making accurate records of past contacts invaluable for continuity.

23.3 SELLING CM SERVICES

Prior to the advent of CM, active selling of services in the construction industry was essentially confined to design firms, consultants, and design–build contractors. General and trade contractors do pursue sales but to a limited extent compared to the others mentioned.

General contractors that enter the CM market place sometimes have difficulty accepting the fact that sales is not a part-time job for those who happen to have time on their hands. Selling CM services is not a casual task, nor can it be considered overhead. Selling is a direct cost of doing business as a construction manager; this reality often takes time to advance from a suggestion to a requirement and in the process wastes time and money on less costly alternatives.

As with successful safety and quality programs, top management must be dedicated, supportive, and actively involved in a firm's sales effort. The person in charge of sales should be elevated to the executive level of the firm if at all practical. This is not a superficial move for status purposes but a strategic business move to enhance the effectiveness and productivity of the firm. Sales should be involved in company management decisions by virtue of its vital role.

23.3.1 Approaches to Selling Services

Selling CM services includes many areas of opportunity. The execution of the Marketing Plan should consider the use of most if not all of them. Participation in some areas is more costly than in others, but the cost of many is absorbed through the extra efforts of the firm's executive, operations, resource, and support personnel and not a burden on the sales budget.

Each area of opportunity contributes to the success of the sales venture in its own way. A comprehensive list of the areas follows.

Advertising
- Direct mail
- Trade publications
- News media
- Yellow pages
- Web sites

Public relations
- News releases
- Sponsorships
- Appearances
- Award competitions

Conferences/Seminars
- Participants
- Exhibitors
- Presenters

Associations
- Members
- Committees
- Officers

Publications
- Articles
- Technical papers

Networking
- Social
- Professional
- Community
- Internet

Personal contacts
- Cold calls
- Appointments

23.3.2 Firm-Wide Participation

Of the areas listed above, the designated salesperson must be exclusively involved in three of them: advertising, public relations, and personal contacts. Advertising and public relations will occupy a relatively small portion of the salesperson's time. Personal contacts generally require more time than the salesperson has available in the normal working day.

The other four areas—conferences/seminars, associations, networking, and publications—are those where responsibility is assigned to other CM personnel, and their efforts are guided and coordinated by the salesperson.

To accomplish significant sales-related results in these areas, as many CM personnel as possible should be active members of at least one appropriate association or organization at a local, state, or national level. Membership status will provide opportunities to attend meetings, conferences, and seminars, write articles and papers, network with other members, and gain access to all corners of the marketplace. Whether the local Chamber of Commerce or the National Society of Professional Engineers, involvement in structured groups will precipitate information that will contribute to the CM firm's overall sales effort.

23.3.3 Advertising

Although advertising is relatively expensive, a certain amount is necessary. The most productive of those mentioned is the Yellow Pages. Experience has shown that many owners simply look to the Yellow Pages when seeking CM candidates. Smaller CM firms operating in a wide geographic market area should list a toll-free number in community telephone directories to facilitate direct access.

Direct mail advertising should be limited to a controlled list of potential clients compiled by the salesperson. Experience has demonstrated that arbitrary and purchased lists are not worth the expense of mailing. The text provided in the mailing should be usable information that will be of interest to the recipient (perhaps a CM newsletter that covers the details of a specific CM responsibility or a CM project case study).

Newspaper advertising should be restricted to listings in the traditional "congratulations to the client" panel when a project is completed and occupied. Experience shows that few clients consult the news media for CM service requirements.

Advertising in trade publications provides two functions. It shows a spirit of cooperation with the organization sponsoring the publication and maintains the CM firm's name in the minds of the readers. A CM firm should demonstrate favorable rapport wherever practical. Publications selected should be those that are part of the construction industry and read by potential client groups.

Web sites can be effective, providing the site is regularly maintained and updated. The very large number of sites on the Internet is cause to question its effectiveness for small CM firms that operate in limited geographic areas. However, for medium and large firms operating in wide geographic areas, a web site can enhance sales.

23.3.4 Public Relations

Salespersons should review the appropriate media to ascertain the types of news releases to publish and when an opportunity arises, write and issue a release in a concise and timely way. Local and national news media and construction industry professional and trade media should be considered. If nothing else, news releases keep the firm's name in front of a targeted audience.

Sponsorships take many forms, from financially backing a local peewee league hockey team to providing lunch on the exhibit floor of a conference or funding college scholarships for high school students. In addition to providing a material benefit to recipients, sponsorships connect the CM firm's name with a positive contribution. As with news releases, the audience should be targeted. Sponsorships are easy to find, so it is up to the salesperson to select only those that best meet the goals of the marketing plan.

Simply attending significant events maintains the visibility of the firm, indicates its interest, and improves its image. These opportunities can be scheduled to permit other persons in the organization to be involved and ease the burden on the salesperson. There will be times when an officer of the firm is better suited to make an appearance, but the salesperson should control the list. Appearance opportunities abound. They range from local zoning board hearings, weddings, and funerals to dinners honoring outstanding persons.

As many competitions as are appropriate and available should be entered. Participation alone can provide a public relations benefit if deftly handled. Entry in award competitions connects the firm's name with excellence.

23.3.5 Conferences and Seminars

As a participant in a conference or seminar, a CM firm's representative will be exposed to a captive audience with a common interest in the topics being discussed. The function doesn't have to be an essential learning experience for the person—the reason for attending is to mingle with other conferees that could become clients. However, only functions that last more than a day should be considered so there is an opportunity for casual contact during breaks and evening hours. An attendance list that can be used for direct mail or cold call contacts is usually distributed at events.

The ultimate productive situation at a seminar or conference is to be a presenter and do a good job of it. Rather than having to seek out conferees, conferees will seek out the presenter. Unless instructed otherwise by the event sponsors, a presenter can only subliminally sell the services of the CM firm during sessions. However, there is never a doubt in the minds of the conferees as to who the presenter is and the firm she/he represents.

A more direct way to sell services at conferences and seminars is to participate as an exhibitor. It is more expensive than being a participant or presenter but should be used where an obvious presence is expected by the sponsorship or the competition for CM services warrants it. The quality and content of an exhibit must match or exceed those that surround it. Salespersons should staff the booth supported by operations personnel that can handle specific questions about the firm's approach to services, and selected handouts should be available to give visitors a lasting contact with the firm.

23.3.6 Organizations and Associations

Membership in trade and professional organizations affiliated with a CM firm's complement of service offerings to clients provides access to current trends. Each of the twelve areas of knowledge are represented in whole or in part by one or more organizations that welcome membership. It is suggested that all persons at the CM firm's executive and management levels, as well as some from the administrative level, maintain active memberships in at least one trade or professional organization. The firm should pay all expenses of membership, including expenses to attend meetings.

Most associations operate through a committee structure. The CM firm should urge and reward personnel who achieve chairperson status in selected organizations. Attaining chairperson status indicates the quality of the person's involvement in the association's activities and assures the firm that time devoted to the group is well spent. The return to the CM firm is increased recognition by peer groups (and evidence to potential clients) that the firm has a vested interest in the progress of the construction industry.

Becoming an officer of an association or organization is a large step up from being a chairperson. Exposure to a broader segment of the construction industry and to some public sectors is assured. Officers of one association have access to their counterparts in other associations and can spread their message and influence to the local, state, and national levels in the public sector. A firm's reputation within and beyond the industry is expanded and enhanced, opening more doors for sales opportunities. It is relatively expensive for a firm to support an association officer, not only in dollars, but in the time the person must devote to association business. Fortunately, the term of an association's president (the position that requires the most support) rarely exceeds one year.

23.3.7 Publications

Well-written, timely articles about construction and CM projects, especially in the form of case studies, are usually welcomed by editors of construction-oriented publications. Salespersons should stay in contact with the firm's projects and identify those that have publishing potential. Articles need not be long; in fact, they are generally more acceptable if they are short but not brief. They should provide a complete account of the topic and be accompanied with photographs, if appropriate.

The salesperson should identify those in the CM firm who have a propensity for writing and enlist their aid in producing articles or in writing them themselves. Salespersons should also target appropriate publications to ascertain their publishing policies and requirements.

Published articles contribute to the firm's public relations efforts. Reprints of articles, obtained from the publisher, are excellent handouts to distribute to potential clients during direct-selling activities.

Articles relate facts and information based on experience rather than research. The credibility of articles need not be established beyond the extent of the information presented. Articles are subject to opinion and criticism but not to the point where stated facts should be challenged.

Technical papers are scholarly dissertations that present arguable conclusions and are based on research and unquestionable data. Status as a paper implies that the information provided is literal. Papers are subject to review and challenge by the author's peers. Papers require scholarly knowledge and considerably more effort and time to write than articles. The majority of papers are theoretical; they are almost exclusively products of academia, where their writing provides promotion credits for authors. Their practical value to CM practitioners is limited.

Technical papers published by CM personnel contribute little to a firm's marketing effort; articles are far more effective. However, the status of CM as an emerging profession should be argument enough for CM practitioners to write papers, if for no other reason than to convey the practical aspects of the CM system on which academia can build.

23.3.8 Networking

Because CM services are usually engaged through a qualification-based selection process that is often initiated without prior advertisement, both the presence of the CM firm and its reputation must be highly visible to the right people at all times. Networking is by far the best tactic to use.

Networking is establishing relationships with a wide but select group of individuals who have a mutual inclination to assist each other in the pursuit of an individual's business enterprise (the cliche, "if you scratch my back, I'll scratch yours" applies). Networking is a framework of persons who have mutual respect for one another by reputation, previous business dealings, or simply casual friendship.

The glue that keeps the framework in tact and functioning is recurrent personal contact. This can occur at conferences and seminars; meetings of associations and organizations; community, social and professional functions; via the Internet, correspondence and telephone; and most impressive of all, by one-on-one personal contacts. Everyone in the firm should maintain a network, and the salesperson should be kept current with the status of each.

23.3.9 Personal Contacts

Approximately 80% of a salesperson's time should be spent on the road, contacting potential clients personally and visiting current and past clients. The selling seeds sown in the other six areas of new business opportunity are difficult to cultivate into sales, unless face-to-face contact is made.

Two types of contacts—cold calls and calls made by appointment—can be made. Cold calls are drop-in calls where no appointment has been made. The assumption is that the party the salesperson wants to see is in the office and has both the time and inclination to meet. Calls made by appointment determine the trip's itinerary. In essence, cold calls are subordinate.

Preparation for sales trips should include meeting with other CM firm personnel to determine if they have received network information about potential clients in the salesperson's assigned area. The time, effort, and money put into a sales trip warrants close scrutiny of the possibilities.

Regardless of which type of call is to be made, the salesperson must be prepared to make it as successful as possible. As much information about the person to be visited and the company he/she represents must be uncovered before the call takes place. Networking may provide some information; a few well-placed phone calls could provide more.

Being highly familiarized with the company and the person representing the company during an initial contact could make the difference between a "goodbye" or a "by the way, the city is planning a new facility" or a "I would like to learn more about your services" response. Preparation for the call is one of the keys to selling success.

Whether a call was considered a success or not, a follow-up mailing is imperative. It is also an opportunity to contact the person and the company one more time. A letter or simple thank you note should be sent within a few days of the visit. The style of the follow-up should be congenial, not aggressive. If implied or requested at the meeting, additional information on the firm and CM should be included; if not, the salesperson should determine what material should be included, if any.

23.4 NETWORKING/PROSPECT RECORDS

Records of sales contacts with potential clients should be kept in a computerized file. The files should be kept in a database for quick access and easy updating. If the records are kept in a simple format, any person in the CM firm that needs information can access the appropriate file by using the name of the company or the name of the network contact person.

Firm-wide networking produces new information on potential clients on an uncontrolled basis. It would be time consuming and problematic for the salesperson to gather and enter all information received. Therefore, personnel actively involved in the firm's networking activities should be able to update the file on a provisional basis. Provisional information can only become part of the prospect's permanent file after the salesperson has clarified and entered it as such.

Records can be used to keep in contact with prospects on a consistent basis. By entering contact types and the dates contacts are made, the salesperson can periodically scan the file to determine which prospects to recontact. Selling CM services to some prospects often takes several years; a relationship of trust and confidence in the salesperson and the firm must be established. Indicating an ongoing interest through low-key periodic contacts at appropriate intervals can help create this rapport.

23.5 SALESPERSONS

Selecting a salesperson for the CM firm is not an easy task. Selling services is not the same as selling products that can be seen. Services are intangible, and among services that are sold, such as design services, CM services are more intangible.

When architects and engineers sell their services, they can emulate the product by providing comparable visual examples of their work. Through photographs and visits to completed projects, the A/E's aesthetic and functional design capability can be clearly demonstrated to prospects. Along with references from past clients, architects and engineers can closely follow selling strategies used by product salespersons.

The CM service salesperson cannot use these strategies. Although a picture of a past project can be shown, it does not demonstrate anything about the CM's management ability, except that the project was completed. If a picture is used, it should serve as the background for a list of the CM's achievements on the project (i.e., the project was completed within budget, on time, with specified quality and no business interruption to the owner).

The only substantiation of a CM firm's ability is references from past clients. The salesperson could provide the prospect with a handful of endorsements and references but it is doubtful that such an approach would lead to an eventual sale. A CM salesman must sell him/herself before selling CM services to the prospect. To bring this about requires time, patience, and talent.

When searching for a qualified salesperson, the question that usually arises is whether to teach a competent CM person the art of selling, or to train a competent salesperson in the practice of construction management. In the first instance, competent CM persons are valuable to the firm when executing CM services. To subtract one from the operations roster could be a problem, especially if the firm is small. Additionally, persons with the necessary level of CM competence are often those with college degrees who will question whether selling CM services is an appropriate way to put their education to use. Although the importance of selling CM services is obvious, status as a salesperson (in spite of the title "Vice President, Business Development") does not always meet personal aspirations.

If the decision to teach a competent CM person the art of selling is made, care must be exercised to select a person who exhibits the attributes of a good salesperson; personality, organization, communication, optimism, tenacity and dedication, in copious amounts. A person with these attributes who is already a member of the CM firm will very clearly stand out.

Optimism and tenacity are of signal importance; especially when making cold calls, salespersons experience rejection much more often than acceptance. However, each rejection can be viewed as a step closer to a sale.

Teaching a technically-competent person how to sell CM services should be left to experienced teachers. There are professional associations that offer seminars to teach selling techniques and books on the subject for those who prefer this approach to learning. Becoming a member of marketing and sales associations will provide opportunities to attend conferences and meet with other salespersons to gain confidence and additional knowledge. The person should select media that relate to selling services, not products.

The strict tenet to follow is that no one should sell CM services without being properly prepared. Although much will be learned from experience, a premature start could prove demoralizing to the firm as well as to the salesperson. Sufficient money, time, and effort should be budgeted to develop a confidently competent salesperson before exposing her or him to the CM marketplace.

If it is decided to hire and train a competent salesperson in the practice of construction management, the CM philosophy of the firm and the CM format used by the firm can be taught in-house. However, care must be taken to select a candidate who thoroughly understands the construction industry and has previously successfully sold services (preferably engineering, architectural, or construction services if not CM services for another company).

Although this training seems straightforward, all three areas must be covered and conveyed. Teaching the science of CM should be straightforward to all candidates;

however, the firm's CM philosophy and format is unique with each company. These two aspects of services are often the deciding factors when owners make their choice of a CM firm. They both must be meticulously transferred and thoroughly understood by salespersons-in-training, especially those who have previously sold services for another CM firm.

When an in-house CM person is learning to be a salesperson, the firm's CM philosophy and format is indelibly fixed in the salesperson's mind. Only the selling aspects must be acquired. Although learning to sell is by no means an easy task, sometimes replacing preconceived ideas with new ones is more difficult. When training persons with previous selling experience for architects, engineers, and CM firms, extra care should be exercised to see that the firm's CM philosophy and format is transferred and indelibly fixed.

CHAPTER 24

Acquiring CM Services

A construction manager's qualifications are significantly different from those of architects, engineers, and general contractors. Owners considering using CM should be well aware that the most important point is that selection should be qualification based, not based on cost as one might select a contractor.

24.1 CM FIRM CONSIDERATIONS

A prototype construction management organization does not exist. A CM can be any organization that has the unique combined resources necessary to proficiently execute the form and variation of CM best suited for or selected by the owner. All organizations that call themselves construction managers are not equally equipped to provide services. Additionally, unlike engineering and architecture, construction management services are not universally governed by law. CM standards of practice are in their formative stages, and the professional status of the CM and those that the CM employs remains for the future to determine.

The financial strength of the CM firm is only important when using a form of construction management that requires the CM to provide explicit services as a constructor or contractor. When using a GMPCM or XCM form (other than Design–XCM), construction managers should be evaluated financially as one would evaluate general contractors. When using Agency–CM or Design–XCM, construction managers should be financially evaluated as one would evaluate design professionals when considering services.

A record of proficient performance as a design professional or as a contractor is not an automatic indication of proficiency as a construction manager. Although many CM's are also architects, engineers, and general contractors (or unique organizations spawned by one of these), successful construction management execution requires much broader disciplines than those inherent to the sole practice of architecture, engineering, or contracting. A proficient CM firm must have the basic resources and qualities of all three; a multi-discipline organization that assimilates the compound expertise of a design–build firm.

CM's have flexible geographic mobility. Their performance is insignificantly affected when functioning in locations other than where their headquarters are

---

Portions of this chapter are reprinted with permission from C. E. Haltenhoff, "Qualification and Selection of Construction Managers With Suggested Guidelines For Selection Process," *Journal of Construction Engineering and Management* 113 (March 1987):51–89. Copyright 1987 by the American Society of Civil Engineers.

located. In fact, CM performance is often enhanced by the CM's investigations in a new area. Familiarity with local conditions and contractors may mitigate the CM's objectivity rather than enhance it. Objective decision making and an inquisitional approach to project delivery are essential to successful CM operations. Owners should not overvalue the "home-town advantage" and select a CM firm simply because it is locally based.

A CM firm with extensive CM experience on a variety of projects is often a better choice than a firm whose experience is limited to a specific project type. The CM's prime contribution to the success of the project is vested in his ability to manage. The technical aspects of the project are essentially vested in the expertise of the A/E and the owner. The CM system of checks and balances in decision making provides sound direction so long as all the expertise is available from team members. In practice, a CM with varied project type experience is the best choice. That CM can close the experience circle through her/his ability to relate solutions from one project type to another. A CM should be considered a manager first and a project consultant second.

24.2 WHEN TO HIRE THE CM

The owner should hire the construction management firm as early in the project as possible. It is recommended that the A/E and CM be hired approximately at the same time. This permits exposure to one another (before agreements are signed) and compatibility checks. It permits the meshing of their respective agreements with the owner and provides an opportunity for a collective start. The nature of the CM's preconstruction services make an early team start very desirable. The CM's expertise in conceptual estimating can be a valuable assist to the owner and the design professional during feasibility.

24.3 CONTRACTS AND SERVICES

The services to be provided by the construction manager are determined by the CM form and variation selected by the owner, and these services should be prescribed in the owner/CM agreement. Standard document series are available. It is advisable to review the series to ensure that those selected conform as closely as possible to the CM form and variation selected. All standard documents must be amended to reflect the services that are unique to each project. Once selected, it is recommended that a single document series be used. Most documents address this consideration, pointing out that mixing could create problems between contractually interfacing parties. The use of proprietary documents should be approached with similar care. It is not advisable to use amended non-CM documents.

A construction manager's services can be identified with the phases of project delivery. Broadly stated, these phases are: (1) Pre-design, (2) Design, (3) Construction, and (4) Occupancy.

During Pre-design and Design, the CM provides services unique to traditional industry practices. In the GC system, input to these phases is via the A/E and almost exclusively reflects a design perspective. The contributions of the construction manager competently broadens that perspective by incorporating comprehensive construction and contracting input into the design.

During the Construction phase, the CM performs and accentuates the management services provided by the general contractor in the GC system. (They are reoriented to benefit the owner rather than the GC.) Most CM service contracts extend into the Occupancy phase at least one year from the date of occupancy or coincide with the duration of the warranties, guarantees, and surety bonds stipulated in owner/contractor agreements, providing the owner with call-back coordination services during occupancy.

24.4 GUIDELINES FOR SELECTING A CONSTRUCTION MANAGER

The format for selecting a construction manager closely resembles that of selecting a design professional or any other consultant an owner is considering. However, some forms and variations of CM, as well as some local laws and regulations, mandate a selection process that more closely resembles that of selecting a contractor. Before initiating a selection process, the owner should seek guidance to ensure propriety, legality, and protocol.

24.4.1 The Process

The selection process should proceed as follows:

1. Determine the form and/or variation of the CM System to be used
2. Determine if any specific laws or regulations govern the selection process for the project
3. Draft a concise physical description of the project, indicating time constraints and budget
4. Draft a brief description of the services required based on the CM form and/or variation to be used
5. Issue a request for interest in the project by CM firms in the form of an Advertisement or Invitation
6. Screen responses based on the preliminary information provided by interested construction managers
7. Issue an Initial Request for Proposal (IRFP) from between five and ten of the most promising respondants
8. Screen respondants to the IRFP and develop the "long list" of those worthy of further consideration
9. Issue a second Request for Proposal (RFP) to those on the Long List requesting uniform and specific information
10. Screen respondants to the second RFP and develop a "short list" of those worthy of an interview
11. Request final information from those selected from the Short List and conduct individual interviews with that group only.

24.5 STEPS IN THE SELECTION PROCESS

24.5.1 Steps 1 & 2

Determining the form and variation of CM to be used is not part of the selection process but is a necessary issue that requires resolution before proceeding. It is also necessary to determine which contract document series is appropriate for the form and variation of CM contemplated. The same issues apply to determining the laws and regulations that govern the conduct of the selection process.

24.5.2 Steps 3 & 4

Before contacting construction managers, it is necessary to define the project to the extent possible at the stage of its development. This information provides criteria with which to match up CM firms with the requirements of the proposed project. The intent is to provide CM firms with an opportunity to quickly decide whether or not to show further interest in the project. The more complete the information, the more definite their reaction can be. Minimum information should include the project's type, location, size, budget, schedule, unique characteristics, and design commitments, if any.

It is mandatory that the selected CM form and variation be explicitly stated, the documents identified, and a clear description of the services being sought from construction managers provided.

There is no advantage in withholding information about the project. The amount of information provided by the owner should only be controlled by practical publishing considerations.

24.5.3 Step 5

There are several ways to inform construction managers about a project. One is to publish a notice in a construction periodical or other media source such as a daily newspaper. Another is to obtain a listing of CM firms from a consultant or CM practitioner and contact as many as necessary, by telephone or by letter, to obtain the recommended numerical response. On public projects, laws often dictate the manner in which advertisement and solicitation must be done. Without regulation, each owner can use his/her own judgment to determine both the conduct and extent of initial contact with construction managers.

Figure 24.1 is an example of a typical advertisement for CM services.

24.5.4 Step 6

The goal of the advertisement is to locate between five and ten responders who clearly state their genuine interest in the project and indicate CM form and variation compatibility. The less certain the owner is about each respondant's stand on these important conditions, the larger the pool of respondants becomes.

During the screening process, it is important to recognize certain conditions and criteria that will help in the final selection of the construction manager.

1. Fees for CM services should not be addressed in any way until the interview stage has been reached. CM services are professional in the business sense of the

Advertisement for Construction Management Services

The (Pine Creek Building Authority) is seeking the services of a Construction Manager for an (office tower) project in (Pine Creek, Iowa).

1 The project consists of a (4-level service and parking structure and an 18-story office tower), located in (the downtown area).
2 The work consists of the (demolition of 7 existing 5-story masonry structures; excavation; underpinning/protection of 3 adjacent structures; construction of the 220,000 sq. ft. facility including tenant improvements and 2 skywalk connections to neighboring buildings).
3 The Architect selection process is to be conducted concurrently with the CM selection process. The project is funded and will proceed immediately, with full occupancy scheduled for (month, year).
4 The Construction Manager will be required to provide a Guaranteed Maximum Price at approximately 60% completion of design, and is not permitted to bid or perform any of the construction work involved.
5 Construction managers with experience on (urban high-rise) projects are invited to express preliminary interest by sending a maximum length five-page qualification letter to:

(Pine Creek Building Authority
1111 West Branch Street
Pine Creek, Iowa 00000)

The Authority will review all properly submitted responses postmarked on or before (day, date, year) and advise respondents of their status for further consideration no later than (day, date, year). The date of this notice is (day, date, year).

Notes to User:
1 A brief, general description of the project.
2 A brief explanation of the work involved.
3 An insight to the current status of the project
4 A description of the form and variation of CM services required.
5 The overall qualifications required.

Words in parentheses are project-specific.

FIGURE 24.1 A typical advertisement for Construction Management services.

word. The efforts of the selection process should be totally oriented toward the scope and quality of services and not prematurely influenced by fee discussions.

2. All communications with all respondants and potential respondants should be kept uniform to create a fair and equitable professional atmosphere throughout the selection process. It is important that none of the construction managers are given an unfair advantage or that any get the impression of unfair advantage.
3. Communication initiated by construction managers included or not included in the selection process should be ignored by the owner without exception. It is advantageous for the owner to remain in total control of the selection process at all times and not be influenced by pressures from those with special interests.

The practice of CM has attracted different types of practitioners with a wide variety of business styles and backgrounds. While CM is a professional service, in many cases individual practitioners are not professionals and may not sell their services at an assumed level of professional ethics. Precautions should be taken to preclude potential problems in the overall selection process.

Figure 24.2 is a selection process schedule for use where the owner is not familiar with CM. When owners are familiar with CM, the time can be substantially reduced.

Activity		Wk 1	Wk 2	Wk 3	Wk 4	Wk 5	Wk 6	Wk 7	Wk 8	Wk 9	Wk 10
Determine Project Budget	?										
Determine CM Suitability	?										
Determine CM Form and Variation	?										
Search Laws and Regulations	?										
Draft Project Description	–	X X									
Determine Time Constraints	–	X X									
Issue Invitation/Advertisement	–	0 0 0 0 X									
Screen Inv./Adv. Responses	–	– – – – –	– – – – –	X X X X X							
Issue IRFP	–	– – 0 0 0	0 0 0 0 0	0 0 0 0 0	X						
Screen IRFP Responders	–	– – – – –	– – – – –	– – – – –	– – – – –	– – – – –	X X X				
Telephone References	–	– – – – –	– – – – –	– – – – –	– – – – –	– – – – –	X X X				
Issue RFP to Long Listed CM's	–	– – – – –	– – – – –	– – – – –	0 0 0 0 0	0 0 0 0 0	0 0 0 0 X				
Screen RFP Responders	–	– – – – –	– – – – –	– – – – –	– – – – –	– – – – –	– – – – –	– – – – –	X X		
Develop Short List of Firms	–	– – – – –	– – – – –	– – – – –	– – – – –	– – – – –	– – – – –	– – – – –	– – – X X		
Request Final Information	–	– – – – –	– – – – –	– – – – –	– – – – –	– – – – –	– – – – –	– – – – –	– – 0 0 X		
Issue Interview Invitations	–	– – – – –	– – – – –	– – – – –	– – – – –	– – – – –	– – – – –	– – – – –	– – – – X		
Conduct Interviews	–	– – – – –	– – – – –	– – – – –	– – – – –	– – – – –	– – – – –	– – – – –	– – – – –	– – – X X	
Telephone References	–	– – – – –	– – – – –	– – – – –	– – – – –	– – – – –	– – – – –	– – – – –	– – – – –	– – – – X	X – – – –
Rate Those Interviewed	–	– – – – –	– – – – –	– – – – –	– – – – –	– – – – –	– – – – –	– – – – –	– – – – –	– – – X X	
Open Fee Envelopes	–	– – – – –	– – – – –	– – – – –	– – – – –	– – – – –	– – – – –	– – – – –	– – – – –	– – – – X	X – – – –
Negotiate with Choice CM Firm	–	– – – – –	– – – – –	– – – – –	– – – – –	– – – – –	– – – – –	– – – – –	– – – – –	– – – – –	– X – – –
Agree on Contract Terms	–	– – – – –	– – – – –	– – – – –	– – – – –	– – – – –	0 0 0 0 0	0 0 0 0 0	0 0 0 0 0	0 0 0 0 0	– X X – –
Sign Contract	–	– – – – –	– – – – –	– – – – –	– – – – –	– – – – –	– – – – –	– – – – –	– – – – –	– – – – –	– X – – –
		Wk 1	Wk 2	Wk 3	Wk 4	Wk 5	Wk 6	Wk 7	Wk 8	Wk 9	Wk 10

Legend: (–) weekdays; (?) indeterminate; (0) preparation days; (X) action days

FIGURE 24.2 A comprehensive selection process time schedule.

24.5.5 Step 7

The IRFP is the first screening device used in the selection process. It should be strategically prepared to effectively surface prime candidates for further scrutiny. It is suggested that the IRFP be in questionnaire form to preclude respondants from providing a lot of unwanted prose. The IRFP should essentially formalize the original information provided by the respondants to the owner solicitation (Figure 24.5 on page 384 is an example of an IRFP).

When forwarding the IRFP to the group of five to ten respondants, it is important to enclose a time schedule for the selection process. Specific dates should be dedicated to the remaining steps to be taken and the due-dates selected should be met if at all possible.

24.5.6 Step 8

A screening process should be established to uniformly compare the responses to each question in the IRFP. In situations where competition is a legal requisite, a rating system can be devised that will satisfy this requirement. It is suggested that owners create a similar rating system even if they are not subject to competitive constraints. The information can best be sorted and evaluated if it is systematically handled. Without a formal system, the information will become unwieldy and hinder the results of a well-intended selection process. Figure 24.3 on page 377 is an example of a simple comparison rating system.

All CM firms that responded to the IRFP should be notified of the owner's "long list" decision by the date specified in the selection process schedule.

24.5.7 Step 9

This second RFP complements the IRFP by requesting information from those on the "long list" that was not previously required. It is important that all remaining evaluation information required for selection be accumulated from this questionnaire.

Specific responses pertaining to handling the project (such as a proposed management plan and staffing plan) should be requested. Resumes of personnel to be assigned to the project, the proposed contract documents to be used, and the construction manager-owner agreement should be requested for review unless the documents to be used have already been specified by the owner.

Essentially all information deemed to be necessary to make a decision on a single firm should be included in the RFP, so that the final interview consists mainly of questions and answers pertaining to information already provided. (Figure 24.6 on page 388 is an example of an RFP.)

24.5.8 Step 10

As soon as it is determined which firms are to be interviewed, all of the firms on the long list should be contacted and advised of the decision. Interview dates for those on the short list should be made as soon as possible.

A procedure similar to that suggested in Step 8 should be used in this final screening process.

	Rating Factors	Score	Weight	Total
1	Longevity of Business Organization	______	______	______
2	Longevity of CM Service Organization	______	______	______
3	Versatility of Available Services	______	______	______
4	Depth of In-house CM Organization	______	______	______
5	Technical Level of Organization	______	______	______
6	Industry Association Involvement	______	______	______
7	Project TYPE Match-up	______	______	______
8	Project SIZE Match-up	______	______	______
9	Experience Based on VOLUME of Work	______	______	______
10	Experience Based on PROJECT NUMBERS	______	______	______
11	Demonstrated Geographic Versatility	______	______	______
12	Project SIZE Versatility	______	______	______
13	Project TYPE Versatility	______	______	______
14	Experience with Different Architects	______	______	______
15	Versatility as a Construction Manager	______	______	______
16	Dispute Avoidance Record	______	______	______
17	Capacity to Absorb New Projects	______	______	______
18	Capacity to Accept THIS Project	______	______	______
19	Stability of Personnel Turnover	______	______	______
20	Quality of Listed Owner References	______	______	______
		Total Score	——→	______

Notes to user:

- Each owner can determine the weight of each Rating Factor based on the unique conditions of the project at hand.
- The Factors listed above generally coincide to the questions asked in the Initial Request For Proposal (IRFP).
- Scoring should be on an even number scale. One to four is valid.
- Multiply the Score by the Weight to get the Total. Highest score indicates the most desirable CM firm.

FIGURE 24.3 A typical CM firm rating system for responders.

The next step is interviews and selection. One of the CM firms on the short list will be the construction manager. It is important that impressions relating to potential team compatibility and confidence in each firm's ability to perform be carefully noted.

If additional information is required from those on the short list, contact should be made and the information obtained. As recommended in all other steps, care must be exercised not to favor any of the contenders. A fair and successful selection process is highly dependent on an arms-length relationship with all firms under consideration. Maintaining this posture in the late stages of the process is sometimes difficult but very advantageous to the owner.

Interview dates should be established to easily accommodate the number of firms on the short list. There are advantages and disadvantages to firms being interviewed relative to their position in the interview schedule. Do not solicit or accept requests from the CM firms for preferred interview times; use an unbiased schedule selection method and hold the firms to the selected dates and times.

It is best not to schedule more than two interviews on any one day. If it is necessary to speed the process up, spread three or four interviews over a full day. No more than four interviews should be scheduled in a day. (This prevents confusing one firm's information with another's.)

Select interview dates and times that allow full attendance by the group responsible for selection. Absenteeism, in whole or in part during the interviews, is unfair to those being interviewed and detrimental to the selection process. An accurate comparison of the competing firms cannot be based on partial information. It is to the owner's advantage to participate fully.

24.5.9 Step 11

The interview process will be a success if the owner is properly prepared for the task. It has already been established that the CM process is not simplistic and not readily understood by many users of the system. It is assumed that those who are to conduct the interviews have educated themselves by reading the information submitted in the IRFP's and RFP's. A comprehensive Request For Proposal process allows owners to learn more about the CM system as well as more about the competing CM firms before final selection occurs.

24.5.10 Interview Format

The interview for each competing CM firm should take approximately an hour and a half. The first thirty minutes should be allocated to an unrestricted presentation by the CM firm. During this time, the CM should be allowed to "sell" his candidacy as he sees fit. The next fifteen minutes should be devoted to the CM's response to the inquiry forwarded to each firm prior to the interview. The CM should also address specific issues directly applicable to the project.

The next fifteen minutes should be used by the owner to question the CM on his/her presentation or anything connected with the information previously provided in the RFP's. The CM should be excused for the next fifteen minutes to permit the interviewers to compare notes and formulate any further questions that should be asked in the final scheduled fifteen minutes of the interview.

Time allocations need not be rigid within the allotted time, but each interview should be held to an hour and a half and used consistently from firm to firm. Every effort should be made to provide a credible comparative process during the interviews.

24.6 HANDLING THE CM FEE

Each CM should present his fee for the project in a sealed envelope before beginning its presentation. It is strongly suggested that the owner not open these envelopes until all interviews have been completed. This ensures the opportunity to rate candidates on their performance potential without considering fees. The fee issue should only be discussed during the final phase of the selection process.

24.7 QUESTIONING THE CM

It is assumed that the process as outlined will provide the interviewers with pertinent and meaningful questions during the interviews. It is generally useful to frame a specific series of questions before starting the interview process. These questions should be addressed to each firm to generate specific comparative criteria for final evaluation.

24.7.1 Typical Owner-Construction Manager Interview Questions

The interview is primarily an opportunity to assess owner, CM, and A/E compatibility. The team must have mutual confidence and trust in order to extract the maximum benefit from the CM system. The information on the construction manager's ability and capacity is obtained from the IRFP and RFP, and the interview is the time to see and talk to the people that will actually provide the services.

The CM spokesperson should be asked to introduce him- or herself and the principle staff members who will be assigned to the project at the site and in the CM's managing office. These are the individuals with whom the owner and A/E will be working throughout the project.

Several days prior to the interview, the owner should obtain resumes of proposed principle staff members from each CM as Final Information. The resumes should be screened by all owner personnel who plan to participate in the interview process. The owner should require the CM to bring key project personnel to the interview.

It is helpful for the owner to match up the CM's principle staff members, as they are introduced, with the Organization Chart provided by each CM as part of the RFP. It is probable that each CM's project organization and the titles of the positions will differ. Matching the names to the chart will facilitate identification during and after the interview.

Some questions should be addressed to each of the CM contingent. Questions should be framed to clarify or substantiate the facts contained in the IRFP the RFP and the Final Information provided prior to the interview.

For example, it would be appropriate to ask the person assigned to lead the project at the site how she/he intends to carry out responsibilities during construction; to ask the person who will lead the total CM effort how she/he will remain involved in the project during design and construction and provide visibility in day-to-day project activities; to inquire of a managing office staff person how accurate the estimating has been on the CM firm's other projects.

Many CM firms have excellent marketing and sales personnel who understand their services and recite them eloquently. The interview should not concentrate on a sales pitch, and it may be advisable to exclude the CM's salesperson from the interview. It is best to spend the time talking to the people who will be directly involved in project activity, and the CM should be made aware of this when the interview is arranged.

The interview is a good time to request the names, addresses and phone numbers of additional references. (Those previously listed in the IRFP were understandably the best references the CM firm could find.) The people at the interview should be asked to provide references that can attest to their performances on other projects.

Not all key personnel need be asked to provide additional references, but it is advisable to get one or two from the designated field leader and the person to be in overall charge of the project.

If the owner is using a formal rating system (such as the one in Figure 24.3) to evaluate the CM firms, it is advisable to maintain the consistency of the questions at each interview so that easy comparisons can be made. A list of core questions can be prepared ahead of time and used, along with random additional questions, during each interview. The core questions should address the seven major service functions and the estimated person-hours the CM indicated in the RFP.

The interview can best be concluded by allowing five or ten minutes for a CM spokesperson to express anything that was not covered or that might assist in the goal to be selected. This provides an opportunity for the CM to point out attributes not previously mentioned. The limited timeframe of the interviews does not permit all the time the CM would like to have, but it is adequate for its purpose if the CM organizes it effectively.

As the interview is concluded, the owner should ask the CM for his Fee Envelope. All fee envelopes should remain unopened until all interviews have been completed.

The CM firms should be given the opportunity to insert the fee amounts after the interview in case information affecting the fee is revealed during the interview. The sealed envelope containing the fee should be given to the owner before the conclusion of business on the day of the interview. A preprinted fee form provided by the owner should be used by each CM firm. Only the information requested should be entered on the form.

24.8 INTERVIEW DECORUM

It is important that the owner remain in charge of each interview and conduct each one in an effective and efficient manner. The groundrules should be formulated and made known well ahead of the interview dates. On public projects, it is mandatory that all competitors be treated with an even hand. This practice benefits the interviewers and should also be practiced on private projects.

24.9 MAKING THE DECISION

Discussions relative to a decision on the construction manager should not be initiated until the final interview has been completed and a day or so for reflection has elapsed. During this time, no contact should occur between those interviewed, but a free exchange of comparative data between the interviewers is encouraged.

When decision time is at hand, the interviewers and others involved in the project should meet to compare notes, score sheets, and impressions and come to agreement on ranking the CM firms. Once this is accomplished, fees should be revealed and further discussions of the best candidates (based on fee information) should occur. The firms should then be re-ranked if necessary.

As soon as possible, and within the schedule set for the selection process, the owner should notify all CM firms interviewed and inform them of the ranking. The owner should let them know that he/she intends to enter into an agreement with the first choice, thank them for their interest and time, and answer nominal queries concerning the decision.

24.10 CM OFFICE VISITATION

Before discussing the terms of an agreement, it is recommended that the owner visit the offices of the CM being considered. If the owner would be dealing with a branch office operation, she/he should visit both the branch and the main offices to see the entire operation. Knowing how each office would be involved, who would be responsible in each case, and how intercommunication would occur is also important. The owner should review the staffing plan, if one was previously submitted, be satisfied that it will serve the purpose, feel free to re-question the CM on unclear responses provided during the interview, and make sure the CM's participation is as it was perceived at that time.

24.11 AGREEMENT NEGOTIATIONS

In addition, the owner should thoroughly review the terms of the Owner-CM agreement with the construction manager. She/he should have present those who can help clarify any discussions of the provisions or any modifications to them. The document should cover the necessary interactions of the owner, design professional, and construction manager in sufficient detail to avert future confusion.

It is suggested that a Memorandum of Understanding or a Responsibility Chart be developed to provide clear definitions of the team members' interacting responsibilities. All team members should contribute to its formulation. This document can be an attachment to each team members' agreement with the owner. Figure 24.4 on page 382 is a sample page of a Responsibility Chart.

24.11.1 Fee Negotiations

It is assumed that the first ranked CM firm was chosen with due consideration of both ability and fee. Consequently, negotiation of fee amount is not normally expected. However, if the owner deems the fee amount to be excessive or deficient, negotiations are definitely in order. The time for fee discussions is after a clear, mutual understanding of services and requirements has been achieved.

24.12 CM FEE STRUCTURES

There are several ways to arrange fees for CM services. All are derived from similar arrangements common to the industry, and essentially they include Lump Sum, Cost Plus, and combinations of both. Some variations of CM have propagated incentive and merit fee structures.

24.12.1 Typical CM-Owner Fee Arrangements

The fee arrangements between owners and construction managers are more varied than the forms and variations under which CM is performed. Certain forms and variations of CM require the use of unique fee arrangements to accommodate dual responsibilities. The following represent the more common fee arrangements and are provided as a guide only.

24.13 AGENCY CM FEE FORMS

Lump Sum, with or without an added cost clause: This is used on small- to medium-size, straightforward, new projects where the required involvement and responsibilities of the owner, A/E, and contractors are readily predictable.

Owner-A/E-CM Responsibility	Owner	A/E	CM
addenda—procedural	review/approve	review/analyze	generate/write
addenda—technical	review	generate/write	review/analyze
advertisement—bids	issue/pay	review/comment	generate/write
bid opening	attend/witness	open/read	check/record
bid reviews	review	review/comment	analyze
bidders lists	review/approve	review/comment	generate
bidding allowances	review/approve	analyze/write	generate
bidding alternates	review/approve	generate/write	generate
bidding documents	approve/pay	assemble/print	distribute
bonding requirements	establish	review/comment	analyze
budget updates	review/approve	analyze	provide
budget, project	establish	analyze	analyze/codify
contract awards	review/approve	recommend	review/comment
contract drawings	review/approve	generate	review/comment
contracting methods	review/approve	review/comment	generate
contracts, construction	provide/sign	review/comment	review/comment
costs, contruction	review/spprove	analyze	determine
design development documents	review/approve	generate	review/comment
design parameters	establish	analyze/codify	analyze
insurance values	establish	review/comment	analyze
letters of intent	issue	write	review
meetings, prebid	attend/comment	participate	conduct
notice to proceed	issue	review	write
project manual	approve	assemble	review/comment
proposal forms	review/approve	analyze	generate/write
bidder qualification	review/approve	analyze	generate
quality standards	establish	analyze/codify	generate
schedule updates	review/approve	analyze	provide
schedule, design	review/approve	provide	comment/codify
scheduling, construction	review/approve	analyze	generate
schematic drawings	review/approve	generate	review/comment
specs., front end	review/approve	generate/write	generate
specs., outline	review/approve	generate/write	analyze
specs., technical	review/approve	generate/write	review/analyze
value engineering	review/approve	analyze design	generate
work-scope description	review	analyze	generate
work-scopes	approve	review/comment	generate

FIGURE 24.4 A sample page of an Owner-A/E-CM Responsibility Chart.

Projects such as warehouses, light industrial plants, local shopping malls, office buildings, minor civil works, medical offices, clinics, and elementary and high schools can effectively use this type of fee arrangement.

A contingency clause is generally included to reimburse the CM for extended project time resulting from acts or occurrences not under his direct control. It should stipulate a daily or weekly lump sum amount, that does not include overhead or profit, for field personnel required after a specific completion date named in the contract.

Lump Sum plus Reimbursables: This is used on any of the projects that accommodate a lump sum fee and larger or more complex projects where the involvement/responsibilities of the owner, A/E, or contractors are not readily predictable.

Hospitals, process plants, high rise buildings, university facilities, jails, prisons, luxury hotels, regional malls, heavy industrial plants, and civil engineering projects fit well under this fee arrangement.

The lump sum portion of the fee arrangement usually covers the cost of the construction manager's services prior to the start of on site construction, including the fee, general overhead, and profit for the total project. The owner reimburses the CM for field costs incurred during construction. Variations to this format can change the phases covered by lump sum and reimbursables.

The reimbursable portion is based on predetermined, listed rates for personnel and equipment and for specific services listed in the agreement. The reimbursable portion is usually limited to a not-to-exceed amount and consists of costs only (the fee, general overhead, and profit having been paid for in the lump sum portion of the total fee).

When using this CM fee form, it is important to define what will be included in each portion. This is especially true when requesting and analyzing fees during the CM selection process. Requiring the CM to provide person-hour requirements will help sort out potential confusion when comparing one candidate with another (Figure 24.6; 4.0 on page 388).

24.13.1 Reimbursables

The total flexibility of this CM fee arrangement makes it applicable to any project regardless of the CM form or variation used. The construction manager is reimbursed for the services provided on an hourly or per diem rate basis plus all expenses incurred in the providing of services. This fee arrangement is characterized as cost expended plus a percentage of cost expended or cost expended plus a stipulated lump sum.

Owners should be aware that this type of fee arrangement is not tied to predetermined performance by the construction manager regarding staffing usage. Due to its open-ended character, this fee should only be used when none other will fit.

24.13.2 Fee Enhancement Provisions

Occasionally the construction manager's fee arrangement is materialistically enhanced by the owner in hopes of improving the CM's performance. Two forms of fee enhancement are sometimes used: Incentive and Performance.

Incentive arrangements: These are usually connected to either the cost or the time schedule for the project, and sometimes both. Additional fees, specifically named in the CM-Owner agreement as to amount and payment conditions, may be earned by the CM for completing the project below a budgeted amount or in a shorter time than originally scheduled.

When considering incentive fees, owners should realize that establishing both budgets and schedules is the normal prerogative of the construction manager. In establishing both the budget and the schedule, caution should be taken that economic returns under an incentive clauses are not inadvertently or strategically included in the budget or the schedule.

Continued on page 389

INITIAL REQUEST FOR PROPOSAL

Submitted To: ______________________________
____
____

Submitted By: ______________________________ a: Corporation ______
______________________________ Partnership ______
______________________________ Individual ______
______________________________ Other ______

1.0 BUSINESS ORGANIZATION

1.1 How many years has your firm been in business under the name stated above? __________

1.2 How many years has your firm provided Construction Management services? __________

1.3 What other services does your firm provide under the name stated above? __________

1.4 If a Corporation, please answer the following:

1.4.1 Date of incorporation __________

1.4.2 State in which incorporated __________

1.5 If a Partnership, please answer the following:

1.5.1 Date of organization __________

1.5.2 Names of partners and percent of participation:
____
____

1.6 If Individual, please answer the following:

1.6.1 Date of organization __________

1.6.2 Name of owner __________

1.7 If "Other," please explain. ______________________________
____
____

1.8 Address of Main Office ______________________________
____
____
____

1.9 Location of branch offices, if any.
____
____
____
____

2.0 CM ORGANIZATION

2.1 The number of permanent employees __________

2.2 How many are permanently assigned to CM services? __________

2.3 List "in-house" personnel in the following categories.

2.3.1 Field Managers or Superintendents __________

2.3.2 Project Administrators or Managers __________

2.3.3 Resource Personnel __________

2.3.4 Support or Administrative Persons __________

2.3.5 Principals or officers __________

2.3.6 Total (must equal 2.1) __________

2.4 How many of the persons in 2.1 have been hired in the past two years? __________

FIGURE 24.5 A typical Initial Request For Proposal (IRFP).

2.5 What has been your turnover number (layoffs *or* replacements) during the past five years? ______

2.6 Designate the technical level of all in-house personnel by category.

		Registered	Degreed Non-Reg.	Non-Degreed
2.6.1	Engineers, Civil	________	________	________
2.6.1	Engineers, Mechanical	________	________	________
2.6.3	Engineers, Electrical	________	________	________
2.6.4	Architects	________	________	________
2.6.5	Value Managers	________	________	________
2.6.6	Planners	________	________	________
2.6.7	Estimators	________	________	________
2.6.8	Attorneys	________	________	________
2.6.9	Business Admin.	________	________	________
2.6.10	Computer Science	________	________	________
2.6.11	All other	________	________	________

2.7 List the professional and trade associations that are represented by in-house personal memberships.

____

____

____

____

____

____

2.8 Of all in-house employees listed in 2.1, how many are classified as minority employees? _____

2.9 When performing CM services, what portions of your CM services do you normally contract out?

____

____

____

3.0 EXPERIENCE

3.1 List your specific experience as a *Construction Manager* in the following project categories.

		Number of Projects	Largest ($) Project
3.1.1	Commercial	________	________
3.1.2	Industrial	________	________
3.1.3	Health Care	________	________
3.1.4	Educational	________	________
3.1.5	Correctional	________	________
3.1.6	Hotels	________	________
3.1.7	Process	________	________
3.1.8	Services	________	________
3.1.9	Civil Works	________	________
3.1.10	________________	________	________
3.1.11	________________	________	________
3.1.12	________________	________	________
3.1.13	________________	________	________
3.1.14	________________	________	________
3.1.15	TOTAL	________	________

FIGURE 24.5 Continued

3.2 What had been your annual volume of CM projects during the past five (5) years, based on construction costs?
$ ________ $ ________ $ ________ $ ________ $ ________

3.3 How many CM projects have you either completed or have underway in the following areas of the United States?

		Number of Projects Completed	Number of Projects Underway
3.3.1	Northeast	________	________
3.3.2	Southeast	________	________
3.3.3	North Central	________	________
3.3.4	South Central	________	________
3.3.5	Northwest	________	________
3.3.6	Southwest	________	________
3.3.7	Alaska and Hawaii	________	________
3.3.8	TOTALS	________	________

3.4 How many different Architectural/Engineering firms have you worked with on the CM projects listed in 3.1?
____

3.5 On how many of the CM projects listed in 3.1 have you provided the following services in whole or in part?

3.5.1 Held contracts for construction work ________

3.5.2 Performed construction work ________

3.5.3 Provided a guaranteed maximum price for work ________

3.5.4 Performed construction support work ________

3.5.5 Provided design services for the work ________

3.6 On how many of the CM projects listed in 3.1 did you enter into litigation/arbitration with the owner? _____

3.7 On how many of the CM projects listed in 3.1 did contractors enter into litigation/arbitration with the owner?
____

4.0 CURRENT WORK LOAD

4.1 Express the current work load of your CM organization on a project basis, broken down as follows:

		Number of Projects	Dollar Value
4.1.1	Feasibility/Planning	________	________
4.1.2	Schematics/Prelim. Design	________	________
4.1.3	Design Devlp./Final Design	________	________
4.1.4	Contract Documents	________	________
4.1.5	Out for Bids	________	________
4.1.6	Construction	________	________
4.1.7	Occupancy/Start-up	________	________
4.1.8	Warranty-Guarantee	________	________
4.1.9	Complete, but not closed out	________	________
4.1.10	TOTALS	________	________

4.2 Based on your current staff listed in 2.1, and average projects from your list in 3.1, what is the estimate annual volume capacity of your CM firm? $____________

4.3 If the project which is the subject of this inquiry put your work load above the estimated capacity stated in 4.2, how would you sufficiently increase your capacity to handle it properly?

		Yes	No
4.3.1	Add staff by direct hire	________	________
4.3.2	Contract services out	________	________

FIGURE 24.5 Continued

4.4 Based on your current staff listed in 2.1, and assuming an average project, what is the largest single project your CM organization can effectively handle? $ ______________

4.5 If the project which is the subject of this inquiry was larger than the maximum size stated in 4.4, how would you sufficiently increase your staff to handle it properly?

		Yes	No
4.5.1	Add staff by direct hire	________	________
4.5.2	Contract services out	________	________
4.5.3	Both of the above	________	________

4.6 Provide the following information on no more than five (5) of your *largest* current projects.

	Project Type (Ques. 3.1)	Location (Ques. 3.3)	$ Size	Completion Date
4.6.1	________________	________	________	________
4.6.2	________________	________	________	________
4.6.3	________________	________	________	________
4.6.4	________________	________	________	________
4.6.5	________________	________	________	________

5.0 REFERENCES

5.1 Provide the following information on owners that have used your services more than once. (Attach a separate sheet)

5.1.1 Owner's Name

5.1.2 Owner's Address

5.1.3 Owner's Phone Number

5.1.4 Projects Involved

5.2 List the following information on three (3) of the five (5) projects listed in 4.6. (Attach a separate sheet) *Do not use projects listed in 5.1.*

5.2.1 Project

5.2.2 Owner's Name

5.2.3 Owner's Address

5.2.4 Owner's Phone Number

5.2.5 Project's Architect/Engineer

5.2.6 A/E's Address

5.2.7 A/E's Phone Number

5.3 Provide the following information on project Architects or Engineers that you have worked with on more than one CM project, for the same owner or different owners. (Attach a separate sheet)

5.3.1 Architect's/Engineer's Name

5.3.2 Architect's/Engineer's Address

5.3.3 Architect's/Engineer's Phone Number

5.3.4 Projects Involved

6.0 CERTIFICATION OF INFORMATION PROVIDED

6.1 All of the information provided herein is to the best of my knowledge complete and accurate, and can be accepted by the solicitor as a valid response to the questions asked.

6.1.1 Firm ______________________________________

6.1.2 Signed: ______________________________________

6.1.3 Signed by: ______________________________________

6.1.4 Title: ______________________________________

6.1.5 Date: ______________________________________

FIGURE 24.5 Continued

REQUEST FOR PROPOSAL

Submitted To: ____
____
____

Submitted By: ____ a: Corporation _____
____ Partnership _______
____ Individual _______
____ Other _______

The information provided by the Construction Manager must address the specifics of the Owner's Project as far as possible. Some of the interpretive project conditions have been simulated in order to obtain comparative answers from responders. It is requested that all questions be answered on the basis of the information as provided herein.

1.0 Provide your proposed Management Plan for the project based on:

a) a design time of ___*___ months (weeks)
b) a construction time of ___*___ months (years)
c) occupancy of area (building) ___*___ on ___*___, ___*___, ___*___
d) occupancy of area (building) ___*___ on ___*___, ___*___, ___*___
e) occupancy of the entire project on ___*___, ___*___, ___*___

1.1 Submit your Proposed organization in the following format.
1.1.1 On-Site organization Chart with position Titles identified.
1.1.2 Off-Site organization Chart with position Titles identified.
1.1.3 Job Descriptions of all positions shown on Both charts.

1.2 Provide a 2-Page narrative of how the organization will Function during Design and during Construction.

1.3 Provide a Schedule (bar chart or other) showing how the Major element of the project will meet the time requirements.

2.0 Using no more than One Page Each, explain how you propose to accomplish the following functions:

2.1 Cost Management—Estimating, Budgeting, Cost Reporting

2.2 Value Management—Value Engineering, Life Cycle Costs

2.3 Decision Management—Checks and Balances, Expertise

2.4 Schedule Management—Systems, Types, Applications

2.5 Information Management—Communications, Documentation

2.6 Risk Management—Insurance, Bonding, Decisions

2.7 Contract Management—Dividing, Bidding, Coordination

3.0 Using no more than 1 Page, identify any unique aspects of this project and briefly explain your approach to each.

4.0 Provide an Estimate of the Person-Hours you would expect to use on this project to accomplish:

4.1 Preconstruction activities ______________________________ PHrs

4.2 Construction activities ______________________________ PHrs

______________X______________

4.3 The total project ______________________________ PHrs

* Pre-entered information by Owner.

FIGURE 24.6 A typical Request For Proposal (RFP).

5.0 Certification of Information Provided

5.1 All information provided herein can be considered by the solicitor, to be provided in good faith and attainable on the project as proposed, by the CM firm named below.

5.1.1 Firm ____

5.1.2 Signed: ____

5.1.3 Name: ____

5.1.4 Title: ____

5.1.5 Date: ____

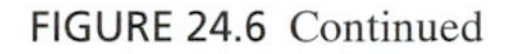

FIGURE 24.6 Continued

Performance rewards: These can be connected to many aspects of the construction manager's performance and consequently are not as simple to measure as time and cost incentives. Additional fees are contractually set aside as a reward for effective performance (as seen from the owner's perspective).

Due to the nebulous criteria, the CM must depend on the owner's good faith for reward. Having made the commitment, owners must retain sufficient budget funds to cover the eventuality and not be tempted to expend them for other purposes.

24.13.3 Fee Incentive Ethics

An ethical question arises regarding the dual level of performance inferred by performance rewards. Does an owner have to pay for improved performance, or is a construction manager obligated to provide the best performance when the original agreement is made? Owners should seriously question the eventual worth of performance incentives from both an ethical and value-received perspective before including them in a CM fee arrangement.

24.14 EXTENDED SERVICES AND GUARANTEED MAXIMUM PRICE CM FEES

Fees for the forms and variations of the CM system are generally combinations of an ACM fee arrangement and one that reflects the particular form or variation being provided. (Review Chapter 6, CM Under Dual Services Agreements.) The arrangements include:

Lump sum, with or without a contingency clause to cover any extended services or increased time resulting from acts or occurrences caused by other than the construction manager;

Lump sum plus reimbursables, where the reimbursable portion usually covers the construction manager's cost and expenses during either the preconstruction or construction phase of the project, and sometimes both;

Fee enhancement provisions may or may not be included, however, their inclusion is more common with CM forms and variations than they are with the ACM form.

Financial risk enhancement: construction managers assume varying degrees of financial risk when providing services beyond those required by the ACM format. The assumption of risk entitles the CM to proportionally higher fees.

24.15 EXTENDED SERVICES CM FORMS

24.15.1 Design–XCM

If the Owner-A/E agreement for design services was entered into without consideration for CM services, a second agreement or an amendment to the design agreement is required. The fee for each CM service is not related in terms of amount, and it should never be assumed that combining A/E and CM services will produce total fee savings. Design–XCM is a combination of standard A/E services and standard ACM services, and fees should be negotiated within that divided context.

If Design–XCM is selected at the outset of the project, a combined A/E-CM agreement with a single fee covering design and construction management services is negotiated. The fee amount for design services and the fee amount for CM services should be generated and analyzed separately, as if the services were not to be provided in combination. It is important that the concept of separate services prevail even though one firm is to provide them under a single agreement. There is no overlap of services between the two.

24.15.2 Contractor–XCM

This variation combines ACM services and the holding of trade contracts by the construction manager. A single fee, covering ACM services increased to compensate the construction manager for the financial risks of holding contracts, facilitates the fee agreement.

Determining risk value for fee purposes is based on the unique terms of the agreement with respect to the type, size, and number of contracts to be held by the construction manager.

24.15.3 Constructor–XCM

This CM variation combines ACM services and the construction of part of the project by the construction manager's own forces. The fee must reflect the ACM services plus the CM's exposure to the loss or gain connected with performance of the construction requirements. The fee amount reflects the CM's exposure to the unique project conditions and, particularly, the fee arrangement under which the CM is to be paid for services.

If the CM's exposure to construction risk is eliminated (as it would be if a cost-plus percentage, or lump sum fee arrangement were used for the constructor portion of the agreement), no CM fee increase should be expected. In this situation, the CM/Constructor has no loss exposure from constructor services; he/she will in fact earn a profit from the "plus" provision of the cost-plus fee arrangement.

24.15.4 Contractor/Constructor–XCM

This variation combines ACM services with contracting and construction services. The CM fee is established to reflect the risks involved in providing the contracting and construction functions as explained for Contractor–XCM and Constructor–XCM.

24.16 GMPCM FEE FORMS

The GMP form of CM, and its three variations, includes a risk element that usually warrants an increase in the CM's fee. In addition to the fee for ACM involvement, the construction manager is entitled to a premium for assuming financial responsibility for the construction budget. The premium amount is based on the obligations placed on the construction manager by the provisions of the Owner-CM agreement.

The GMPCM form obligates the construction manager to pay for the cost of construction overruns on a total-project basis. Some GMPCM agreements (those involving fast-track or phased contracting techniques) obligate the CM to pay for cost overruns on a phase-budget basis. Other GMPCM agreements obligate the CM to pay overruns on a budget line item basis.

The two latter agreements provide little performance flexibility and are very demanding on the CM's budgeting and estimating skills. The fees for providing a GMP on the total budget can be expected to be considerably less than the fees for providing a GMP on either a phase or line item basis.

24.16.1 Contractor, Constructor, and Contractor-Constructor GMPCM Variations

These variations further complicate determination of fair and equitable fees for construction management services. It is appropriate to individually consider each element of the total services with respect to CM's exposure to financial risk. Determining fees for the GMP variations follows the steps prescribed for ACM and XCM, as previously stated.

The CM variation obviously deserving of the highest fee is Contractor/Constructor–GMPCM, which in analysis is a combination of Contractor/Constructor–XCM and GMPCM and represents the maximum degree of CM services available that are directly related to construction. Fee development must consider ACM services, contracting services, construction services, and the provision of a guaranteed maximum price for the total project.

24.17 COMMENTS ON THE INTERPRETATION OF CM FEES

When reviewing proposed fees, it is important to recognize that CM services vary in quality and quantity from one construction manager to another. These variables must be properly evaluated when comparing fees during the final steps of the selection process.

When requesting fee amounts, owners should require a breakdown of the hours to be expended in each of the obvious CM service categories. Each CM firm should be requested to provide hour/cost information for the personnel to be utilized in providing services. From the fee amount, hours, and personnel wages, the owner can realistically compare the real value of each CM firm's proposed services.

The value of each CM's services should be rated on the number of hours to be devoted to the project and/or to each category of service and the average hour cost of services in total or by category. Analysis of this information, along with prior client reference information, will provide a quantitative/qualitative evaluation base for each construction manager and provide valuable support during fee negotiations. Estimated hours are required in the RFP, Section 4.0, but may be expanded by using the categories in Section 2.0 of the RFP and submitted with the sealed fee proposals.

24.18 THE IRFP AND RFP

The IRFP was designed to determine the overall suitability of responding CM organizations. The RFP is designed to determine specific suitability for the project at hand. The project information previously supplied to interested firms for their response to the IRFP may or may not need enhancement, depending upon the completeness of the initial information.

APPENDICES

APPENDIX A

A Suggested Technical Knowledge Base for CM Operations Personnel

CM personnel closely interact with architects, engineers, owners, and various consultants during the course of the project. Interaction occurs at meetings, via phone and computers, during teleconferences, in correspondence and meeting minutes, and in one-on-one conversations.

All these communications deal with the technical facets of design and construction as well as the management aspects of project delivery. Many interactions are at a high technical level (i.e., those with architects and engineers in the civil, mechanical, and electrical disciplines). These professional persons have completed a course of learning at the university level or gained the equivalent from experience and have proven minimum competence by becoming legally registered or licensed.

As stated in Chapter 8, The CM Organization, a construction manager is an organization, not an individual—an organization comprised of specialists and generalists who collectively possess the expertise to provide CM services. The specialists are the resource persons whose backgrounds and education put them on a par with the specialists with whom they must interact at high technical levels. The generalists are the operations persons who must have, in addition to their proficiency in the management aspects of project delivery, a level of competence in the technical areas to function effectively.

Throughout a project, CM operations persons interact constantly with specialists, both in their own organization and the organizations that comprise the project team. Their management roles as communicators and decision facilitators mandate that they have the required technical knowledge.

To identify the level of competence that CM operations personnel should have, the following questionnaire is included. The list is representative only of the technical knowledge CM operations persons should have, and it should not be construed as either complete or as an academic measure of overall CM competence.

Referring to the CM operations personnel identified in Chapter 8, Level 1 (executive) persons should be able to answer *most* questions correctly, Level 2 (management) persons should be able to answer *almost all* questions correctly, and Level 3 persons (on-site administration) should be able to answer *some* questions correctly.

It should be noted that simply finding the correct answers to the questions will not demonstrate competence; it is necessary to understand the subject matter behind the questions. When using the questionnaire, the "I Don't Know" option should be selected if there is any doubt concerning a choice of "True" or "False." The object is not a passing score, it is to provide a self-assessment of your comfort level as an operations person in a CM firm.

A.1 REPRESENTATIVE TECHNICAL KNOWLEDGE FOR CM OPERATIONS PERSONS

001 In some cities, zoning ordinances establish fire zones, where all buildings must be constructed of noncombustible materials. T F ?

002 The intent of building codes is to protect public health and safety by establishing a minimum standard of construction quality. T F ?

003 Fire resistance ratings for all construction components and materials can be found in the BOCA Code. T F ?

004 Most building codes in the United States are based on either the BOCA Code, Uniform Building Code, or the Standard Building Code. T F ?

005 The dollar amount of fire insurance premiums influences the construction standards for a building. T F ?

006 The CSI Masterformat divides construction components and materials into 16 primary divisions. T F ?

007 The CSI Masterformat provides example technical specifications for the 16 primary divisions. T F ?

008 The total live load of a building is the sum of furnishings, occupants, moveable equipment, snow, ice, and water on the roof. T F ?

009 Wind can exert forces on a building or structure in a lateral, upward, or downward direction. T F ?

010 Clay is generally referred to as noncohesive or cohesionless soil. T F ?

011 Poured concrete footings should be placed on undisturbed soil, not on engineered fill. T F ?

012 Steel H piles driven to refusal and caissons drilled in to bedrock are comparable structural solutions for foundation construction. T F ?

013 Framing lumber is considered "seasoned" if its moisture content is less than 19%. T F ?

014 The actual dimensions of a seasoned 2 × 12 southern pine plank is $1^1/_2'' \times 11^1/_2''$. T F ?

015 A 2 × 8 wood joist, 12 feet long, contains approximately 14 feet board measure. T F ?

016 Medium strength mortar (Type N) is satisfactory for masonry work above grade. T F ?

017 If the color of a brick is very dark, it probably has a low rate of water absorption. T F ?

018 When storing concrete blocks on site prior to use, they should be protected from the weather. T F ?

019	A 12″ thick, solid masonry wall, laid up of modular bricks, would be 2 wythes wide.	T F ?
020	A rowlock modular brick is laid on its end with its face parallel to the wall.	T F ?
021	When constructing a rectangular masonry foundation wall, the mason first lays up the corners and then fills in between them.	T F ?
022	Mortar used in masonry construction consists of Portland cement, lime, sand and water.	T F ?
023	An 8″ × 8″ × 16″ standard concrete masonry unit would weigh approximately 30 to 35 pounds.	T F ?
024	The standard for determining the allowable bearing capacity of reinforced masonry walls is to build a section of wall and test it.	T F ?
025	Movement joints in masonry walls should be placed at discontinuities (changes in thickness or height) in the wall.	T F ?
026	Effluorescence on masonry walls can be avoided by selecting masonry units that do not contain water-soluble salts.	T F ?
027	The fire resistance of an 8″ modular brick wall is approximately twice that of a standard 8″ concrete block wall.	T F ?
028	The extra 18#/lf in a W10×30, compared to a W10×12, is more in web thickness than in the width and thickness of the flanges.	T F ?
029	If corrosion is not a factor, and deflection controls the design of a simple beam, the substitution of A242 high strength steel for A36 carbon steel would not be justified.	T F ?
030	Load indicator washers are used in structural steel bolted connections to indicate if a bolt has been properly tensioned.	T F ?
031	Type llA Portland cement is used when it is necessary to obtain a fast initial set in a concrete pour.	T F ?
032	When used in concrete, fly ash increases strength, decreases permeability, reduces mixing water, and improves pumpability.	T F ?
033	The workability of concrete with a low water-cement ratio can be improved by replacing Type lA cement with Type l cement.	T F ?
034	The same concrete mix poured under water (by tremie) will reach a higher eventual compressive strength than if poured in the dry.	T F ?
035	The external controlling factors in concrete form design are the rate of pour and the ambient and concrete temperatures.	T F ?
036	Post-tensioned concrete beams have higher bending strength when the tensioning strands are draped than when they are straight.	T F ?
037	The configuration of the structural reinforcing steel in a one-way slab precludes the need for temperature steel.	T F ?
038	In a thermally-insulated office building envelope, the vapor barrier is always placed on the cold face of the envelope system.	T F ?

039	The lower the temperature of the air, the more water vapor it is capable of containing.	T	F	?
040	The "U" value of a specific material is the reciprocal of its "R" value.	T	F	?
041	One of the safety features of tempered glass is that, when broken, it reduces to small granules rather than large sharp shards.	T	F	?
042	A full-scale mock-up of a curtain wall systems is usually laboratory tested before the installed system is tested in the field.	T	F	?
043	The three ratings for interior finishes specifically covered by code are flame spread, fuel contribution, and smoke development.	T	F	?
044	A fire separation wall and a fire wall serve the same purpose but have different physical requirements.	T	F	?
045	An axial load is one that is applied parallel to the long axis of a structural member.	T	F	?
046	A 1/4″ thick steel plate measuring 12″ × 12″ would weigh close to 10 pounds.	T	F	?
047	An S4S, softwood, 2 × 12, 8 feet long would weigh ± 30 pounds.	T	F	?
048	Internal compressive/tensile bending stresses result from the application of external non-axial forces to structural members.	T	F	?
049	A penetrometer is a measuring device commonly used to approximate the compressive strength of a material.	T	F	?
050	Relative humidity is the amount of water vapor in an air mass at a specific temperature and pressure.	T	F	?
051	A beam-to-column shear connection resists the tendency of the beam to slide down or up the column but does not resist rotation.	T	F	?
052	A shear panel is a wall, floor, or roof surface that acts as a diaphragm to stabilize a structure against external forces.	T	F	?
053	A block of concrete which prevents the lateral movement of a water main at a 90E change in direction is called a thrust block.	T	F	?
054	A difference between CPM and PERT scheduling is that CPM selects one duration per activity and PERT picks three durations (optimistic, most probable, and pessimistic) and relies on statistical selection.	T	F	?
055	The straight-line depreciation method writes off equipment faster than either sum-of-the-years or declining-balance methods.	T	F	?
056	When using present worth to calculate life-cycle costs of alternatives, the higher present worth is the economical choice.	T	F	?
057	An item's life-cycle expense includes its purchase, installation, operating, maintenance, replacement, and finance costs.	T	F	?
058	A four-cycle engine's horsepower, produced under restraint of a governor, is called that engine's brake horsepower.	T	F	?
059	Four-cycle engines lose power when working at high elevations and gain power at higher temperatures.	T	F	?

060 A 30 ton PCSA Class 12-105 crane can lift 30 tons at a radius of 12′ with its basic boom and 10,500# at 40′ with a 50 foot boom. T F ?

061 The radius, used to calculate the lifting capacity of a crane, is measured from the base of the boom. T F ?

062 The practical dewatering depth of a single stage wellpoint system becomes less at higher altitudes. T F ?

063 The cubic yards of sand excavated from an undisturbed borrow pit will provide an equal volume of compacted sand at the fill site. T F ?

064 The LCY capacity of a 5 CY front-end loader is found by multiplying 5 CY by the bucket's fill factor in a given material. T F ?

065 Of all excavated material to be hauled, solid rock swells the most from BCY to LCY. T F ?

066 Factors that increase the load on vertical concrete forms are: rate of placement, slump, vibration and temperatures. T F ?

067 The maximum design pressure for a concrete wall-form design need not exceed the unit weight of concrete times the height of the pour. T F ?

068 The design of shores supporting concrete deck and beam pours falls into three length classes; short, medium, and long. T F ?

069 The two major factors to consider when calculating a shore's supporting capacity are its un-braced length, and least sectional dimension. T F ?

070 The three basic forms of heat involved in HVAC calculations are: sensible heat, latent heat, and radiant heat. T F ?

071 The British thermal unit is the unit used in the United States as the measure of heat gain and heat loss in HVAC calculations. T F ?

072 The percent difference between the two readings of a sling psychrometer in a space is the relative humidity in that space. T F ?

073 Undersizing or on-the-nose sizing of HVAC equipment not only lowers initial cost but also reduces life-cycle cost. T F ?

074 Optimum comfort for occupants of a building is approximately 70°F DB and 40% RH. T F ?

075 An adiabatic change in air characteristics occurs when water vapor is added without an accompanying change in air temperature. T F ?

076 To obtain desired supply air quality, air is often cooled and then reheated. T F ?

077 Infiltration of outside air through a building's envelope is not counted as part of a building's make-up air requirements. T F ?

078 One air-change-per-hour (ACH) means that all current air in a ventilated space is replaced within an hour. T F ?

079 The temperature of the water entering the cooling coil of an HVAC chiller must be less than 35°F. T F ?

080 In HVAC systems, the transport characteristics and capabilities of round ducts are better in all respects than rectangular ducts. T F ?

081 Low-velocity duct-work transports air at less than 2500 fpm. T F ?

082 The design of HVAC systems is heavily influenced by the provisions of fire codes. T F ?

083 A fire-tube boiler is one where fire and hot gasses circulate around the tubes and water flows through the tubes. T F ?

084 Enthalpy is synonymous with total heat; the sum of latent heat and sensible heat. T F ?

085 Absorption refrigeration machines depend on the use of dry lithium bromide and a heat source for their operation. T F ?

086 The water pressure in municipal water systems is generally between 70 psi and 80 psi at a building's meter. T F ?

087 The water pressure required for the normal operation of the common building plumbing fixtures is 40 psi at the fixture. T F ?

088 In tall buildings, it is not generally permissible to use the same tank to store domestic water and fire protection water. T F ?

089 A criteria for stored water in sprinkled buildings is sufficient volume to control the fire until fire fighting units arrive. T F ?

090 The purpose of ground-level siamese fire connections is to allow fire-fighting equipment to tap into the building's fire protection system. T F ?

091 Storm water and sanitary waste can use one common building drainage system if the common drain is properly vented. T F ?

092 Roof drainage design is usually based on the maximum rainfall, in inches per hour, for the area in which the building is located. T F ?

093 Electrical power transmission lines carry high voltage electricity because high voltages are cheaper to transmit than lower voltages. T F ?

094 An AWG 18 gage conductor is physically larger than an AWG 10 gage conductor. T F ?

095 The use of bus-ducts is popular in industrial construction because of their flexibility and ease in hooking up electrical equipment. T F ?

096 The colors produced by light fixture sources is an important factor when the architect selects interior finishes and colors. T F ?

097 The design of a vertical transportation system is accomplished by using a performance specification issued by the architect. T F ?

098 An important factor in the design of elevator systems is the lobby waiting time. T F ?

099 The architect or engineer hired by the owner to provide design services has the legal responsibility for all design decisions. T F ?

APPENDIX B

A Model Program for the Certification of Construction Managers

In 1984, the CMAA installed a committee to study the subject of "the certification and registration of construction managers." The report, in the form of a certification model, was completed in 1986. It was determined that registration (legal licensing) is a consideration of state governments and may occur in the future, but the current effort should be certification.

Two certifications options were investigated, one for individuals who made CM their career and another for firms that provide CM services to clients. Firm certification was dependent on individual certification; individual certification stood on its own. A model was created for each option.

It was suggested that an individual's proficiency be tested in the twelve areas of knowledge (Chapter 9, The CM Body of Knowledge) with certification granted in as many areas as the individual successfully pursued. Firm certification was mainly dependent on having certified individuals on staff in all twelve areas.

The following is a review of the model proposed for CM firm certification. It provides additional insight into the practice of construction management as presented in this text.

B.1 MODEL OVERVIEW

The model provides certification of individuals who have chosen construction management as their career and firms that practice construction management by providing CM services to owners.

It is assumed that individuals who have chosen CM as a career will be employed by the firms that practice CM. The model bases certification of a CM firm on the mix and number of certified personnel it employs. This stacking arrangement fosters the concept that a CM is an organization and provides the key to the practical success of a CM certification program—a market is created for certified individuals.

Unlike most current construction industry certification programs where a framed certificate is the only tangible reward for becoming certified, certification is a path to employment by CM firms.

The model recognizes that construction management practice requires expertise in several areas of knowledge, and that it is possible for expertise in all areas to be vested in a single individual. However, the model assumes that under most project conditions the required construction management function cannot be physically accomplished by an individual. The model strongly supports the accepted criteria that a CM firm be a multi-discipline organization with all of the required disciplines in its employ.

Twelve areas of knowledge were established to define the CM body of knowledge. Each area of knowledge was evaluated for its contribution to the performance of credible CM services. The areas of knowledge are established in a bibliography of appropriate texts, papers, and articles, and a database pool of test questions are available to measure expertise in each area.

The test questions are compartmented into the twelve areas, and each question is assigned a specific degree of difficulty (A, B and C, with A questions the most difficult and C questions the least difficult). Tests are formulated randomly by computer selection for each area of knowledge tested and for the appropriate degree of difficulty required for a particular certification.

The content of each area of knowledge is established specifically for CM performance. However, other construction industry organizations/associations that certify individuals have knowledge areas that are similar or overlapping. Consequently, provisions are included in the model to accept certification granted by other organizations/associations as credit toward CM certification.

An individual interested in becoming certified can choose from three levels of achievement: Certification as a Specialist (resource person) in one or more of the twelve areas of knowledge; a Construction Manager (operations persons) spanning all areas of knowledge, or a Construction Manager-in-Training with an option to elevate to a higher level.

The criteria for individual certification includes education, experience, and expertise. Education and experience are demonstrated by documentation, expertise by examination.

Experience and education are prerequisites to examination. Formal education in construction-related programs above high school and experience in construction and construction-related positions combine to satisfy this requirement.

Individual certification requires periodic updating through subsequent examination in additional areas of knowledge and continuing education.

The goal of the certification program was to improve construction managers' performance by stipulating minimum requirements for individuals and firms involved in the providing of CM services and by upgrading CM firms through the continuing education of their personnel.

The model accommodates CM practice at all levels and in all forms by requiring conformance to existing standards, not a change in the standards themselves. It establishes the certification of CM firms as the key to credibility in CM practice by assuring clients that the firms they hire have competent personnel on staff.

The certification of CM firms is practical because it is based on the certification of the firm's employees. The model's "statement" is that a CM firm's capabilities to provide services can be confidently measured by the capabilities of the individuals it employs. The strength of the model is in the dual-certification concept.

Certification is awarded to CM firms based on their ability to match current construction industry criteria. The intent is to establish a firm's potential performance based on experience and available expertise. Experience takes into consideration the types and sizes of projects completed; expertise, the size, diversity, and demonstrated ability of the firm's personnel.

The awarding of certification neither guarantees nor promises that a certified firm will perform to industry standards, but rather that a firm:

- meets the criteria of certification
- has a demonstrated capability to perform to industry standards
- has the abilities listed in a data sheet that evidences a firm's certification.

When certification is granted, a certificate and a data sheet is issued to the firm. As a condition of acceptance, the firm is required to sign an agreement binding it to compliance with the certification code of ethics. Prior to each firm's certification anniversary, the firm must submit an updated self-audit form for review. The certificate is either renewed and the data sheet updated or certification is withdrawn.

Firms that achieve certified status are designated as Certified Construction Managers and can place the letters CCM after their company name. The term "construction manager" currently applies to both individual practitioners and practicing firms. The model promotes the exclusive use of the term construction manager for the *firm* providing services rather than the individuals employed by the firm. This selected usage reinforces the fact that construction management services are provided by CM firms, not individuals.

B.2 THE CERTIFICATION OF INDIVIDUALS

B.2.1 Goals and Objectives

Construction management practice is a multi-discipline service, the success of which is almost entirely dependent upon the capabilities of the individuals engaged by the construction manager. There are three factors that contribute to the individual's capabilities: expertise, experience, and education. To become certified, an individual must meet or exceed the minimum prescribed requirements in these three areas.

The areas of knowledge that contribute to the complete practice of construction management in the form of resource personnel are:

Budget Management
Contract Management
Decision Management
Information Management
Material/Equipment Management
Project Management
Quality Management
Resource Management
Risk Management
Safety Management
Schedule Management
Value Management

Effective CM practice is highly dependent on the adequacy, availability, and proper linking of all areas. Consequently, operations personnel who coordinate the output of resource personnel are a key ingredient in a CM firm.

B.2.2 Certification Categories and Titles

Fourteen separate certification categories are recognized in two equal professional titles and one internship title.

- **A "Construction Management Specialist"(CMS)** certification indicates expert capabilities in one or more of the twelve listed areas of knowledge and general capabilities in all areas.
- **A "Professional Construction Manager" (PCM)** certification indicates general capabilities in all twelve areas of knowledge as an administrator or in the field.
- **A "Construction Manager-in-Training" (CMIT)** certification indicates internship status with general knowledge of construction management practice and the capability to advance to the PCMS or PCM level.

To qualify for CMS certification, an individual must:

- Meet the minimum combined education and experience requirement
- Receive a passing grade in the "C" Level written examination testing the individual's expertise in overall CM practice or hold CMIT certification
- Receive a passing grade in the "A" Level written examination specifically testing the individual's expertise in his/her chosen area(s) of specialization.

To qualify for PCM certification, an individual must:

- Meet the minimum combined education and experience requirement
- Receive a passing grade in a "B" Level written examination testing the individual's expertise in overall CM practice.

To qualify for CMIT certification, an individual must:

- Receive a passing grade in the "C" level written examination testing the individual's expertise in overall CM practice.

B.2.3 Reciprocity

Other construction industry groups have certification programs that are equal to and compatible with some of the disciplines and areas of knowledge involved in individual certifications. These included the Society of Value Engineers, the Construction Specifications Institute, and the Project Management Institute.

Certification duly received from these and other approved organizations is accepted as conclusive evidence of expertise in comparable areas of certification. Consequently, CMS status is awarded to applicants without the "A" level examination requirement in the specialization. However, the "C" Level examination must be taken and a passing grade received.

Applicants seeking CMS status through reciprocity with SAVE, PMI, or CSI certification must submit a written request with their application, including supporting evidence of certification. Other certification reciprocity requests must be supported by documentation establishing the credibility of the certification program to the satisfaction of a Certification Administration Unit.

B.2.4 Current Reciprocity Certification Equivalents

The Project Management Professional (PMP) certification qualifies for Construction Management Specialist (CMS) status in the specialty areas of Information Management, Material/Equipment Management, and Project Management.

The Certified Value Specialist (CVS) certification qualifies for Professional Construction Management Specialist (PCMS) status in the specialty areas of Value Management and Budget Management.

The Certified Construction Specifier (CCS) certification qualifies for Professional Construction Management Specialist (PCMS) status in the specialty of Material/Equipment Management and Quality Management.

B.3 OVERVIEW OF THE PROCEDURE

Individuals seeking certification must complete an application form. Individuals can determine eligibility levels by reviewing the requirements section of the form.

Eligible applicants must complete the examination request portion of the application and forward it to the administration unit. Upon receipt, review, and approval, examination information is forwarded to the applicant, including the appropriate areas of knowledge with bibliographies and information on the next examination date and location.

After the examination(s) have been taken and processed, individuals are notified of the results. Those who pass are required to sign the conditions of certification statement committing him or her to the code of ethics and other professional obligations before receiving a certificate of certification

Individual certification as CMS and PCM is subject to renewal every three years. CMIT certification remains valid, requiring no renewal.

B.4 CERTIFICATION CRITERIA

To be eligible for the CMS or PCM and CMIT examinations, a combination of experience and education is required.

Education must be in a construction or construction-related two or four year college level program, accredited by either the ACCE or by ABET and evidenced by a valid Associate or Bachelor diploma or Technician certificate.

Education credit will not be given for incomplete programs or programs less than two years in length. Experience credit will be given in lieu of education credit for time spent in incomplete and less than two year construction or construction-related programs.

Experience must be in the construction industry or (upon exception) in a closely related field. Experience with a construction management firm, contractor, architect, engineer, or as an owner's representative on construction projects equally qualify.

Experience may substitute for all or part of the minimum education requirement. The experience equivalent of a two-year associate degree program is three years; a two-year technician program, two years; and a four-year undergraduate bachelor degree program, six years.

The minimum experience requirement for certification is one year at the time application is made, education and experience equivalents not withstanding.

As examples, a four-year bachelor degree in an accredited construction program and one year of acceptable experience meets the minimum requirements. Seven years of acceptable experience without applicable construction education, or a two-year associate degree and four years experience, also meet minimum requirements.

B.5 THE INDIVIDUAL'S CERTIFICATION ALGORITHM

Enter the points applicable to your education and experience and enter the total. Education points only apply to ONE of the four education programs listed.

Four-year Bachelor degree:	6 points
Four-year Associate degree:	5 points
Two-year Associate degree:	3 points
Two-year Technician Certificate:	2 points
Experience, per 12 month period:	1 point

Education Points: ____________

Experience Points: ____________

Total Points: ____________

Requirement for CMS and PCM Application: 7 or more total points.

Requirement for CMIT Application: 1 or more total points.

B.6 INDIVIDUAL CERTIFICATION RENEWAL

Certification must be renewed every three years on the anniversary date of original certification. A renewal form is automatically sent to the individual by the administration unit. The person completes the form by providing the requested information and returns it to the unit.

The requirement for renewal is evidence of professional improvement and updating during the three previous years. The requirement can be satisfied in any one of four ways:

- By becoming certified in one or more additional specialist areas of knowledge
- By receiving completion certificates from four or more short courses
- By earning 4.8 continuing education units (C.E.U.'s) in a course or courses approved for credit

- By formally participating in a construction or construction-related higher education program for at least one semester or quarter.

Individuals that fail to renew their certification as stipulated above can renew retroactively upon evidence that the professional improvement requirement for the elapsed period of time was met. Certifications not renewed within one year of the individual's renewal date shall be noted as inactive on the official records.

B.7 THE CERTIFICATION OF CONSTRUCTION MANAGEMENT FIRMS

B.7.1 Goals and Objectives

Firms that practice CM depend on four attributes to provide high performance levels of service.

1. Abilities of individuals that provide services
2. Management systems used to provide services
3. Experience gained on past projects
4. Basic philosophy of services provided.

A CM firm must establish its capabilities through the abilities of the individuals it employs to become certified. The firm's term of certification is then dependent on its capacity and expertise in the three remaining areas. Certification establishes that the firm:

1. has met the criteria for certification
2. has the potential to perform to established standards
3. has specific performance capacities and abilities.

Certification is neither a guarantee nor promise that a certified firm will perform to established standards. Membership in the certifying organization is not a prerequisite for certification, nor does it influence the granting of certification.

B.8 OVERVIEW OF THE PROCEDURE

Firms applying for certification must request and complete an application form. The criteria and requirements are a part of the application. When completed, the form decisively determines certification status.

Continuing toward certification evaluation is at the applicant's option. If the applicant meets the requirements and desires certification, the application is forwarded to the Certification Administration Unit. If upon review all is in order, an audit date and itinerary are set for a visit to the firm. Discrepancies found during the visit of the audit team must be resolved.

An audit report is sent to the firm shortly after the visit, indicating the approved term of certification and offering suggestions for possible improvement. Certification terms extend from two to five years after which they require renewal.

Certification is evidenced by a certificate with an expiration date, a certification symbol that may be used on the firm's printed materials, and a standard detailed data sheet attesting to the firm's capabilities. As a condition of acceptance, the firm must subscribe to the code of ethics and the conditions of certification.

B.8.1 Certification Criteria

1. The Firm's Organizational Expertise

   The firm must have at its immediate disposal sufficient personnel, that collectively possess the spectrum of expertise specified in the prescribed areas of knowledge, to execute projects of the type, size, and complexity they undertake under the terms of the CM services contracts they enter into.

   The expertise of a firm shall be measured in terms of the combined qualifications of its entire complement of personnel with credit given for certified individuals, professional status, and advanced education.

2. The Quality of the Firm's Management Systems

   The firm must have a full complement of well-developed management systems and procedures available to its personnel to permit the efficient and effective accomplishment of its responsibilities, in accordance with the form(s) and variation(s) of construction management the firm provides to owners.

   The level of development of the management systems and procedures shall be measured by assessing: (a) documentation in procedure manuals or in other descriptive materials, (b) actual examples of system outputs, and (c) general descriptions of the firm's systems and procedures.

   The adequacy of the management systems and procedures shall be measured by correlating the firm's available systems with the systems appropriate to the CM service(s) provided by the firm.

3. The Firm's Applicable Experience

   The firm must be able to document its applicable experience in sufficient detail to adequately establish its expertise in the form(s) and variation(s) of construction management in which it requests certification.

   Applicable experience shall be restricted to that which applies to the firm's specific certification request but shall also reflect the basic disciplines that contribute to the overall expertise of construction management practice.

   Experience details shall include project numbers, types, sizes and complexities, and any unique services in the experience of the firm.

4. The Firm's Basic CM Philosophy

   The firm must be able to clearly state its philosophy with regard to providing construction management services within the context of the form(s) and variation(s) of CM it practices.

The firm shall describe its approach to the various aspects of CM services within the context of how specific contractual responsibilities are shared, assigned, accepted, and carried out by the project team.

The firm must have sufficient personnel to execute its philosophy as measured by proportioning the number, size, and complexity of the firm's ongoing projects.

B.9 THE CM FIRM CERTIFICATION ALGORITHM

ORGANIZATION: Certified Personnel

(Enter the number of currently active certified personnel in each specialization under the applicable column.)

Areas of Knowledge	Directly Employed		Otherwise Available *		
1. Budget Management	______		______		
2. Contract Management	______		______		
3. Decision Management	______		______		
4. Information Management	______		______		
5. Material/Equipment Management	______		______		
6. Project Management	______		______		
7. Quality Management	______		______		
8. Resource Management	______		______		
9. Risk Management	______		______		
10. Schedule Management	______		______		
11. Safety Management	______		______		
12. Value Management	______		______		
13. PCM certified	______		______		
14. CMIT certified	______		______		
Total Certified Personnel:	______	+	______	=	______

* Personnel listed in the "Otherwise Available" column must be under a written formal agreement for services with the Applicant to be considered for credit. Evidence of the arrangement must be provided to the Audit Team at the time of their visit.

REQUIREMENTS:

(1) $\dfrac{\text{Total Certified Personnel:}}{\text{Total Operations And Resource Personnel:}}$ ______ = ______ (Must be greater than 0.80)

(2) Areas of Knowledge 1 through 12 *must be represented* by *certified* personnel for a firm to be certified.

ALL PERSONNEL

Account for all personnel in the firm's employ by listing them in the appropriate category. List each person *once* even though he/she qualifies in more than one category. Select the most advanced level of each person when listing them in a category.

OPERATIONS and RESOURCE (O&R) Personnel: (those directly involved in providing CM services to clients).

Registered Arch. or Engr.	______	× 10 =	______
Masters Degree, Arch. or Engr.	______	× 6 =	______
4yr Undergraduate Degree, Arch/Engr.	______	× 4 =	______

4yr Undergraduate Degree, Construction	______	× 5 =	______
4yr Assoc. Degree, Arch/Engr/Const.	______	× 4 =	______
2yr Assoc. Degree, Arch/Engr/Const.	______	× 2 =	______
Masters Degree, Business Admin.	______	× 4 =	______
4yr Undergraduate Degree, Bus. Admin	______	× 2 =	______
Other O&R personnel not included above	______	× 1 =	______
Total Operations and Resource Personnel Points:		=	______

Support and Administrative (S&A) Personnel: (those providing support for O&R personnel in the business aspects of the firm).

Total Support and Administrative Personnel: = ______

TOTAL PERSONNEL = ______

REQUIREMENT:

(3) $\frac{\text{Total Operations and Resource Points:}}{\text{Total Operations and Resource Personnel:}}$ ______ = ______

(Must be greater than 4.0)

B.10 FIRM CERTIFICATION RENEWAL

The term of each firm's certification varies according to the audit team's evaluation of the firm's in-place management systems, experience, and philosophy. The maximum certification term before renewal is five years; the minimum term before renewal is two years.

A certified firm must apply for renewal no later than the expiration date indicated on the firm's certificate. A renewal application is sent to the firm by the Certification Administration Unit about six months prior to the expiration date. The firm is to complete the form and return it to the unit.

The certification renewal form for firms previously certified for a term less than the maximum will specifically address the status of the suggested upgrading noted by the audit team during their visit to the firm. An indication of progress in these areas and compliance with the mandatory certified personnel requirements is necessary for renewal.

Firms previously certified for less than a five year term, and firms not visited by an audit team within the previous five years, will require an audit visit as part of any current certification renewal process.

Firms that fail to renew their certification within one year of the expiration date shall be listed as uncertified on the records of the certifying organization. A firm that allows certification to lapse for a period of one year or more must seek reinstatement through the established application procedure for new firms.

B.11 EVIDENCE AND USE OF CERTIFIED STATUS

Evidence of a firm's certified status would be in the form of a Data Sheet verifying that the firm meets the staffing requirements and has the necessary systems and philosophy in place to provide CM services. The Data Sheet would also include a list of completed projects to support experience requirements.

The Data Sheet would be a standard form issued by the certifying body based on information provided by the firm and substantiated during the audit process. It would include:

- Firm identification information
- Business organization information
- Personnel data listing certified persons
- Experience information
- CM services philosophy
- CM systems information
- Work load capacity information.

Data Sheets can be sent in response to requests for proposals from owners to establish CM capability and provide credibility to the other information requested from the owner.

In addition to the data sheet, the firm would be furnished a "certified firm" logo for its letterheads and business cards to quickly establish the firm's credibility as a CM practitioner. The logo could only be used during a firm's certified tenure.

B.12 THE SOURCE OF THE MODEL PROGRAM

The model produced in 1987 and issued in an abridged form in 1988 is essentially the product of the Construction Management Studies and Research Group (CMSRG), a part of the Construction Engineering and Management program in the Civil and Environmental Engineering Department at Michigan Technological University. Contributing to the effort were members of the CMAA Registration/Certification Committee. The combined group included Kristine Sanders, David Still, David McAlvey, Linda Phillips, Merlin Kirschenman, Derek Calomeni, and Craig Marriner. C. E. Haltenhoff chaired the work of the CMAA committee and the combined committee.

APPENDIX C

An Example of Innovative Contracting on a Complex Project

C.1 THE SITUATION

A manufacturing corporation based in Europe needs to extend its operations in the United States. A 120 acre site in the Midwest has been selected for the project. The firm already has a raw materials plant in the Southwest that was put into operation two years ago and now wants to build a new raw materials facility and a finished product facility on the midwest site.

Financing is not a problem and budget limitations appear liberal. However, time is of the essence. It is important that the facility be in operation within 22 months to gain a competitive edge in the U.S. market.

This is a tight schedule, based on the 24 months it took to design and build the raw materials plant in the Southwest and the 17 months it took just to construct the most recently finished product plant the company built in Europe. Additionally, several existing structures on the site must be demolished, and considerable earth moving, road building, and site utilities work is required on the 120 acre site.

The company hired a construction manager to perform a contractability study and provide management services for the contract management plan that resulted from the study.

The initial constructability meeting between the company (owner) and CM produced the following information.

C.1.1 General

Owner

1. The owner has competent process design personnel in-house at their European headquarters who will be available for design review on the U.S. project.
2. The owner will only dedicate three employees full-time to the project at the site: an executive person (Level 1) with authority to approve expenditures for originals, modifications and change orders up to $500,000 without prior approval from Europe; a management person (Level 2) experienced in company policy

(public relations/publicity/legal) who can make owner decisions; a technical person (Level 2) who is experienced in design and construction of the facilities being constructed.

3. The owner will provide a technical start-up team to inspect and commission the facility and has all necessary use-permits in hand.
4. The owner has no objection to multiple prime contracts, fast-track or phased construction, or any combination thereof.
5. The owner prefers to make a single payment on the 15th of each month to cover all billings for the project received at the owner's office at the site prior to the 10th of that month. Billings must be supported by invoices.
6. The owner will purchase and deliver all major, special, and proprietary process equipment for the project to the site but wants to use local contractors to install it.
7. The owner requires on-site inspection of construction in addition to installation permits and certifications and laboratory testing.
8. The owner requires complete standardization of mechanical and electrical equipment on this facility and with the existing raw materials plant in the Southwest.
9. The owner has no existing affiliation with building trades labor in the U.S. and has no preference regarding the use of union or merit shop labor on the project.

CM

1. The CM has found that there is a significant labor pool (both union and merit shop) to meet construction needs and that competent general contractors and trade contractors (both union and merit shop) are available locally to meet the schedule and provide competition.
2. The CM has found that there are no major trade contracts that will expire during the next two years and that unions are willing to sign project agreements with the owner if necessary.
3. The CM reports that local authorities will grant building permits on partial contract documents in support of phased and fast-track construction.

C.1.2 Raw Materials Plant

1. The owner explained that the design and construction of the existing raw materials plant was a combined effort between the owner's engineers, a local engineering firm, and a local union general contractor. It was paid for under separate cost-plus-fixed fee arrangements. Design was completed and approved before construction began. Design required three months, construction 20 months, and commissioning one month, for a total of 24 months from the start of design to full operation.
2. All drawings and specifications are the sole property of the owner and are protected by a confidentiality agreement with both the engineer and the contractor.

The drawings and specifications are filed at the owner's headquarters in Europe and are available for reference purposes on this project.

3. The new raw materials plant will be similar but not identical to the existing plant. The size will be the same, however, the location of the new plant will require redesigned electrical and mechanical systems, and code differences will change some architectural and structural components.

C.1.3 Finished Product Plant

1. The finished product plant will be a duplicate of the last one built by the owner in Europe.
2. The plans and specifications are available, but the dimensions, sizes, and capacities are metric, and some building materials, products, and systems are not available or compatible with construction in the U.S.

C.2 THE INITIAL CONTRACTING STRATEGY

C.2.1 For the Raw Materials Plant

Use Design–Build. This contracting system can produce a project faster than any other. As well, there are other advantages: (1) the southwest raw materials plant was a "path-finding" form of design-build, (2) the new plant will be similar, and (3) the existing plans are available for reference. The overall delivery time of the Southwest plant can probably be reduced by three months or more.

Contact the contractor who constructed the Southwest plant. Inquire if there is an interest in designing and building the raw materials plant in the Midwest. The contract arrangement would be a design–build joint venture, involving the original engineer and contractor. If the joint venture wants to involve a local engineer or general contractor, there would be no objection.

Design would be reimbursed on a cost-plus-fixed-fee basis. When design is sufficiently complete, or at the end of 30 days, whichever occurs first, the joint venture would negotiate a GMP for the plant's construction. If the owner approves the GMP, on-site construction should start within ten days of notification.

If the GMP is not satisfactory, the owner would have the right to contract construction to others. The joint venture would be obligated to complete the design and provide complete drawings and technical specifications suitable for competitive bidding by the owner within 60 days after notification of rejection.

C.2.2 Finished Product Plant and Site Development

Use modified General Contracting. Adapting the design documents used in Europe should take no more than three months. This would permit one month for bidding and allocate 18 months for construction. Using modified general contracting (separately bidding general, mechanical, and electrical contracts) instead of CM multiple contracting would reduce the number of contracts held by the owner, simplify contract administration, and make better use of CM personnel and the owner's limited cadre of on-site personnel. In the absence of budget restrictions, general contracting is an acceptable trade-off.

Contract with a full service architectural/engineering firm (one with in-house architectural, structural, electrical, and mechanical design capabilities) to modify the metric documents and produce bid packages for the general work, the mechanical work, and the electrical work.

This firm should also have responsibility for site design including demolition, mass excavation, roads, parking and drainage, and produce separate bid packages for demolition, mass excavation, roads, parking and drainage structures.

Proposals for these work-scopes should be competitively bid by contractors selected on their qualifications. Dollar proposals should be lump sum for each work-scope contract. The contracts for the mechanical and electrical work should be assigned by contract to the selected general contractor for construction coordination but not contract administration purposes. The CM should coordinate the demolition, mass excavation and roads, parking and drainage contractors.

C.2.3 Design Coordination and Site Utilities

The owner's requirement for standardization of mechanical and electrical equipment, on this facility, and with the existing raw materials plant in the Southwest, will require comprehensive and timely design coordination of the finished product plant engineering firm and the design partner of the design–build contractor.

To effect this, a local mechanical/electrical engineering firm, or a mechanical and electrical joint venture, should be hired whose responsibility will be to ensure standardization. This firm should also design the electrical and site mechanical needs and produce bid packages for electrical site utilities and mechanical site utilities.

C.2.4 Testing and Inspection

The nature of the projects—processing plants—requires considerable high-tech, on-site testing. To fill this need, an independent testing and inspection company will be hired by the owner.

C.2.5 Summary of Contracts

	Construction Contracts	**Design Agreements**
Raw Materials Plant	1 Design–build	X
Finished Product Plant	3 Lump sum GC/TC	1 Architect/Engineer
Site Development	5 Lump sum TC	1 Architect/Engineer
Design Coordination	X	1 Mechanical/Electrical Engineer
Construction Inspection	X	1 Inspection Firm
Construction Management	X	1 Agency CM Firm

C.2.6 The Role of the Construction Manager

Provide:

Budget	Quality
Contract	Resource
Decision	Risk
Information	Safety
Material/Equipment	Schedule, and
Project	Value Management, appropriate for a multiple contract project.

C.2.7 CM On-Site Personnel

1 Executive Level Person
1 Management Level Person
2 Administrative Level Persons
2 Scheduling Persons
1 Receptionist/Word Processor
1 Support Person/Computer Operator

C.2.8 The On-Site Organization Chart

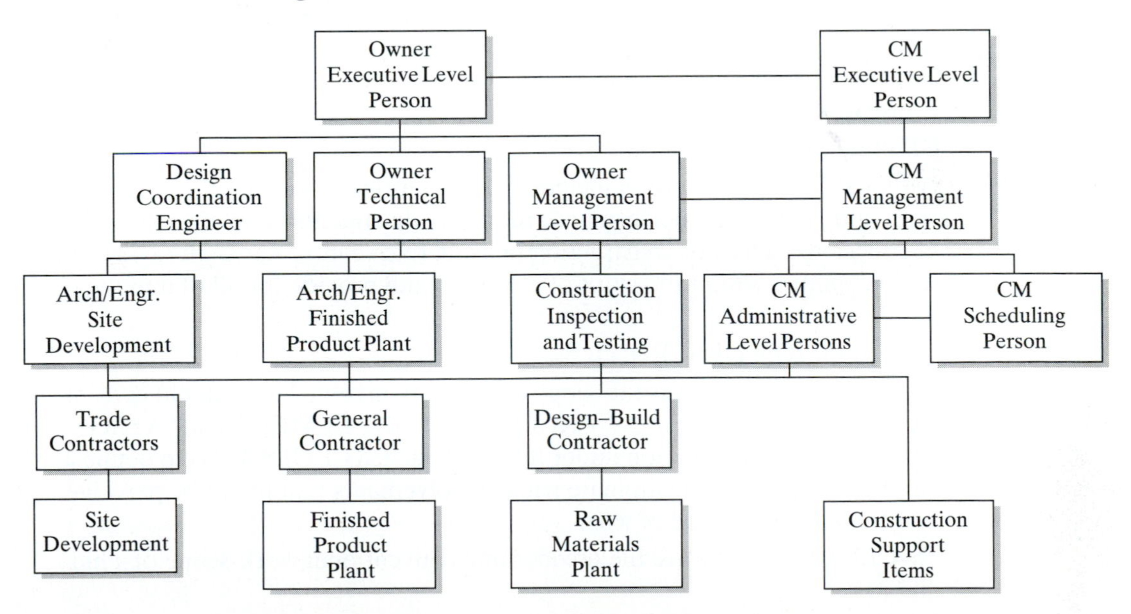

FIGURE C.1 The On-Site Organization Chart.

APPENDIX D

Some Contract Provisions Used to Convert Single Prime to Multiple Prime Contracts

The following is an example of contract provisions that were used to convert standard contract documents for a GC project into a multiple bid CM project before standard CM documents were published in 1975. These provisions vividly demonstrate the significant differences between the GC system and the CM system.

D.1 PROPOSAL SECTION

A. ATTENTION ALL BIDDERS

A.1 This is a Construction Management Project. There is no General Contractor. All contractors on this Project are considered prime contractors. The Owner will award separate contracts for all Bid Divisions (defined work-scopes) involved in the Project. The Project will be controlled and coordinated by a Construction Manager.

B. BID DIVISION UNIT

B.1 Although each Bid Division includes a conventional segment of subcontracting, multiple-contract performance requires adjustments be made so that each bid division is a unit of construction in itself. Each contractor shall review the responsibilities within the work of a division and provide for all of it in the proposal.

C. BID DIVISION DESCRIPTION

C.1 For clarification purposes the scope of the work involved in each bid division is specified in three categories: "EXCLUDED," "INCLUDED," and "ALSO INCLUDED." Information under the heading "EXCLUDED" is to define a point of beginning and eliminate fringe involvements that might be perceived as included in the scope of work.

C.2 "INCLUDED" items are the obvious and conventional work-scope of a bid division.

C.3 Information under "ALSO INCLUDED" identifies unconventional and less obvious items of work included in the division, as well as the fringe involvements that could inadvertently be missed in evaluating the scope of the work.

D. MANDATORY INTERFACES

D.1 The scope of each contractor's work is defined in the Division(s) selected to be bid. Each contractor shall become familiar with the requirements of all Bid Divisions that interface with the one(s) selected to be bid. Consideration shall be given to the fact that the work of one contractor follows the work of another and that the work of several contractors could interface with one Division.

E. WORK ASSIGNMENTS

E.1 Nothing contained in the contract documents, and especially in the work-scope of a Bid Division, shall be construed as a work assignment to any trade. Each Bid Division contractor is responsible for work assignments and shall make them in accordance with the practice in the area of the project; and in such a way that scheduled progress shall not be adversely affected by the decisions.

E.2 Disputes which arise from improper assignments or assignments claimed by more than one trade shall be immediately settled by the Bid Division contractor and shall in no case result in a slow-down or stoppage of work of any contractor on the project.

F. PRE-BID MEETINGS

F.1 Meetings with all interested bidders shall be held after all documents have been distributed and before the bid date. The purpose of the meetings is to answer all questions generated by the contractor's initial review of the bidding documents. The Owner, Architect, Engineer, and Construction Manager shall be on hand to answer questions. The schedule of meetings will be sent out by the CM. All Bidders are urged to attend.

G. BIDDING AND AWARDS

G.1 All bidders must submit their proposals on the form provided. Failure to do so will jeopardize the offerers' chances of receiving an award.

G.2 There is no limit as to the number of Bid Divisions any one contractor can bid. However, each contractor is required to enter a figure for each and every Bid Division being bid in order to be considered for award in that division. Space is provided in the Proposal Form for a combined proposal deduct if a contractor bidding more than one Division wishes only to be considered on a combined bid basis.

G.3 The award of contracts will be based on the dollar value of the proposal, the qualifications of the contractor, and his ability to perform. Bidders are cautioned to fill in all blanks on the pages of the proposal he is submitting by noting "N/A" in those blanks not applicable to their particular proposal.

H. ACCEPTANCE OR REJECTION OF BIDS

H.1 The bidder acknowledges the right of the Owner to accept or reject any and all bids and to waive any informality or irregularity in any bid received.

H.2 The bidder acknowledges the right of the Owner to accept any combination of Bid Division proposals.

H.3 The bidder, by submitting a bid, represents that award will be accept if offered, regardless of who the other Bid Division contractors may be.

H.4 The bidder further represents that neither his Work nor the Work of other Bid Division contractors will be prejudiced because of sex, race, color, creed or labor affiliation of other contractors under Contract to the Owner on this project.

I. MILESTONE CONSTRUCTION SCHEDULE

I.1 Proposed Milestone dates for the Project have been developed by the Construction Manager, using a computerized Critical Path Method (CPM) approach. Each bidder is required to review the dates and their implied time spans and endorse or amend them for work being bid. *A space is provided on the Proposal Form for the bidder's endorsement or amendment.*

J SCHEDULE ENDORSEMENT OR AMENDMENT

J.1 The Milestone Schedule as endorsed and/or amended by the successful bidders and accepted by the Owner will be used as the basis for the construction schedule by which the project will be built.

J.2 The effect of any amendment to the schedule shall be considered when selecting a contractor for performance of the work. Bidders are obligated to comment on the proposed schedule if in their opinion it does not realistically depict the sequence and/or time interval for performance of the work in their Bid Division(s).

K. OWNER PURCHASED MATERIAL AND/OR EQUIPMENT

K.1 As an expedient, the Owner will purchase certain material and equipment to be incorporated into the work by designated Bid Division contractors. The designated contractor shall accept delivery, unload, handle, store, and install the items.

K.2 Upon delivery, the designated contractor shall verify product suitability, quantity, quality, and condition and shall accept full responsibility for care, custody, and control. Delivery status shall be reported to the Field CM.

L. RETAINAGE ON OWNER PURCHASED ITEMS

L.1 The Owner may retain five thousand dollars ($5,000.00) or ten percent (10%) whichever is the lessor, on material/equipment purchased from suppliers for inclusion in the work, until it is satisfactorily installed. The purpose of this provision is to ensure proper conformance to the plans and specifications.

M. PROMPTNESS OF EXECUTION

M.1 It is the intention of the Owner to complete the Project in the shortest practical time. The conditions inherent to construction notwithstanding, it is the intent of this contract that each contractor maintain the progress established in the Milestone Schedule and Short Term CAP.

N. PERFORMANCE OF THE WORK

N.1 All contractors shall diligently provide input to the Short Term Contractor Activity Plan (CAP); the day-to-day schedule that moves the project from milestone to milestone. It is the obligation of all contractors to provide resources and manpower to meet their CAP commitments.

O. PROGRESS PAYMENTS

O.1 The Owner intends to favor contractors who adhere to the CAP. Contractors who maintain committed CAP progress shall be eligible for Progress Payments without a retainage deduction.

O.2 Contractors who fail to maintain committed progress shall be subject to retainage, on the full amount of progress payments due, at the date the Progress Payment is scheduled to be submitted by the contractor. The CM shall determine a contractor's schedule status.

P. PAYMENT FOR STORED MATERIALS

P.1 To offset escalation on short-lead material/equipment items purchased by contractors, and ensure the receipt of competitive bids, the Owner will provide payment for contract items stored off-site as well as at the site of the work.

P.2 To qualify for payment, material/equipment shall be properly stored, protected and insured against loss or damage, inspected by the CM, and dedicated to this project only. The cost of off-site storage shall be included in Bids Division proposal.

P.3 Material/equipment stored on-site, within project property limits, shall be in the area designated by the CM. The Owner takes no responsibility for material/equipment lost through theft or mishandling; they shall be replaced by the contractor without cost to the Owner.

Q. CONTRACTOR'S COST BREAKDOWN

Q.1 Each contractor's Schedule of Values shall include three mandatory line items; Housekeeping, Final Clean-Up, and Punch List.

Q.2 Progress Payment percentages shall be applied to Housekeeping in accordance with the contractor's carrying out of this contract responsibility.

Q.3 Progress Payments shall not be requested for Final Clean-Up until this item has been 100% accomplished in terms of the contract.

Q.4 The value of the Final Punch List line item shall be 10% of the contract value, or for certain contractors, an amount between 2% and 10% as determined to by the Owner.

R. QUALITY CONFORMANCE

R.1 Quality Conformance is the responsibility of the contractor performing the work. Each contractor shall inspect his work daily. Inaccurate, faulty, defective, and uncompleted work shall be corrected or completed by the contractor when it is observed to be so.

R.2 Contractor questions pertaining to quality shall be addressed to the A/E either directly or through the Field CM. Neither the CM nor A/E is responsible for determining quality conformance, except during each contractor's Punch List process.

S. INTERFACE QUALITY CONTROL

S.1 When a contractor determines that the quality of his work is in jeopardy as a result of the schedule or the coordination of the project, or for any other reason, the contractor shall immediately stop work, and just as immediately, inform the CM of his action and the reasons therefore.

S.2 If the situation is not satisfactorily corrected by immediate action of the CM, the contractor shall reduce the situation to writing on the same day the stop-work action occurred and submit it to the CM with a copy to the A/E. Upon subsequent investigation by the CM and A/E, a decision shall be made and the contractor shall proceed with his work under the terms of the decision.

S.3 If the contractor finds no satisfaction in the decision, he is entitled to pursue satisfaction under the dispute resolution terms of the contract.

T. MATERIAL AND EQUIPMENT EXPEDITING

T.1 The CM will initiate and coordinate an expediting program in cooperation with each contractor that incorporates all critical items of material/equipment provided by Bid Division contractors.

T.2 To ensure timeliness and accuracy, each Contractor shall cooperate by providing order and acknowledgment documentation (without pricing) as required by the CM.

T.3 Contractors shall further cooperate by keeping the CM informed of any and all changes in the delivery commitments incorporated into the expediting program, and when deemed necessary by the CM, provide material/equipment supplier contacts for direct use in expediting by the CM.

T.4 The CM expediting program is a back-up program and shall not relieve Contractors from their performance responsibilities prescribed elsewhere in the contract documents.

U. PROTECTION OF THE WORK OF OTHERS

U.1 When interfacing with work of others, contractors shall protect other contractors' work as if it were their own. Care shall be taken not to damage work of others in any way.

U.2 When it is necessary for a contractor to move personnel or transport materials or equipment across floors, grades, roofs, or other surfaces, that contractor shall provide appropriate surface protection to prevent damage to those surfaces.

U.3 The contractor that damages the work of others to the point where, in the opinion of the A/E, repair is required, shall fully compensate the offended contractor for repairing the damage.

V. LAYOUT AND MEASUREMENTS

V.1 Layout and measurements pertaining to the work of a Bid Division is the Bid Division contractor's responsibility.

V.2 Each contractor shall examine the condition and dimensional accuracy of the work his work is reliant upon before beginning work. Flaws and inaccuracies shall be reported to the CM the same day they are discovered and the contractor shall not proceed with his work until corrections are made or the situation resolved by the A/E.

V.3 If a contractor inadvertently or knowingly proceeds with his work on a contractor's work that is flawed or inaccurate, the contractor who proceeds shall be liable for all costs of correction of his work when the situation is resolved.

W. MANDATORY ATTENDANCE AT MEETINGS

W.1 It is the responsibility of each Contractor to be appropriately represented at every Project Meeting, Progress Meeting, and special meetings when requested to do so by the CM, A/E, or Owner.

X. PUNCH LIST PROCEDURE

X.1 When a contractor's work is 95% complete, a blank Certificate of Substantial Completion will be forwarded to the contractor by the CM. The contractor shall fill out the Certificate and designate a future date, the Date of Substantial Completion. The contractor shall attach (1) a list of work that requires correction, (2) a list of work that is in complete, and (3) a Punch List Schedule that demonstrates that the lists can be completed in 30 calendar days from the Date of Substantial Completion and return it to the CM.

X.2 The CM and A/E will verify the lists and schedule, amend them if necessary, and return them to the contractor. When the CM and A/E agree with the contractor's lists of punch list work, the Certificate of Substantial Completion will be forwarded to the Owner for signature and returned to the contractor.

X.3 The contractor shall begin punch list work no later than the Date of Substantial Completion. Contractors failing to meet progress as defined in their Punch List schedule are subject to termination with two working days notice, to permit another contractor to finish the contractor's work, as well as other contractors to finish their work, on schedule.

Y. PRE-ON SITE ACTIVITY MEETINGS

Y.1 Every Contractor shall meet with the CM at the site, at least one week prior to beginning work on-site. The meeting is to (1) review the technical specifications pertaining to the contractor's work, (2) review the contractor's Quality and Safety Plans, (3) discuss staffing and the construction means, methods, and techniques to be employed, (4) introduce and integrate the contractor into the CAP, and (5) generally orient the contractor.

Z. RETURN ACTIVITIES

Z.1 Each Contractor is required to report to the Field CM before starting a working day after an absence from the site of three or more working days. The purpose of reporting is to alert the Field CM to the contractor's reinvolvement and to provide an update of any conditions that could affect the work of the contractor since last at the site.

APPENDIX E

Activities During Project-Delivery Phases for the GC, D–B, and CM Systems

E.1 ACTIVITIES DURING THE NINE PHASES OF A GC BUILDING PROJECT

The sequence of activities required to deliver a building project using the General Contracting System follows the GC standard contract documents for an Architectural Project.

E.1.1 GC Feasibility Phase

This is the phase where the project is conceived and analyzed by the owner. It begins when a need for a facility is recognized and ends when a decision is made to construct or not construct. The length of this phase is dictated by the owner's requirements and timing demands.

Project definition gains focus as requirements are considered and discussed. Definition is generally given in broad parameters, such as minimum square footage of space and maximum budget expenditure. The feasibility phase is also called the Conceptual Phase or Programming Phase.

During this phase, Special Consultants will be most helpful. The Basic Services of the A/E-Owner Design Contract do not cover services during this phase. If A/E services are deemed necessary, they can be contracted for separately or included as Additional

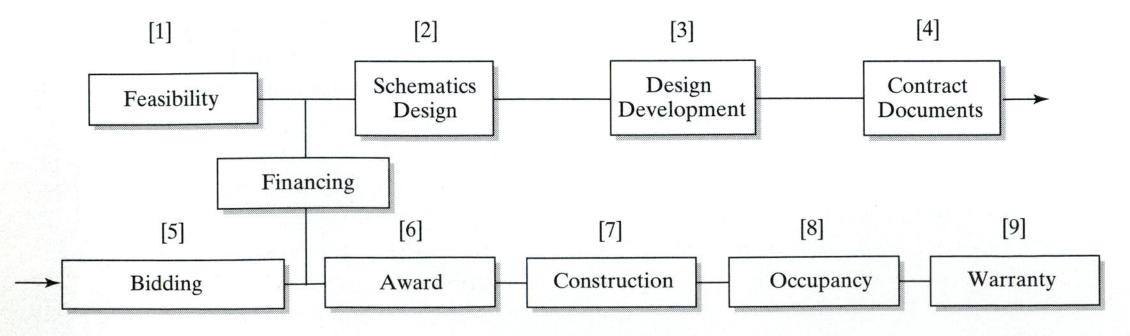

FIGURE E.1 The Nine Phases of a GC Building Project.

Services in the A/E-Owner agreement if the project moves ahead. Engineering project contracts usually include feasibility services as part of basic services due to their unique engineering requirements.

The documentation produced during Feasibility ranges from a Program Statement, with or without Sketches or Rough Drawings, to formal Conceptual Drawings and highly developed Brochures and Conceptual Models. The mandatory documentation requirement should be sufficiently detailed and reliable information from which the A/E can productively begin the Schematic Design Phase.

On many projects in both the Public and Private Sectors, the end of this phase establishes the need to locate a Funding Source for the project. This requires estimates of project costs to the degree of accuracy acceptable to the funding source. Depending on the detail of the documentation, the range of budget accuracy at the end of the feasibility phase may be 30% plus or minus.

E.1.2 GC Design Phases

The design phase of an architectural project is divided into three sub-phases: (1) Schematic Design, (2) Design Development and (3) Contract Documents. The design phase of an engineering project is divided into two sub-phases: (1) Preliminary Design and (2) Final Design. The purpose of sub-phases is to provide an Owner Review point and also a means of determining design progress for A/E payment purposes.

The sub-phases of design are loosely defined in A/E design agreements, making it difficult for anyone to review in-progress design documents and determine exactly which of two adjacent sub-phases the design is actually in. A/Es interpret design sub-phase definitions for their own convenience unless guidelines are set before design begins.

E.1.3 GC Schematic Design

On an architectural project, the Schematic Design Phase begins when the Feasibility Phase ends. Design has a different meaning to architects and engineers. Architectural Design integrates use and space combining form and aesthetics into Occupational Function. Engineering Design deals less with form and aesthetics, dwelling on the effects and forces of nature by producing or injecting design solutions with Physical Performance Credibility. Working together, architects are often involved in engineering projects; engineers are always involved in architectural projects.

Schematic design is the design sub-phase which requires the most demanding Liaison with the owner; the time when Communication between owner and design professional are most critical. The conclusions reached by the owner during the Feasibility Phase must be Successfully Conveyed to the A/E to be Accurately Incorporated into the design.

Schematic design produces Project Configuration—the shapes, sizes, and relationships of required spaces. Alternative Solutions are developed and considered by the A/E. Drawings are very preliminary, small scale (1/16″ and 1/8″) without detail. Architects often provide Sketches to show space locations and relationships. Technical

Specifications are not yet being formulated, but Design Criteria is being recorded as decisions are made on Building Systems, Materials And Equipment.

The Probable Cost Of Construction is determined and compared with the owner's budget as design definition develops in drawings and design criteria is established. The owner's Original Budget, which is often established without drawings, is the A/E's constant concern as Design Delineation and Material Selections are made. The A/E tries to Design To Budget from the beginning of the design phase. It is possible to maintain the Probable Cost Of Construction within 25% plus or minus during schematic design.

The schematic design phase ends when the A/E submits acceptable Schematic Drawings and an acceptable Probable Cost of Construction to the owner.

E.1.4 GC Design Development Phase

The Design Development Phase begins when the owner signs off on the Schematic Design Phase. Working from the approved Schematic Design, the A/E further refines the design through increased delineation and documentation. The Goal of this phase is to complete and coordinate all architectural and engineering design decisions so that final delineation and documentation can be accomplished during the Contract Documents Phase.

Drawings enlarge from 1/16″ and 1/8″, to 1/8″ and 1/4″ scale, and details are drawn in larger scales. Walls and partitions, shown as lines on Schematic Drawings, are shown in section with thickness and scale. Openings are located and dimensioned. Space is allocated to accommodate electrical, mechanical, and structural requirements. Other design disciplines are coordinated.

Building Systems decided upon during schematic design are detailed and fundamentally specified. Outline Specifications begin to take shape, based on design decisions made during the schematic phase. Building Elevations and Sections are developed to assist in dimensioning and Design Coordination.

The Probable Cost of Construction is updated based on increased design definition. Its accuracy should be in the range of 15% by the end of the phase. Designing to Cost remains a goal, although standard design agreements makes this obligation a contract option of the owner rather than an automatic requirement.

The phase is complete when the A/E submits the Design Development Documentation and an updated Probable Cost of Construction to the owner and they are accepted. Any changes the owner makes to Previously Approved Schematics are charged as Additional Services to the owner.

E.1.5 GC Construction Documents Phase

This phase is production intensive, requiring considerable person-hours to fully document the design that evolved during the previous two phases. Bidding Documents, which include Working Drawings, the Project Manual (front end and technical specifications), Instructions to Bidders, and Proposal Forms are developed as flawlessly as possible and distributed to Bidding Contractors. Alternates, if necessary, and Allowances, if necessary, are decided upon. The Advertisement For Bids is drafted by the A/E and issued in the name of the owner.

Final Drawings are 1/4″ scale with larger scales for construction details. The Drawings are sectioned to accommodate Civil Site, Architectural, Structural, Mechanical, and Electrical disciplines, and the instructional requirements for these disciplines are included in their proper section in the Technical Specifications.

Front-End Specifications, including the General Conditions of the Contract for Construction, Supplementary and Special Conditions of the contract are selected and edited for inclusion in the Project Manual.

Interested Bidders are identified and, where possible, Prequalified as to their ability. Sets of bidding documents are printed and distributed to interested GCs in exchange for a Bidding Document Deposit of a sufficient amount to cover costs if they are not returned in good condition.

Complete sets of Bidding Documents are sent to Plan Rooms to facilitate review, quantity take offs, and pricing by Trade Contractors. Usually, only GCs are permitted to obtain individual sets of bidding documents; trade contractors generally rely on Plan Rooms or other named locations, such as the offices of the owner and A/E, to get the information they need to submit proposals.

When the owner is satisfied that Contract Documents and Bidding Documents are complete, and the Probable Cost of Construction is within 5% or less of the budget, the owner signs off on the Contract Document Phase.

E.1.6 GC Bidding Phase

Bidding Phase: Overt, Competitive Bidding is the usual requirement on Public Projects. On Private Projects, owners sometimes negotiate contracts, or use a combination of bidding and negotiation. This phase could be called the Negotiation Phase, if the owner elects to use that process of acquiring a contractor.

The Bidding Phase is the first phase in which contractors become involved in the project. The beginning of the bidding phase is the end of the Design Phase. During this phase the Probable Cost of Construction is converted to the Actual Cost of Construction through the competitive bidding process.

The A/E is not free of responsibilities during this phase. There are Questions From Contractors relative to the Bidding Documents which require responses; Requests For Substitutions to act upon; Pre-Bid Meetings to organize and attend; and Pre-Bid Addenda to issue to Bidders of Record when changes are made in the Contract Documents. (This to ensure that all bidders have the same information on which to develop their proposal.) A final, thorough review of the Contract Documents must be accomplished, before the Bid Date.

General Contractors and Trade Contractors also have plenty to do. Bidding requires Construction Planning, Quantity Surveys, and Estimates of construction costs. In essence, the GC must completely build the project on paper from a construction cost perspective and do it during the brief period (several weeks on average, and dependent on the project's size and complexity) set aside for assembling proposals. The GC quantifies and estimates the cost of the work reserved for his own forces and solicits Trade Contractor Proposals for the work he intends to have done by Subcontracted forces.

Trade Contractors will get bidding and technical information from Plan Rooms (established by local Construction Associations) which have been supplied Bidding

Documents by the owner via the A/E for their use. To complete their proposals, Trade Contractors quantify and price their work and obtain quotations for Material and Equipment from various Suppliers.

Unless specifically defined, Trade Contractors determine their own work-scope, price their work accordingly, and give oral Proposals to GCs, who have the Option of Using or Not Using the quoted prices. GCs receiving trade contractor proposals must Compare them in terms of work-scope and price. Without a GC-prescribed Work-Scope, GCs have difficulty comparing Cost Proposals submitted by the same trade.

Few if any Contract commitments are made by the GCs to Trade Contractors prior to the Owner's bid date. Trade Contractor Proposals are usually Offers made by telephone. Unless explicitly Withdrawn prior to the bid date, trade contractor offers remain Valid until accepted, rejected, or there is a clear indication by the GC that different proposals are being Solicited for the same work.

Both GCs and trade contractors must be alert for Pre-Bid Addenda and, if issued, revise their Proposals accordingly. Each Bidding GC must acknowledge receipt of Addenda on their Proposal Form. If All issued addenda are not acknowledged, the GC's Bid will not be considered by the Owner.

The Location and Time for receiving proposals from GCs is first stated in the Advertisement For Bids and later established in the Instructions To Bidders. GC proposals Must be received at the appointed place before the designated time. Late Bids are not accepted. Only Responsive Proposals, those submitted in accordance with the Instructions To Bidders, will receive consideration for award.

At the Bid Opening, GC proposals are opened and read publicly, one at a time, usually in the order in which they were received. Any Award Criteria, other than Price, must be clearly stated in the instructions to bidders. Although the owner usually reserves the right to Reject Any and All Bids, Waive Irregularities, and accept the offer which is In The Owner's Best Interests, the low bid is usually accepted. A decision Not to award the project to the Low Bidder usually faces a legal challenge from the low bidder.

After all bids have been publicly read, the owner carefully reviews each one and makes a decision. The options available to the owner are to make an Award, Reject all bids, or allow the stated proposal holding time to Expire.

E.1.7 GC Award Phase

The Award Phase is a suspenseful period for trade contractors and general contractors, especially the Apparent Low Bidder. Questions on proposals may arise which require clarification, and Post-Bid Qualification of the low bidder may be suggested by the A/E. The owner must make the very important decision of whether to accept the low bidder or not. The Owner must take all of the prescribed actions which governing law and statutes specify and consider the advice of the A/E.

When an Award decision is made by the Owner, action can be taken to enter into the Contract For Construction. It is sometimes convenient or expeditious for the owner to issue a Notice of Award letter to the contractor stating any pre-conditions, such as providing Labor and Material/Performance Bonds and Insurance Certificates.

A Notice to Proceed may also be issued by the owner in the event the signing of the Contract For Construction will be delayed due to a technicality. The Notice to

Proceed usually imposes conditions on the contractor to ensure against any potential problems before Insurance, Bonds, or Funding have been arranged.

Bid Document Deposits are exchanged for unmarred sets of drawings returned by unsuccessful bidders, and Bid Bonds (or other forms of Bid Security) are returned to them.

The Awarded Contractor proceeds to Buy Out the project, mobilize forces, and commence construction. The Contract Time starts when the contractor begins operations as required in the Contract For Construction or the Notice to Proceed.

E.1.8 GC Construction Phase

The GC has control of the site and the project according to the terms of the Contract For Construction during the Construction Phase. As an Independent Contractor, the GC is responsible for completing the project On Time, In Conformance with the Technical Specifications, and for the Bid Price stated in the Proposal.

The A/E has responsibility for Contract Administration during the construction phase. This consists of Checking Shop Drawings, Product Data and Samples, Processing Change Orders, Certifying Progress Payments, and serving as the exclusive Liaison between the contractor and the owner.

There are other responsibilities required by the Agreement. The A/E is the Judge of Performance Under the Contract of both the contractor and the owner and is the first level of Dispute Resolution between the two parties.

Inspection of the contractor's work is not a specific A/E responsibility but, by contract, the A/E is to look after the owner's interests and interpret the Contract Documents in performance. The independent contractor status of the GC places Performance Responsibility solely on the shoulders of the GC.

The A/E, however, Judges GC Performance and accordingly makes constructive recommendations to the owner. The Rejection of the GC's work and the Denial of Progress Payment Requests are the prerogatives of the A/E. In essence, the A/E Cannot Stop Work that is not according to specification; the A/E can only tell the contractor that it will not be accepted or paid for by the owner.

There are signal points during the construction phase that demand the attention of the A/E. One is the Date of Substantial Completion (the point in the project where the owner can use the facility, or parts thereof, for its intended purpose). The A/E Certifies this date for the owner's concurrence; it is the date which Terminates the Contract Time.

Another important juncture is the fifty percent completion point where the Retained Amount (money earned but not paid to the GC by the owner as insurance against GC failure to comply) is often reduced, sometimes halved.

The Contract Starts on a documented date and ends when the GC accepts Final Payment under the terms of the contract. The Construction Phase starts when the contract is awarded to the GC and ends when all construction is completed. The nature of construction requires that the Construction Phase and Occupancy Phase run concurrently during a variable period after the Date of Substantial Completion. There are always work items to be corrected or completed by the GC or a Subcontractor and Latent Defects which need attention after the facility is occupied by the owner.

At the Date of Substantial Completion, the A/E and the GC agree on the items which still remain to be done. This list is called a Punch List and consists of large and small items, only some of which will prevent the owner's full utilization of the facility the owner.

In practice, many of the smaller items never get satisfactorily completed from the A/E or owner's perspective, and the Contract Amount is consequently adjusted downward.

E.1.9 GC Warranty/Guarantee Phase

There are criteria which determine the length of this phase and the Owner, A/E, and Contractor involvement during this phase. The phase provides a period of time for latent defects in construction, or more importantly, equipment and materials, to surface before the GC is contractually excused from remedial action.

The GC's contract usually requires the contractor to Warrant his workmanship for a period of one year after the Date of Substantial Completion. The surety is usually kept involved for the same period by the terms of the Labor/Material and Performance Bonds.

Various items of equipment and materials which were incorporated into the construction have Specific Warranty Conditions for varying lengths of time, usually measured from their installation or operational date or from the date of substantial completion.

The standard architect's agreement Does Not extend into the Warranty/Guarantee Phase. It is terminated shortly after the Date of Substantial Completion and relates to issuance of the final Certificate For Payment by the A/E.

The Warranty/Guarantee Phase is not the final opportunity for the GC to correct items which were not constructed according to the Contract Documents. Statute of Limitations laws establish a time period during which an owner can seek redress for Breach of Contract, in the event that a problem involving the GC's performance under the contract arises. The owner can hire an attorney and take legal issue.

E.1.10 Financing GC Projects

Although financing is not a "phase of construction," its timing during the project is an important consideration of the owner. Two realities must coincide: (1) the construction cost of the project to the owner must be justified, and (2) the projected cost of the project must be established as accurately as possible.

The GC system has two points during the project where financing can best be arranged, one at the end of the Feasibility Phase and the other after the Bidding Phase. Each point has its advantages and disadvantages which should be considered before making a selection.

The problem with obtaining financing after the feasibility phase is that the final cost of construction cannot be conclusively established based on the information available. At this point, financing must provide for potentially significant overruns or the scope of the project must be such that it can be reduced to offset cost overruns.

Unless there is a "design to budget" clause in the A/E agreement for design services or a pragmatic "contingency" in the owner's budget, there is no guarantee that the project will be within the owner's budget when final bids from general contractors are received.

The problem with waiting until bids have been received to arrange financing is that the A/E fees for the design and bidding phases (about 80% of an A/E's total fee for the project) must be paid by the owner, even if the project must be abandoned due to excessive construction proposals.

If the owner can absorb A/E design and bidding fees, financing is best arranged after bids have been received. At that point, construction costs are usually 95% established. The remaining 5% can be covered by a contingency.

It is also worth the expense to incorporate a "design to budget" clause in the Owner-A/E design agreement; however, the clause should limit the range and magnitude of the A/E's prerogative to change design to accommodate cost reductions.

E.2 ACTIVITIES DURING THE ELEVEN PHASES OF A D–B BUILDING PROJECT

The progression of activities required to deliver a building project using the D–B Contracting System in the Public Sector uses a Bid–Design–Build sequence of events.

E.2.1 D–B Feasibility Phase

At this phase, the project is conceived and analyzed by the owner. It begins when a need for a facility is recognized and ends when a decision is made to construct or not. The length of this phase is dictated by the owner's requirements and timing demands. Project Definition gains focus as requirements are considered and discussed. Definition is generally given in broad parameters, such as minimum square footage of

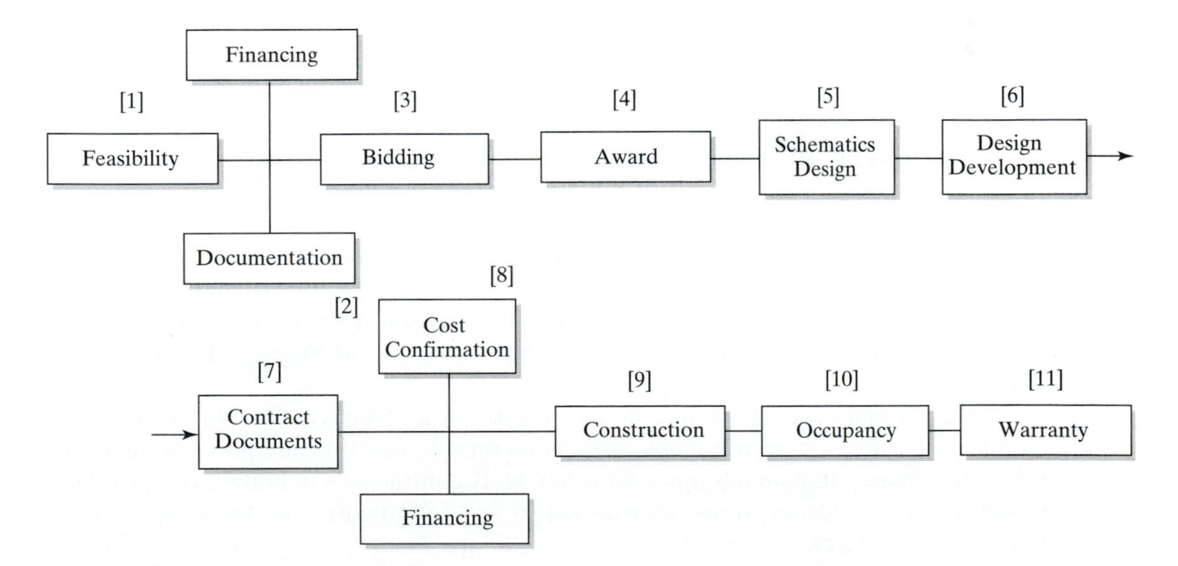

FIGURE E.2 The Eleven Phases of a D–B Building Project.

space and maximum budget expenditure. The feasibility phase is also called the Conceptual Phase or Programming Phase. During this phase, Special Consultants will be most helpful.

The documentation produced during Feasibility ranges from A Program Statement, with or without Sketches, to Conceptual Drawings and highly-developed Brochures and Conceptual Models. Depending on the documentation, the budget accuracy range at the end of the phase should be plus or minus 30%.

Documentation: After the Feasibility Phase has been completed and a decision has been made to proceed with Design and Construction, the Owner's next and exceptionally important step is to produce Documentation that physically describes the Project in sufficient detail to eliminate ambiguity when design–build contractors interpret the project's requirements for pricing purposes.

Project Requirements Documentation usually consists of Performance Specifications, Drawings and Sketches that convey the size, shape, character, and quality of the project to be constructed. Adequate documentation can sometimes be produced by an astute owner. More often than not, owners hire Design Professionals to assist in developing the Project Requirements and Especially the Documentation.

E.2.2 D–B Bidding Phase

On Private Projects, owners have the option to bid or to negotiate contracts or to use a combination of bidding and negotiations. This phase would be called the Negotiation Phase, if the owner elects to use that process to acquire the services of a D–B Contractor.

The Owner has a wide range of responsibilities during this phase and must be familiar with the construction industry to make it successful. If a Documentation Consultant is engaged, that consultant's assistance will beneficially guide the project through this phase.

Interested D–B contractors must be located and prequalified. The Project Requirements Documentation must be distributed to qualified D–B contractors. Questions from contractors relative to the Documentation will require responses; Requests For Substitutions must be act upon; Pre-Bid Meetings must be organized and conducted; and Pre-Bid Addenda may have to be issued to Bidders when changes are made in the Documentation. (All bidders should have the same information from which to develop their proposal.) A final review of the Documentation must be accomplished before the Bid Date.

During the Bidding Period, D–B contractors are engaged in two major activities, (1) developing a Design Proposal, by converting the project requirement documentation into a more Definitive Design, and (2) developing a Cost Proposal based on that definitive design.

The Design Proposal should be in the form of a Definitive Design in sufficient detail to Visually Convey the shape, size, character, and function adequacy of the project to the owner, in competition with other D–B contractor's definitive designs. The design must provide accurate information to the D–B contractor for Construction Estimating purposes.

The Cost Proposal requires construction Scheduling and Planning, Quantity Surveys, and Estimates of construction costs. In essence, the D–B contractor must build the project on paper and do it during the brief period (several weeks on average, depending on the project's size and complexity) set aside for assembling proposals.

The GC quantifies and estimates the cost of the work reserved for his own forces and solicits Trade Contractor Proposals for the work he intends to have done by Subcontracted forces.

Trade Contractors must get bidding and technical information from D–B contractor's offices by invitation. To complete their proposals, Trade Contractors must discuss the requirements with the D–B contractor, quantify and price their work, and obtain quotations for Material and Equipment from appropriate Suppliers.

Trade Contractors determine their own work-scope from the Project Requirements Documentation, price their work accordingly, and give Proposals to the D–B contractor, who has the Option of using or not using the offers. However, it is common for bidding D–B Contractors to give Subcontract commitments to Trade Contractors before the Owner's bid date.

Without finished drawings and definitive technical specifications to bid from, trade contractors have the opportunity to use equipment and materials that are most economical to install yet fall safely within the parameters of the Project Requirements Documentation. Trade contractor proposals must be Compared in terms of the material/equipment Provided as well as work-scope and cost.

D–B contractors must be alert for Pre-Bid Addenda, and if issued, revise their Proposals accordingly with the help of the involved trade contractors. Each Bidding contractor must acknowledge receipt of Addenda in their Proposal. If addenda are not acknowledged, the D–B's bid will require careful review by the Owner.

If D–B proposals are competitively bid, they are usually required to be received by the owner at an appointed place before a designated time. Proposals are usually opened privately by the owner, and they are rarely shared with bidders. The owner has the right to Reject Any and All Bids, Waive Irregularities, and accept the offer which is In The Owner's Best Interests.

After all bids have been received, the owner questions and reviews each one and makes a decision. The owner may make an Award or Reject all bids.

Note: Design–Build Proposals are solicited, using the Project Documentation and Pre-Bid Meetings as the basis for assembling proposals. The overall Quality of the documentation will ultimately determine the Competitive Veracity of the proposals received. The more Complete, Clear, and Concise the project requirements documentation is, the more Accurate and Competitive the bidding will be.

Additionally, the Private nature of the bidding process, and the Absence of the Legal Requirements applicable in the public sector, opens contractor proposals to considerable Negotiations before an award is finally made. The areas and topics of negotiation are such that an inexperienced owner will not benefit from them without the help of a Consultant who is appropriately qualified.

E.2.3 D–B Award Phase

The Award Phase is a suspenseful period for the D–B contractor and the trade contractors. Questions on proposals may arise which require clarification, and additional Qualification questions may be appropriate. Negotiations relating to design, time, cost, and quality usually take place. Several weeks and several meetings will transpire before the owner finally selects the D–B contractor for the award.

When an Award decision is made by the Owner, action can be taken to enter into the Contract For Construction. It is sometimes convenient or expeditious for the owner to issue a Notice of Award letter to the contractor stating any pre-conditions. A Notice to Proceed may also be issued by the owner in the event the signing of the Contract For Construction will be delayed due to a technicality.

E.2.4 D–B Design Phases

The design phase of an architectural project is divided into three sub-phases: (1) Schematic Design, (2) Design Development, and (3) Contract Documents. The design phase of an engineering project is divided into two sub-phases: (1) Preliminary Design and (2) Final Design. Sub-phases provide Owner Review points and a means of determining design progress for payment.

The sub-phases of design are loosely defined, making it difficult for anyone to review in-progress design documents and determine exactly where the design is in the two adjacent sub-phases. D–B contractors interpret design sub-phase definitions for their own convenience unless some guidelines are set before design begins.

E.2.5 Schematic Design Phase

On an architectural project, the Schematic Design Phase begins when the contract is awarded. The term Design has different meanings to architects and to engineers. Architectural Design integrates use and space, combining form and aesthetics into Occupational Function. Engineering Design deals with the effects and forces of nature by producing or injecting design solutions with Physical Performance Credibility. Working together, architects are often involved in engineering projects; engineers are always involved in architectural projects.

Schematic design is the sub-phase which requires the most demanding Liaison with the owner; the time when Communication between owner and design professional is most critical. The Project Requirements Documentation must be Reviewed and Refined and Successfully Conveyed to the D–B contractor for Accurate Incorporation into the design.

Schematic design provides the opportunity to critically review the Design Proposal submitted by the D–B contractor, noting the shapes, sizes, and relationships of the required spaces. Alternative Solutions are developed by the D–B contractor and considered by the owner.

Drawings are still very preliminary; small scale (1/16″ and 1/8″) without detail. Sketches are provided to show space locations and relationships. Technical Specifications are not yet formulated, but Design Criteria are being refined as decisions are made on Building Systems, Materials, and Equipment.

The Cost of Construction is tracked as progress is made and compared with the Cost Proposal. The D–B contractor's highest priority as final Design Delineation and Material Selections are made is to remain within the budget.

The schematic design phase ends when the owner accepts the Schematic Drawings as representative of the Design Proposal.

E.2.6 Design Development Phase

The Design Development Phase begins when the owner signs off on the Schematic Design Phase. Working from the approved Schematic Design, the D–B contractor further develops the design through increased delineation and documentation. The Goal of this phase is to complete and coordinate all architectural and engineering design decisions so that final delineation and documentation can be accomplished during the Contract Documents Phase.

Drawings enlarge from 1/16″ and 1/8″, to 1/8″ and 1/4″ scale, and details are drawn in larger scale. Walls and partitions, shown as lines on Schematic Drawings, are shown in section with thickness and scale. Openings are located and dimensioned. Space is allocated to accommodate electrical, mechanical, and structural requirements. Coordination of other design disciplines is accomplished.

Building Systems decided upon during schematic design are detailed and fundamentally specified. Outline Specifications begin to take shape, based on design decisions made during the schematic phase. Building Elevations and Sections are developed to assist in dimensioning and in Design Coordination.

Designing Within The Cost Proposal remains a goal of the D–B contractor.

The Design Development Phase is complete when the D–B contractor submits the Design Development Documents to the owner and they are accepted as conforming to the Project Requirements Documentation. Any changes the owner makes to Previously Approved Schematics, which result in more or less construction cost, will be incorporated in a plus, minus, or net Change Order.

Note: The contract award is made to the D–B contractor on the basis of a Design Proposal and a Cost Proposal. Both are developed from the Project Requirements Documentation provided by the owner. The cost proposal submitted to the owner was based on the D–B contractor's design proposal and Modified (increased or decreased) to adjust for owner design changes made after bidding or during the schematic and design development design sub-phases.

If the cost proposal is in the form of a Lump Sum or Guaranteed Maximum Price, the current cost proposal as Modified is valid and must be honored by the D–B contractor. It is assumed that the D–B contractor has designed the project within the modified cost proposal and that the lump sum or guaranteed maximum price figure is consequently protected. If the cost proposal is in the form of a Unit Price Schedule or on a Cost Plus basis, the project cost is the owner's responsibility and is monitored during the design sub-phases as the probable cost of construction.

However, it is common for D–B contractors to work under a split compensation arrangement—one amount to be paid for design and another amount to be paid for construction. Design compensation is often on a Cost Plus Fee arrangement and Construction compensation made on a Lump Sum, Guaranteed Maximum Price, Cost Plus Fee or Unit Price Schedule.

When design compensation and construction compensation are separated, the eighth phase of the project, Cost Confirmation, becomes crucial to both the owner and the D–B contractor.

The Cost Conformation phase (shown in figure Appendix E.2 as occurring after the contract document phase) essentially Begins during the Design Development phase (or as soon as the D–B contractor can firmly commit to a lump sum or guaranteed maximum price). Some D–B contracts require the D–B contractor to "provide a lump sum price to the owner as soon as design is sufficiently complete to do so." This has been generally interpreted as after design is approximately 80% complete (sometime early in the contract document sub-phase of the design).

On split compensation projects, the owner often retains the option to terminate the contract with the D–B contractor and use the design documents, when completed by the D–B contractor, to solicit construction proposals from other parties. One reason for this is that the Cost Confirmation number has not been exposed to competitive bidding and, if it seems excessively high, the owner has the right to seek a more economical construction source.

The possibility also exists that the relationship between the owner and the D–B contractor may have deteriorated during the design phase. In this case, it is better for both parties to discontinue the contract arrangement. D–B contracts should provide enough flexibility to accommodate both proceed and not proceed provisions.

E.2.7 Construction Documents Phase

This phase is production-intensive, requiring considerable personpower to fully document the design that evolved during the previous two phases. Construction Documents, which include Working Drawings and the Project Manual (front-end and technical specifications) are developed and provided to the owner.

Working Drawings are 1/4″ scale with larger scales for construction details. The Drawings are sectioned to accommodate Civil Site, Architectural, Structural, Mechanical, and Electrical disciplines, and the instructional requirements for these disciplines are included in their proper section in the Technical Specifications.

Note: The D–B System can accommodate Fast-Track Construction because of its built in, Single Contract, design and build responsibility. Fast-tracking is where construction begins before the project design is completed. As foundation design occurs in the late stages of completion, foundation construction begins based on the foundation design completed. This design–build sequence is repeated as design progresses until the construction is completed.

If fast-tracking is used on a project, the project's Design phase cannot be completed using the Conventional Three Sub-phase procedure outlined here. The completed Working Drawings and Specifications will be made available to the owner at the End of construction in the form of As-Built drawings—drawings and specifications that accurately portray the project as constructed.

E.2.8 D–B Construction Phase

The D–B contractor has control of the site and the project according to the terms of the Contract For Construction during the Construction Phase. As an Independent Contractor, the D–B contractor is responsible for completing the project On Time, In Conformance with the Project Requirements Documentation, the Drawings and Technical Specifications, and within the amount stated in the Cost Proposal as amended by Change Orders or as established in the Cost Conformation sub-phase.

The Owner (with/without the assistance of a Consultant) has responsibility for Contract Administration during the construction phase. This consists of approving Shop Drawings, Product Data and Samples, processing Change Orders, certifying Progress Payments, judging the D–B contractor's Performance, resolving Disputes, and providing an open Liaison with the D–B contractor on day-to-day activities.

Another Owner responsibility is establishing the Date Of Substantial Completion—the point where the owner can use the facility, or parts thereof, for its intended purpose. This date is the date which legally Ends the Contract Time.

The Contract Starts on a documented date and ends when the D–B contractor accepts Final Payment under the terms of the contract. The Construction Phase starts with on-site activity. The nature of construction requires that the Construction Phase and Occupancy Phase run concurrently during a variable period after the Date Of Substantial Completion. There are always work items to be corrected or completed and Latent Defects that need attention after the facility is occupied fully or partially by the owner.

Still another Owner responsibility is issuing Punch Lists. These lists consist of large and small items which are incomplete or not acceptable as completed, only some of which will prevent the owner's full use of the facility. Punch lists are produced by a walk-through inspection of the project, as a condition of certifying substantial completion, and issuing the final payment check to the D–B contractor.

In practice, many of the smaller punch list items never get satisfactorily completed from the owner's perspective, and the Contract Amount is consequently adjusted downward.

E.2.9 D–B Warranty/Guarantee Phase

There are criteria which determine the length of this phase and the involvement of the Owner and the D–B Contractor. The phase provides time for latent defects in construction, or more importantly, equipment and materials, to surface before the D–B contractor is contractually excused from remedial action.

The D–B contract usually requires the contractor to Warrant his workmanship for a period of one year after the Date Of Substantial Completion. If Surety Bonds have been provided by the D–B contractor, the surety is usually kept involved for the same period by the terms of the Labor/Material and Performance Bonds.

Various items of equipment and materials which are incorporated into the construction have Specific Warranty Conditions for varying lengths of time, usually measured from their installation or operational date or the date of substantial completion. The Owner should make sure that all of these special warranties have been Cataloged and Filed for future, timely reference.

A complete file of Operating And Maintenance Manuals for mechanical and electrical equipment, and a supply of frequently needed Spare Parts, should be delivered to the Owner by the D–B contractor.

The Warranty/Guarantee Phase is not the final opportunity for the D–B contractor to correct items not constructed according to the D–B Contract. There are Statute Of Limitations laws which establish periods of time during which an owner can seek redress for Breach Of Contract in the event that a problem arises involving the D–B contractor's performance. The owner must then hire an attorney and take legal issue.

E.2.10 Financing

Although financing is not a "phase of construction," its arrangement during the project is an important consideration for the owner. Two realities must coincide: (1) the construction cost of the project to the owner must be justified, and (2) the projected cost of the project must be established as accurately as possible.

The D–B system, Without Fast-Tracking, has two points during the project where financing can be arranged, at the end of the Feasibility Phase and before the Construction Phase.

When Using Fast-Tracking, the only opportunity to finance is at the end of The Feasibility Phase, as construction will begin before the design phase is complete. The problem with obtaining construction financing after the feasibility phase is that the final cost of construction may not be established unless a Lump Sum or Guaranteed Maximum Price payment scheme is in effect. Project financing under Unit Price or Cost Plus payment arrangements at this point must provide for potential overruns or an understanding that the scope of the project can be reduced to offset cost overruns.

Obviously, projects that are not Fast-Tracked can be conveniently financed initially for design and subsequently for construction, and they often are.

E.3 ACTIVITIES DURING THE NINE PHASES ON AN ACM BUILDING PROJECT

The progression of activities required to deliver a building project using the ACM Contracting System is according to the standard contract documents for an ACM Architectural Project.

E.4 ACM FEASIBILITY PHASE

This is the phase where a project is conceived and analyzed by the owner. It begins when a need for a facility is recognized and ends when a decision is made to construct or not construct. The length of the feasibility phase is dictated by the owner's requirements and the timing demands of the project.

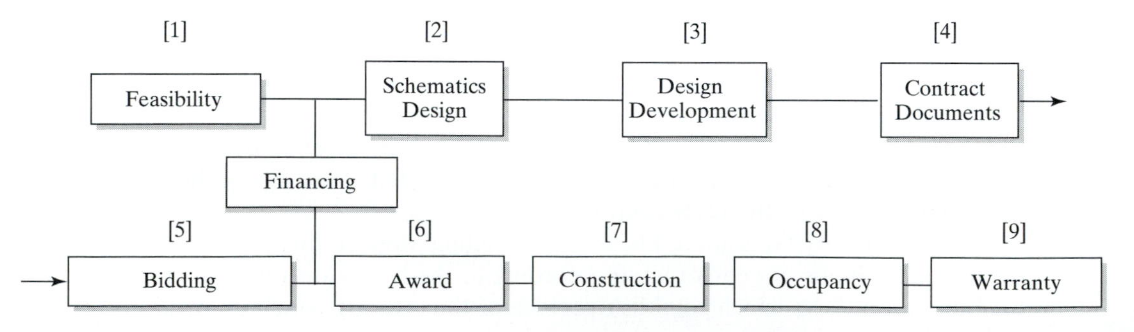

FIGURE E.3 The Nine Phases of an ACM Building Project.

The project gains definition as requirements are considered and discussed. Definition is generally given in broad parameters, such as minimum square footage of space and maximum budget expenditure. The feasibility phase is also called the Conceptual Phase or Facility Programming Phase.

During this phase, Special Consultants will be most helpful. The Basic Services of the A/E-Owner and the CM-Owner Contracts do not Usually cover services during this phase. If A/E or CM services are deemed advantageous, they can be contracted separately or included as Additional Services in the respective agreements if the project moves ahead. (Engineering agreements usually include feasibility services as part of basic services.)

A/E and CM firms sometimes provide limited feasibility phase services without additional fees as an incentive to an owner to engage their services for the project if it moves ahead. For obvious reasons, this practice is frowned upon by some professional associations. (Owners should be aware that feasibility services provided under these conditions can have a self-serving motive—to move the project ahead one way or another.)

The documentation produced during Feasibility ranges from a Program Statement, with or without Sketches or Rough Drawings, to formal Conceptual Drawings and highly developed Brochures and Conceptual Models. Documentation should be sufficiently detailed and reliable to permit the A/E and the CM to productively begin their responsibilities in the Schematic Design Phase.

On many projects in both the Public and Private Sectors, the conclusion of this phase establishes the need to locate a Funding Source for the project. This requires estimates of project costs to the degree of accuracy acceptable to a funding source. Depending on the detail of the documentation, the range of budget accuracy at the end of the feasibility phase should be plus or minus 15%.

E.5 ACM DESIGN PHASES

The design phase of an architectural project is divided into three sub-phases: (1) Schematic Design, (2) Design Development and (3) Contract Documents. The design phase of an engineering project is divided into two sub-phases: (1) Preliminary Design and 2) Final Design. The sub-phases provide an Owner Review Point and establish a means of determining progress for A/E and CM payment purposes.

These sub-phases are loosely defined in design and CM services agreements, making it difficult for anyone to review in-progress design and CM services and make a determination of exactly where the services are in the two adjacent sub-phases. A/Es and CMs interpret design sub-phase definitions to their own benefit unless some guidelines are set before the design phase begins.

E.5.1 Schematic Design Phase

On an architectural project, the Schematic Design Phase begins when the Feasibility Phase ends. Design has a different meaning to architects and to engineers. Architectural Design integrates use and space, combining forms and aesthetics into Occupational Function. Engineering Design must conform to architectural design but additionally deals with the effects and forces of nature on the design, producing solutions with adequate Physical Performance Credibility. Working together, architects are often involved in engineering projects; engineers are always involved in architectural projects.

Schematic design is the sub-phase which requires the most cooperation and demanding Liaison with the owner, the time when Communication between owner, A/E, and CM is critical. The Conclusions reached by the owner during the Feasibility Phase must be Accurately and Thoroughly Conveyed to the A/E and CM and Accurately Incorporated into the design.

The start of Schematic Design is the start of the CM and A/E's Project Management responsibilities and the formation of the Project Team consisting of the Owner, A/E, and CM. The initial activities are team Brainstorming and Organizational meetings to clarify team interactions and establish a clear path toward synergistic results. The expertise and knowledge of each team member provides Checks And Balances during all phases of the project.

There is no designated Team Leader. The Owner has the contractual right to make Most decisions. However, the owner defers to the expertise of the A/E and CM and the common goal of the team by requiring that team leadership shift from one member to the other, depending on the input requirements at the particular point during the project.

During the Schematic Design sub-phase, the A/E leads Design activities, the CM leads Project Management activities, and the owner follows the lead by providing necessary Information and decisions.

Schematic design produces Project Configuration (the shapes, sizes, locations, and relationships of the required spaces). Alternative Solutions are suggested and reviewed by the A/E, CM, and owner. Drawings are very preliminary, small scale (1/16″ and 1/8″) without detail. Sketches are provided to delineate the project configuration. Technical Specifications are not yet being formulated, but Performance Specifications and Design Criteria are being recorded by the Team as decisions are made on Building Systems, Materials, And Equipment. Designability And Constructability studies begin and continue throughout design.

An Estimate Of Construction Cost is determined by the CM and compared with the owner's Feasibility Budget. The Feasibility Budget (often established without drawings) is the constant criterion of the A/E and CM as owner-information is converted to Design and Material/Equipment Selections are made. The CM provides ongoing (not periodic) estimates of construction costs as the A/E proceeds with design.

Economically correct Building Systems decisions are collectively made by the Team through Value Engineering and Life-Cycle Costing studies instituted by the CM. By properly considering alternatives, it is possible to maintain the Estimated Cost Of Construction within plus or minus 5% during schematic design.

During the Schematic Design phase, the CM also develops a Program Schedule with input from the Owner and A/E and begins the Contractability and Risk-Management review processes and the development of the CM Project Manual.

The schematic design phase ends when the A/E submits acceptable Schematic Drawings, and the CM submits an acceptable Cost Of Construction and Program Schedule to the owner.

E.5.2 Design Development Phase

The Design Development Phase begins when the owner Accepts the Schematic Design Phase Documentation from the A/E and CM. Working from the approved Documentation, the A/E further refines the design through increased delineation and documentation. The CM initiates Constructability review procedures and continues other responsibilities. The Goal of this phase is to complete and coordinate all design decisions so that final delineation and documentation can be completed during the next phase, the Contract Documents Phase.

Drawings enlarge from 1/16″ and 1/8″, to 1/8″ and 1/4″ scale, and details are drawn in larger scales. Walls and partitions, shown as lines on Schematic Drawings, are shown in section with thickness and scale. Openings are located and dimensioned. Code requirements and local Building Regulations are incorporated. Ancillary space is allocated and space is designated to accommodate electrical, mechanical, and structural requirements. Design Coordination of the work of other design disciplines is ongoing.

Building Systems determined during schematic design are detailed and fundamentally specified. Outline Specifications begin to take shape, based on design and material/equipment decisions made during the schematic phase. Building Elevations and Sections are developed to refine dimensioning and aid in Design Coordination.

The Estimate Of Construction Cost is updated as design definition improves. Its accuracy should remain in the 5% range. Designing Within The Budget remains the goal of the Team.

The phase is complete when the A/E submits the Design Development Documentation, the CM submits a Cost Of Construction and Schedule to the owner, and they are Accepted. Any changes the owner makes to Previously Approved Schematics are charged as Additional Services to the owner by the A/E and sometimes by the CM.

E.5.3 Construction Documents Phase

This phase is production-intensive, requiring considerable personpower from the A/E and CM to fully document the design and contractability procedures that evolve during the previous two phases.

Bidding Documents (which include Working Drawings), the Project Manual (front-end and technical specifications), and Instructions To Bidders and Proposal Forms are developed for distribution to Bidding Contractors in the next phase. Alternates, if necessary, are selected, and Allowances decided upon. The Advertisement For Bids is drafted by the CM and issued in the name of the owner.

Final Drawings are 1/4″ scale with larger scales for construction details. The Drawings are sectioned to accommodate Civil Site, Architectural, Structural, Mechanical, and Electrical disciplines. The instructional requirements for these disciplines are included in their proper section in the Technical Specifications.

Front-End Specifications, the General Conditions of the Contract for Construction, Supplementary and Special Conditions, and Bid Forms are developed and edited to accommodate specific CM contract requirements and added to the Project Manual.

The CM develops the Construction Schedule (to confirm the project can be constructed within the allotted time) and the construction portion of the Milestone Schedule which provides bidders with a generalized schedule for bidding purposes. The CM writes Work-Scope Descriptions for all Divisions of Work that comprise the total Project and Reviews the Bidding Documents for ambiguities and errors.

Contractor Bidding Procedures are established by the Team and included in the bidding documents. Bidding of Long-Lead Items, previously identified by the Team, is arranged for. The CM identifies interested contractors and, where possible, Prequalifies them based on their performance ability. Contractors on the Bid List are persistently encouraged to submit proposals at bid time.

When the owner is satisfied that Bidding Documents are complete, the Cost Of Construction is within 5% or less of the budget, and a satisfactory Construction Schedule has been produced, the owner signs off on the Contract Document Phase.

E.5.4 ACM Bidding Phase

Overt, Competitive Bidding is the usual requirement on Public Projects. On Private Projects, owners sometimes negotiate contracts or use a combination of bidding and negotiation. This phase could be called the Negotiation Phase, if the owner elects to use that process to acquire a contractor.

The start of the bidding phase ends the Design Phase. During this phase, the Estimated Cost Of Construction will be converted to the Actual Cost Of Construction using the cost information obtained from the competitive bidding process.

Bidding documents are printed; Complete sets are distributed Exclusively to the Contractors who request them in exchange for a Bidding Document Deposit sufficient to cover printing and handling costs. The deposit will be returned if the documents are returned in good condition by the unsuccessful bidders.

To increase bidding interest, complete sets of Bidding Documents are sometimes sent to Plan Rooms to permit review, quantity take offs, and pricing by Trade Contractors. However, contractors must obtain the Bid Forms for the project from the A/E, CM, or owner.

The A/E's and CM's responsibilities during this phase are responding to Questions From Contractors concerning the Bidding Documents which require responses; acting on Requests For Substitutions; organizing and attending Pre-Bid Meetings; issuing Pre-Bid Addenda to Bidders of Record when changes in the Contract Documents are made (to ensure that all bidders have the same information on which to develop their proposal); and a final, thorough review of the Contract Documents before the Bid Date.

Contractors also have plenty to do. Bidding requires Construction Planning, Quantity Surveys, and Estimates of construction costs. Each bidder must "build the project on paper" from a construction-cost perspective and do it during the period (a few weeks on average, dependent on the project's size and complexity) set aside for assembling proposals.

To complete their proposals, Contractors quantify (categorically determine the quantity of work required in the work scope), obtain quotations for Material/ Equipment from Suppliers, and estimate the costs of labor and construction equipment.

Contractors are not usually permitted to define their own work-scope. They must submit their proposals, based precisely on the content of the Work-Scope Descriptions written by the CM and provided in the Bidding Documents and as shown on the Bid Form for the work-scope(s) they are bidding.

Contractors must be alert for Pre-Bid Addenda and, if issued, revise their Proposals accordingly. Each Bidder must acknowledge receipt of Addenda on their Proposal Form. If All issued addenda are not acknowledged, the Contractor's Bid will not be considered by the Owner.

The Location and Time for receiving proposals from Contractors on Public projects is first stated in the Advertisement For Bids and later established in the Instructions to Bidders. Proposals must be received at the appointed place before the stated time. Bids that are received Late are not accepted. Only Responsive Proposals, those submitted in accordance with the Instructions to Bidders, are considered for award.

At the Bid Opening, proposals are opened and publicly read. Due to large number of bidders (sometimes over 100), proposals are opened in Work-Scope groups. This allows bidders to limit their presence to readings of proposals in the group(s) in which they have an interest.

Proposals for each numerically successive Work-Scope group are opened and read, one at a time, in random order. Mandatory proposal enclosures such as Bid Bonds are acknowledged vocally and the Dollar Proposal read. If Award Criteria other than Price is to be considered (such criteria had to be previously stated in the Instructions to Bidders), the bidder's response to that criteria must be Read at the time of opening his/her proposal.

Although the owner usually reserves the right to Reject Any and All Bids, Waive Irregularities, and accept the offer which is In The Owner's Best Interests, the low bid is usually accepted. A decision Not to award the project to the Low Bidder can prompt a legal challenge from the low bidder.

E.5.5 Award Phase

After bids have been publicly read, the owner takes them under Advisement. With assistance from the A/E and CM, the Owner comprehensively reviews each proposal and eventually makes an award decision. The options available to the owner in each Work-Scope category are to make an Award, Reject all bids in a specific work-scope category(s), or allow the stated proposal holding time to Expire.

The Award Phase is a busy time for the CM and the A/E, and a suspenseful period for contractors, especially the Apparent Low Bidder(s). Questions on contractor proposals, regarding Quantities, Pricing, and Schedule may arise which require contractor clarification. Post-Bid Qualification of certain apparent low bidders may be recommended by the Team.

With input from the A/E and CM, the owner must make the final decision to accept a bidder or not. The Owner must take into consideration existing laws and statutes and consider the advice of the A/E and CM. It is common for the CM to recommend awards to the A/E, who in turn recommends awards to the owner.

As each Award decision is made by the Owner, Contracts For Construction are prepared. It is sometimes convenient or expeditious for the owner to issue a Notice of

Award letter to the contractor stating any pre-conditions, such as providing Labor and Material/Performance Bonds and Insurance Certificates. The CM Coordinates this part of Contract Administration.

A Notice to Proceed may also be issued by the owner in the event that the signing of the Contract For Construction is delayed due to a technicality. The notice to proceed usually imposes conditions on the contractor to ensure against any potential problems before Insurance, Bonds, or Funding have been arranged.

Bid Document Deposits are exchanged for unmarred sets of drawings returned by unsuccessful bidders, and Bid Bonds (or other forms of Bid Security) are returned.

Awarded Contractors proceed to Buy Out the project, Mobilize forces/equipment, begin the submittal of Shop Drawings, Product Data and Samples, and commence Construction operations. The Contract Time for the Project begins when the first Start-Up Contractor begins construction operations as required in the Contract For Construction or the Notice To Proceed.

E.5.6 ACM Construction Phase

The CM has control of the site in behalf of the Owner and the responsibility of Coordinating the contractors' work. Using updated schedule input from Start-Up Contractors, the CM produces a Short Term Contractor Activity Plan (CAP)—a construction schedule based on the first portion of the Detailed Construction Schedule and schedule information provided by contractors with their proposals.

Contractors construct the project according to the terms of their Contracts For Construction. As Independent Contractors, they are responsible to complete their Work-Scope(s) in accordance with the Short Term CAP, In Conformance with the Technical Specifications, and for the Dollar Amount stated in the Contract.

The A/E and CM share the responsibility for Contract Administration during the construction phase. The A/E Checks shop drawings, product data and samples, Approves Change Orders, Certifies progress payments, and Determines if specified quality is produced by the contractors.

The CM Expedites shop drawings, product data and samples, Develops and Reviews Change Orders and progress payments, and Determines if contractors are timely in their performance. Both the A/E and CM provide Liaison between the contractors and the owner.

The A/E is the Judge of Performance Under the Contract of both the contractors and the owner and is the first level of Dispute Resolution between them.

Inspection of the contractor's work in-progress is Not a contract responsibility of either the A/E or CM. The independent contractor status of contractors places Performance Responsibility solely on their shoulders. However, as contractual agents of the owner, both have a responsibility to look after the owner's best interests at all times. The A/E and CM are obligated to put procedures in the Contract Documents that ensure to the extent possible contractor performance without comprehensive inspection. However, if either sees work being performed Contrary to the Contract Documents, they are obliged to positively take action.

In essence, neither the A/E nor the CM can Stop Work that is not being performed according to the specifications; they can only inform the contractor that the work will not be accepted or paid for by the owner.

The A/E is the Final interpreter of contractor performance relative to the Drawings and Technical Specifications. However, the CM is the Judge of the contractors' time performance and makes corrective recommendations to the owner when sub-par performance is noted.

There are signal points during the construction phase that demand the combined attention of the CM and A/E:

1. Each contractor's Date of Substantial Completion; that point where a contractor's Work is sufficiently complete to permit Subsequent contractors to begin work without interference from the contractor seeking substantial completion status or when a contractor's work is totally complete. The A/E Certifies these dates upon recommendation of the CM and concurrence of the owner. This date Terminates the Contract Time of the substantially complete or totally complete contractor.
2. The 50% contract completion point where the Retained Amount (money earned but not paid to the contractor by the owner as insurance against contractor failure to complete) is often Reduced, sometimes halved. This date must be determined by the CM, reviewed and approved by the A/E, and agreed to by the owner.

Each Contract Starts on a specified date and ends when the contractor accepts Final Payment under the contract terms. The Construction Phase starts when the contract is awarded to the first Start-Up Contractor and ends when the work of the Last Contractor is declared substantially complete by the Owner.

The nature of construction permits the Construction Phase and the Warranty Phase to run concurrently during a period after Dates of Substantial Completion. After contractors claim their work is complete, there are always work items to be corrected or completed by Contractors and Latent Defects which need attention.

As a condition of Substantial Completion, the A/E, CM, and contractor agree on the items which remain to be completed or need correction. This Punch List consists of items which will not prevent utilization of the contractor's work or prevent subsequent contractors from proceeding with their work. On well-managed projects, only one Punch List needs to be issued. On other projects, two, three or more may be required.

In practice, some items never get satisfactorily completed from the A/E or owner's perspective. In this case, the Contract Amount is adjusted downward by a Change Order and another contractor is used to complete the work.

E.5.7 ACM Warranty/Guarantee Phase

There are criteria which determine the length of this phase and the involvement of the Owner, A/E, and Contractor during this phase. The phase provides time for Latent Defects in construction, equipment, and materials to surface before the Contractor is contractually excused from remedial action.

The contract for construction usually requires the contractor to Warranty the workmanship for a period of one year after the Date of Substantial Completion. The Surety is usually kept involved for the same period of time by the terms of the Labor/Material and Performance Bonds provided to the Owner by the Contractor. Additionally, various items of equipment and materials which were incorporated into the construction have Specific Warranty Conditions for varying lengths of time (usually measured from their Installation or Operational Date or from the Date of Substantial Completion).

The standard architect's agreement Does Not extend services into the Warranty/Guarantee Phase. It is terminated shortly after the date of Project Substantial Completion and relates to the date the final Certificate For Payment is issued to a contractor by the A/E. However, the Owner-CM agreement usually extends to the Completion of the Warranty/Guarantee Phase and requires the CM to arrange for Additional work or Correction of work that falls under warranty/guarantee provisions.

The Warranty/Guarantee Phase is not the final opportunity for Contractors to correct items which were not constructed according to the Contract Documents. There are Statutes of Limitations which establish legal Time Periods during which owners can seek redress for Breach of Contract in the event a problem involving a contractor's performance arises. The owner would have to hire an attorney and take legal action.

E.5.8 Financing

Although financing is not a "phase of construction," its arrangement during the project is an important consideration of the owner. Two realities must coincide: (1) the construction cost of the project must be justified by the owner, and (2) the projected cost of the project must be accurately established.

The CM system provides two points during a project where financing can best be arranged, one at the end of the Feasibility Phase and the other after the Bidding Phase. Although the amount of financing, as determined after the feasibility phase on a CM project, is not as accurate as when determined after bidding, the CM process has inherent cost controls that can maintain both budgets within about 5% of each other.

A P P E N D I X F

Subcontractor Survey Regarding GC Practices

The information in this appendix was extracted from the Trade Subcontractor Attitude Survey conducted by Mark F. Ahlborn, currently a Lecturer in the Civil and Environmental Engineering Department, Michigan Technological University, Houghton, Michigan, as partial fulfillment of the requirements for a Master of Science degree in Civil Engineering at MTU in 1986.

Mark was in the graduate Construction Engineering and Management Program and a member of the Construction Management Studies and Research Group (CMSRG) that conducted construction engineering and management research between 1980 and 1996. The author was the program's mentor.

The survey conducted in 1986 substantiates the assumptions that influenced the direction of the development of the CM system in the 1970s. There were 1219 surveys sent out and 270 usable responses.

F.1 GENERAL

B.3 We are usually aware of all general contractors that are bidding on a particular project.
Agree: **80.1%** Disagree: **10.8%**

B.4 We submit bids to ALL general contractors preparing bids for a project of interest to us.
Agree: **30.2%** Disagree: **3.8%**

C.9 Subcontractors usually submit the same price to all general contractors bidding on the same project.
Agree: **39.5%** Disagree: **54.2%**

C.10 Subcontractors generally submit their "best competitive price" to general contractors prior to the date and time general contractors submit bids to the owner.
Agree: **54.1%** Disagree: **27.1%**

F.2 PRE-BID SHOPPING

C.1 In our experience, we have submitted bids to general contractors that practice "pre-bid shopping."
Agree: **73.3%** Disagree: **13.6%**

C.2 We are aware of certain general contractors that frequently practice "pre-bid shopping."
Agree: **86.6%** Disagree: **4.4%**

C.3 The elimination of "pre-bid shopping" is desirable from the subcontractor's viewpoint.
Agree: **79.2%** Disagree: **6.4%**

F.3 POST-BID SHOPPING

C.4 In our experience we have submitted sub bids to general contractors that practice "post-bid shopping."
Agree: **73.4%** Disagree: **14.8%**

C.5 We are aware of certain general contractors that frequently practice "post-bid shopping."
Agree: **84.8%** Disagree: **6.4%**

C.6 The elimination of "post-bid shopping" is desirable from the subcontractor's viewpoint.
Agree: **84.4%** Disagree: **8.5%**

F.4 REACTIONS TO BID SHOPPING

C.7 We have lowered our bid price to general contractors in response to bid shopping pressures.
Agree: **51.8%** Disagree: **34.6%**

C.8 A subcontractor's price to a general contractor is inflated to compensate for the effect of bid shopping.
Agree: **45.4%** Disagree: **33.8%**

C.11 When lowering a price in response to bid shopping, subcontractors often reduce costs as well as overhead and profit figured in the bid.
Agree: **53.3%** Disagree: **25.1%**

C.12 Realizing that bid shopping is likely to occur, a subcontractor increases the amount of overhead and profit figured in the bid to compensate for any reduction.
Agree: **43.8%** Disagree: **32.7%**

C.13 In order to obtain work (in response to bid shopping) we have lowered our overhead and profit to a level below that which we feel is comfortable.
Agree: **39.1%** Disagree: **44.6%**

F.5 SUBCONTRACTOR PREFERENCES

C.14 It would be desirable if all subcontractors were able to bid a project once on a defined work-scope, with no opportunity to adjust the price after bid submittal.
Agree: **84.5%** Disagree: **8.4%**

C.15 In fair competition, with no potential for bid shopping, we would submit our lowest possible price the first time.
Agree: **91.6%** Disagree: **3.2%**

F.6 WORK-SCOPE DEFINITION

D.1 When submitting bids to general contractors, the scope of work we are to perform is usually explained to us by the general contractor.
Agree: **33.3%** Disagree: **46.6%**

D.2 General contractors usually conduct PRE-BID meetings with subcontractors to clarify bidding requirements.
Agree: **17.9%** Disagree: **66.5%**

D.3 Pre-bid meetings would be beneficial to the subcontractor's understanding of the required work-scope.
Agree: **81.7%** Disagree: **3.6%**

D.4 Vague work-scopes cause subcontractors to add contingency amounts to their proposals.
Agree: **86.1%** Disagree: **5.6%**

D.5 A well-defined, written work-scope would reduce the contingency amounts added to a subcontractor's proposal.
Agree: **92.8%** Disagree: **2.0%**

D.6 A well-defined, written work-scope would enable subcontractors to submit a more accurate proposal.
Agree: **96.0%** Disagree: **4.0%**

F.7 SUBCONTRACTOR SELECTION

E.1 When an awarded general contractor selects a subcontractor for his project, the choice is more often based on price rather than quality performance.
Agree: **88.8%** Disagree: **2.8%**

E.2 General contractors often show favoritism by awarding work to certain subcontractors, regardless of price.
Agree: **43.9%** Disagree: **34.7%**

E.3 General contractors often show favoritism by awarding work to certain subcontractors, regardless of the subcontractors' ABILITIES.
Agree: **50.6%** Disagree: **22.3%**

F.8 CONTRACT ADMINISTRATION AND COORDINATION

F.2 General contractors usually provide efficient processing of subcontractor Change Orders.
Agree: **12.4%** Disagree: **70.9%**

F.4 Submitting subcontractor's claims to an owner through a general contractor is a satisfactory process.
Agree: **28.3%** Disagree: **43.9%**

F.7 It would be beneficial to subcontractors to submit their claims directly to owners instead of through general contractors.
Agree: **55.4%** Disagree: **18.7%**

F.8 General contractors closely monitor the construction quality of their subcontractors.
Agree: **23.5%** Disagree: **47.0%**

F.9 General contractor quality inspections are thorough enough to prevent extensive subcontractor rework.
Agree: **13.2%** Disagree: **58.2%**

F.10 The amount of punch list work by subcontractors could be reduced if general contractors paid closer attention to quality control.
Agree: **76.5%** Disagree: **6.8%**

F.13 Most general contractors provide excellent coordination of subcontractors in the field.
Agree: **13.9%** Disagree: **63.3%**

F.15 We have often been scheduled on a project by the general contractor only to find that we cannot commence our work as promised.
Agree: **86.4%** Disagree: **4.0%**

F.16 Significant interface problems between trade subcontractors arise during the course of most projects.
Agree: **64.1%** Disagree: **13.5%**

F.17 General contractors usually solve trade subcontractor interface problems in a satisfactory manner.
Agree: **23.9%** Disagree: **40.3%**

F.18 At the right price, an ideal job would be one that involves the minimum number of moves on and off the project, and minimal trade subcontractor interface problems.
Agree: **96.8%** Disagree: **0.4%**

F.9 PAYMENT PROCEDURES

G.1 When working for a general contractor, progress payments earned by subcontractors are usually received by subcontractors in a timely manner.
Agree: **15.5%** Disagree: **69.7%**

G.2 Delays in receiving earned progress payments from the general contractor increases the subcontractor's cost of doing business.
Agree: **94.9%** Disagree: **3.6%**

G.3 General contractors often use subcontractor progress payment distribution as a means of leverage to improve subcontractor performance.
Agree: **63.4%** Disagree: **10.0%**

G.4 Delays in receiving earned progress payments from general contractors have a negative effect on the subcontractor's performance on the project.
Agree: **72.9%** Disagree: **13.2%**

G.5 If a subcontractor could be assured of fair, timely progress payments, a lower price would be provided to general contractors for trade contract work.
Agree: **76.4%** Disagree: **6.8%**

G.6 A method of receiving earned progress payments directly from the owner, rather than through the general contractor, would be desirable to subcontractors.
Agree: **86.0%** Disagree: **4.4%**

GLOSSARY OF CONSTRUCTION MANAGEMENT TERMS

Addendum

Supplementary documentation issued prior to the contract award that changes or clarifies information stated in the bidding documents.

Additional Services

Services provided over and above those designated as basic services in owner agreements with A/Es and CMs.

A/E

Architect/Engineer; the design professional hired by the owner to provide design and design-related services.

Agency CM (ACM)

Also: Pure CM; Professional CM; CM Without Risk (Abbr PCM; CM)

A contractual form of the CM system exclusively performed in an agency relationship between the construction manager and owner. ACM is the form from which other CM forms and variations are derived.

Agent (or Agency)

A legal relationship where one party is authorized to act in behalf of, and in the best interest of, another party, as defined in an agreement between the parties.

Agreement

Also: Contract

A legal document that binds two or more parties to specific and implied obligations.

Ancillary Benefits

Subordinate secondary benefits that automatically accrue from the performance of an unassociated prime responsibility.

Apparent Low Bidder

The bidder who has submitted the lowest competitive proposal as determined by a cursory examination of the bids submitted.

Approved Bidders List

The list of contractors that have survived prequalification tests.

Approved Changes

Changes of any nature in contract requirements which have been agreed upon through a change approval process and approved by the owner.

Arrow Diagram

Check: Critical Path Schedule; Critical Path(s); Precedence Diagram

Also known as the *i-j* method or *activity-on-arrow* method. This method uses arrows, pointing in the direction of schedule flow, to represent activities with durations and circles (nodes) at each end of the arrow designating the start and finish events (dates) of each activity.

As-Built Drawings

Also: As-Builts; *Check:* Record Drawings

Drawings produced during or after construction and amended to show the exact location, geometry, and dimensions of the constructed project. As-Built Drawings are not the same as Record Drawings.

Basic Services

Check: Additional Services, Reimbursables

The services specifically listed in the services agreement as basic services.

Beneficial Occupancy
See: Start-up; Commissioning
The point of project completion when the owner can use the constructed facility in whole or in part for its intended purpose even though final completion may not be achieved.

Bid
A binding offer, usually expressed in dollars, to provide specific services within clearly stated requirements.

Bid Bond
A pledge from a third party (usually a surety company) to pay liquidated damages to the owner to the extent of the difference between the bonded contractor bid and the next highest bidder but not to exceed the face value of the bond; if the bonded contractor declines an award offered by the owner.

Bid Depository
A physical location where trade contractor proposals are filed the day before general contractor bids are to be received by an owner for pick-up, opening, acceptance, or rejection by general contractors bidding the owner's project.

Bid Division
Also: Division of Work; Work-Scope
A portion of the total project reserved for contractors for bidding and performance purposes.

Bid Division Description
Also: Division of Work Description; Work-Scope Description
A narrative description of the concise work-scope to be bid and performed by a contractor.

Bid(ding) Documents
The documents distributed to contractors by the owner for bidding purposes. They include drawings, specifications, form of contract, general and supplementary conditions, proposal forms, and other information including addenda.

Bid Shopping
Check: Pre-bid Shopping; Post-bid Shopping, Scope Enhancement
Negotiations to obtain lower costs and prices both prior to submitting proposals and after signing contracts.

Bonus-Penalty Clause
Check: Penalty-Bonus Clause
A positive/negative incentive to comply with a schedule. A bonus is paid for timely performance; a penalty is assessed for untimely performance. The dollar amount of the bonus and penalty must be equal.

Budget Estimate
An estimate of cost based on rough or incomplete information, with a stated degree of accuracy. The more information available, the more accurate the estimate. Loosely called a "ballpark" estimate.

Bulletin
A delineation, narrative or both describing a proposed change for pricing by a contractor(s) and for consideration as a change by the owner.

Change Order
Also: Contract Modification
The document that alters the contract amount, contract time, or contract requirements of the original contract entered into by the owner and a contractor.

Changed Conditions
Also: Concealed Conditions; Latent Conditions
Conditions or circumstances, physical or otherwise, which surface after a contract has been signed and which alter the circumstances or conditions on which the contract is based.

Chart of Accounts
Also: Account Codes; Codes of Accounts
An alpha/numeric identification system for budget line items that ensures that project expenditures are properly debited/credited in the project budget as payments are made in behalf of the project.

Checks and Balances
The term used to describe the use of the overlapping expertise of each team member during team decision making.

Claim
A formal notice sent by a contractor to an owner asserting the fact that the terms of the contract have been breached and compensation is being sought by the contractor from the owner.

Clerk-of-the-Work

An individual employed by an owner to represent him on a project at the site of the work. The clerk-of-the-work's abilities, credentials, and responsibilities vary at the discretion of the owner.

cm

Check: CM; Construction Manager; Construction Management

The abbreviation for construction manager (a person employed by a CM to provide CM services on a project). Not commonly used.

CM

Check: cm; Construction Manager; Construction Management

The abbreviation for Construction Management (the project-delivery system) and Construction Manager (a firm that provides CM services or persons who work for a CM firm).

CM Fee Plus Reimbursables

A form of payment for CM services where the construction manager is paid a fixed or percentage fee for CM expertise, plus preestablished hourly, daily, weekly, or monthly costs for field personnel and equipment.

CM Format

The interactive contracting approach to providing a project's needs, used by the CM project team to manage a project.

CM Partnering

A contractual commitment by the Owner, A/E, and CM to achieve a common goal, and doing so without a stakeholder's exposure to a potential for conflict of interest in pursuit of that goal.

CM Philosophy

An enlightened approach to accomplishing an owner-oriented end result using a system of motivating concepts and principles for achievement.

CM Procedures Manual

The depository for the proprietary micromanagement procedures used by the CM to detail and facilitate the service obligations owed to a client.

CM Project Manual

Check: Project Manual

The common depository for the micromanagement procedures to be used on the project by the team to accomplish project requirements.

CM Services

The scope of services provided by a construction manager and available to owners in whole or in part. CM services are not consistent in scope or performance from one CM firm to another.

Collateral Information

Information of value that is unexpectedly made available through the routine performance of another activity or activities.

Commissioning

See: Start-up; Beneficial Occupancy

The process at or near construction completion when a facility is tried out (put into use) to see if it functions as designed. Usually applied to manufacturing type projects.

Completion Schedule

Also: Occupancy Schedule

A schedule of the activities and events required to effect occupancy or the use of a facility for its intended purpose. It is used to determine if construction progress will meet the occupancy date.

Conditions of the Contract

Terms that refers to the General Conditions and the Supplementary and Special Conditions of the contract for construction.

Conflict of Interest

A situation where it is difficult for a party to clearly choose a direction or make a decision because of a self-interest in the outcome of the situation.

Constructability

Check: Contractability

The optimizing of cost, time, and quality factors with the material, equipment, construction means, methods, and techniques used on a project; accomplished by matching owner values with available construction industry practices.

Construction Budget

The target cost figure covering the construction phase of a project. It includes the cost of contracts with trade contractors, construction support items, other purchased labor, material and equipment, and the construction manager's cost but not the cost of land, A/E fees, or consultant fees.

Construction Coordination
The orchestration or interfacing of performing contractors on-site.

Construction Cost
See: Cost of Construction

Construction Documents
Drawings, technical specifications, and addenda; the contract documents that refer to the physical construction requirements established by the A/E.

Construction Management
A project-delivery system that uses a construction manager to facilitate the design and construction of a project.

Construction Manager
A firm or business organization with the expertise and resources to manage the design, contracting, and construction aspects of project delivery. Individuals who work for a CM firm are also referred to as Construction Managers.

Construction Schedule
Also: Detailed Construction Schedule
A graphic, tabular or narrative representation or depiction of the construction portion of the project-delivery process, showing activities and durations of activities in sequential order.

Construction Support Items
Also: General Condition Items
Purchases, services, or materials required to facilitate construction at the site. As part of the construction budget, these are financial obligations of the owner and the logistic responsibility of the CM.

Construction Team
Check: Project Team
The designated leaders of each trade contractor plus the Level 2 and 3 Managers (Persons) of the owner, A/E, and CM.

Constructor–XCM
A variation of the extended services form of CM where the construction manager self-performs some of the construction on the project.

Contingencies
Line-item amounts in the project budget, dedicated to specific cost areas where oversight is an inherent problem in project delivery.

Contract Administration
Servicing the interactive provisions in the contracts for construction between the owner and the contractor(s).

Contract Document Phase
The final phase of design on an architectural project when construction documents are completed and bidding documents formulated.

Contract Document Review
Also: Biddability Review
A review of Bidding and Contract Documents on a continuing basis, or at short intervals during the preconstruction phase, to preclude errors, ambiguities, and omissions.

Contract Documents
The documents which collectively form the contract between the contractor and the owner. They consist of the bidding documents less bidding information plus pre-award addenda and post-award Change Orders.

Contractability
Also: Contract*ibility*; *Check:* Constructability
The optimizing of cost, time, and quality factors with the contracting structures and techniques used on a project; accomplished by matching owner contracting requirements with available construction industry practices.

Contractor
Check: Trade Contractor; General Contractor
A business entity that contracts to perform a defined scope of work on a construction project.

Contractor/Constructor–XCM
A variation of the extended services form of CM where the construction manager holds construction contracts and self-performs construction on the project.

Contractor–XCM
A variation of the extended services form of CM where the construction manager holds construction contracts for the project.

Control CM
Also: Project Manager; Level 2 Manager; Level 2 Person
A person designated by the CM firm to interface with the owner's and A/E's representatives on the project team at the second management level.

Coordination Meeting
See: Progress Meeting; Field Meeting

Coordinator
Also: CM Coordinator; Project Manager's Assistant
A person designated to assist a Control CM, Project Manager, or Level 2 Manager in executing the CM format.

Cost Control
Also: Cost Management
Deliberations, actions, and reactions to project cost fluctuations during a project to maintain the project cost within the project budget.

Cost(s) of Construction
See: Construction Budget

Credibility
The quality of something that makes it believable.

Critical Date Schedule
See: Milestone Schedule

Critical Path(s)
The continuous chain(s) of activities from project-start to project-finish, whose durations cannot be exceeded if the project is to be completed on the project-finish date. A sequence of activities that collectively require the longest duration to complete (the duration of the sequence is the shortest possible time from the start event to the finish event).

Critical Path Schedule
Check: Arrow Diagram; Precedence Diagram
A schedule that utilizes the Critical Path scheduling technique using either the arrow or precedence diagramming method.

Designability
A pragmatic, value-based assessment of the design in comparison with the stated physical and aesthetic needs of the owner.

Design–Build (D–B) Contracting
A contract structure where both design and construction responsibility are vested in a single contractor.

Design–Build (D–B) Contractor
A contractor that provides design and construction services under a single responsibility contract to an owner.

Detailed Construction Schedule
See: Construction Schedule

Design Development Phase
The term used on architectural projects to describe the transitional phase from the Schematic Phase to the Contract Document Phase during design.

Design–XCM
A variation of the extended services form of CM, where the A/E also provides the CM function.

Direct Costs
The costs directly attributed to a work-scope, such as labor, material, equipment, and subcontracts but not the cost of operations overhead and the labor, material, equipment, and subcontracts expended in support of the undertaking.

Direct Labor Costs
Costs accruing from expended labor excluding the bonus portion of overtime, insurances, and payroll taxes.

Direct Material Costs
Costs accruing from material acquisition including purchase price, freight, and taxes.

Division of Work
Also: Work Division; Work-Scope; Bid Division; *See:* Bid Division

Division of Work Description
Also: Work Division Description; Work-Scope Description; Bid Division Description; *See:* Bid Division Description

Drawings
Graphic representations showing location, geometry, and dimensions of a project or its elements in sufficient detail to facilitate construction.

Dual Services
The providing of more than one principal service under a single contract or another contract(s).

Dynamic Decisions
Decisions that are made without team deliberations. Autonomous or bilateral decisions based on policy, procedures, or experience.

Dynamic Risk
The risk inherent to a speculative decision. The risk-taker can either gain, lose, or break even from the risk.

Employment Agreement
A contract binding an employee to an employer for a specific length of time and for disclosed compensation.

Escrow Account

Money put into the custody of the third party by the first party for disbursement to the second party. A brief temporary depository for progress payments until authorized for release according to the depositor's explicit instructions.

Estimated Cost to Complete

An estimate of the cost still to be expended on a work-scope in order to complete it. The difference between the Cost to Date and the Estimated Final Cost.

Estimated Final Cost

An estimate of the final cost of a work item based on its Cost to Date and the estimated cost to complete it. The sum of the Cost to Date and the Estimated Cost to Complete.

Ethics

Self-imposed rules or standards of performance for professionals set by the organization or association to which the professional belongs or by the public trust.

Extended Services

See: Additional Services

Dissimilar services included in a contract to be performed over and above those that are included as the principal services of the contract.

Extended Services–CM (XCM)

A form of CM where other services such as design, construction, and contracting are included with ACM services provided by the construction manager.

Facilitator

A person who leads by logic, suggestion and example more so than by direction.

Fast-Track(ing)

The process of designing portions of a project while portions already designed are under construction. A series of controlled design–build sequences that collectively constitute a complete project.

Feasibility Phase

Also: Predesign Phase; Conceptual Phase

The phase of a project preceding the Design Phase used to determine from various perspectives whether a project should be constructed or not.

Fee Enhancement

The awarding of an additional fee, over and above the basic fee for services, based on the performance quality of the party providing the basic service.

Fiduciary

One who stands in a special relationship of trust, confidence, and responsibility regarding contracted obligations.

Field-Based CM Field Organization

Check: Office-Based CM Field Organization

A project organization structure that bases the CM's 2nd Level representative and certain resource persons in the field rather than in the office.

Field Construction Manager

Also: Field CM; Level 3 Manager; Level 3 Person; Superintendent, Site Manager

A person designated by the CM firm to interface with the owner's and A/E's representatives on the project team at the third management level. A person located at the site and charged to administer the procedures established by the team's Level 2 Manager for the construction of the project.

Field Management

The coordination and management of owner-contracted resources on-site during construction.

Field Order

An order issued to a contractor by an authorized team member to perform work not included in the contract for construction. The work will become a Change Order. It is an expedient used in an emergency or need situation.

Field Schedule

Also: Construction Schedule

See: Short Term Construction Activity Plan (CAP)

Final Completion

The point at which both parties to a contract declare the other has satisfactorily completed its responsibilities under the contract.

Final Design Phase

The designation used by engineers for the last portion of the design process prior to bidding.

Financial and Management Control System (FMCS)
Also: Management Control System (MCS)
A manual or computerized system used by the team to guide the course of a project and record its progress.

Financial Stakeholder
A party involved by contract to perform a prescribed definitive physical work-scope for a sum of money, who stands to lose or gain money from the eventual outcome of the project or how it is performed.

Float
A scheduling term indicating that an activity or a sequence of activities does not necessarily have to start or end on the scheduled date(s) to maintain the schedule.

Force Account Work
Work done and paid for on an expended time and material basis.

General Condition Items
See: Construction Support Items

General Conditions (of the Contract for Construction)
See: Front-End Specifications
The part of the contract that prescribes the rights, responsibilities, and relationships of the parties signing the agreement and outlines the administration of the contract for construction.

General Contracting System
The traditional project-delivery system that utilizes the services of a general contractor; the GC assembles and submits a proposal for the work on a project and then contracts directly with the owner to construct the project as an independent contractor.

General Contractor
A business entity that provides independent contractor services to owners through the use of subcontractors when using the general contracting system.

Guarantee
An agreement by which a party accepts responsibility for fulfilling an obligation.

Guaranteed Maximum Price CM (GMPCM)
A form of the CM system where the construction manager guarantees, in addition to providing ACM services, a ceiling price to the owner for the cost of construction.

Human Resources
Persons who have inherent and acquired abilities to function to the benefit of an employer.

IRFP (Initial Request For Proposal)
The first request for uniform detailed information from prospective CM practitioners being screened for a project.

Indirect Costs
Costs for items and activities other than those directly incorporated into the building or structure but considered necessary to complete the project.

In-House Resources
Resources, physical, monetary, or human, available within an organization for providing contracted services.

Interview Decorum
The rules and procedures established by the owner for interfacing with firms competing for providing services on a QBS basis.

Job Description
A broad-scope explanation of a position's requirements indicating the duties of the position and the expertise and capabilities required of a person to adequately perform in that position.

Job-Site Overhead
Supportive and necessary on-site construction expense, such as construction support costs, supervision, bonus labor, field personnel, and office expense.

Joint Venture Partner
A party that contracts with another similar party on a project basis to provide greater financial strength, improved services or more acceptable performance qualifications as a combined organization to design, bid, and/or construct a specific project.

Labor and Material Bond
Check: Surety; Performance Bond
A guarantee provided by a surety to pay claims against the owner from contractors and suppliers who have not been paid for labor, material, and equipment incorporated into the project.

Letter of Intent
Check: Notice of Award; Notice to Proceed
A notice from an owner to a contractor stating that a contract will be awarded to the contrac-

tor providing certain events occur or specific conditions are met by the contractor.

Lien
Also: Mechanic's Lien
The right to take, hold, or sell the property of a debtor as security or payment for a debt.

Life-Cycle Cost
The cost of purchasing, installing, owning, operating, and maintaining a construction element over the life of the facility.

Long-Lead Items
Material and equipment required for construction which have delivery dates too far in the future to be included in a contractor's contract at bid time. They are prepurchased directly by the owner.

Long-Lead Time
The time between the purchase date and delivery date of long-lead items.

Lump Sum Fee
A fixed dollar amount which includes all costs of services including overhead and profit.

Management Information and Control System
See: Financial and Management Control System

Management Plan
A microscheme to produce project requirements in terms of policies, procedures and timing developed from the management strategy.

Management Strategy
A macro- and micro-approach to structuring contracts and managing a project; based on owner policies, project demands, and contracting practices in the project area.

Master Schedule
See: Program Schedule

Mechanic's Lien
Also: Lien
A legal claim against an owner's property by a project participant to the value of monies earned but not paid for by the owner or an employing contractor.

Milestone
Selected strategic events of signal importance to progress used in the milestone schedule.

Milestone Schedule
A schedule of milestones spanning from the start of construction to occupancy, used as the main measure of progress to keep the project on schedule.

Multiple Bidding
Soliciting and receiving bids from trade or work-scope contractors when using a multiple-contracting format.

Multiple Contracting
A contracting format that separates the project's single work-scope into a number of interfacing smaller work-scopes, to be individually and competitively bid or negotiated.

Multiple Prime Contracts
Contracts with work-scope contractors, individually awarded by the owner under a multiple contracting format.

Negative Attributes
A quality, character, procedure, or practice inherent to a system (of contracting) that impairs the system's performance.

Notice of Award
Check: Letter of Intent; Notice to Proceed
A letter from an owner to a contractor stating that a contract has been awarded to the contractor and a contract will be forthcoming.

Notice to Proceed
Check: Notice of Award; Letter of Intent
A notice from an owner directing a contractor to begin work on a contract, subject to specific stated conditions.

Occupancy Phase
Also: Warranty/Guarantee Phase; Warranty Phase; Guarantee Phase
A stipulated length of time following the construction phase, during which contractors are bonded to ensure that materials, equipment, and workmanship meet the requirements of their contracts, and that supplier- and manufacturer-provided warranties and guarantees remain in force.

Occupancy Schedule
See: Completion Schedule

Office-Based CM Field Organization
Check: Field-Based CM Field Organization
A project organization structure that bases the CM's 2nd Level representative and resource persons in the home office rather than in the field.

On-Site Supervision
Site-based personnel with supervisory responsibilities.

Owner–CM (OCM)
A form of the CM system that utilizes the owner as the construction manager.

Owner's Representative
A person assigned to a project to represent the owner on the project team at one of the three management levels.

Partnering
Check: CM Partnering
A recommitment to the fundamental terms of a construction contract brought about by the humanistic interactions of the parties to the contract; the stakeholders.

Penalty-Bonus Clause
See: Bonus-Penalty Clause

Percentage Fee
A fee for services, expressed as a percent of the final cost of construction.

Performance Bond
Check: Labor and Material Bond; Surety
A guarantee provided by a surety to complete a project according to the terms of the contract documents in the event that the bonded contractor defaults on the contract or to pay the owner the face value of the bonded amount.

Performance Coordination
See: Construction Coordination

Phased Bidding
Also: Stage Bidding
The process of receiving proposals from contractors on projects that are constructed as more than one total work-scope.

Phased Construction
Also: Stage Construction
A unitized approach to constructing a facility by designing and constructing separate project elements. Each element is a complete project in itself.

Plans
See: Drawings

Positive Attributes
A quality, character, procedure, or practice inherent to a system (of contracting) that elevates the system's performance.

Post-Bid Shopping
Negotiations between prime contractors (buyers) and trade contractors (sellers) to obtain lower prices after signing a prime contract with an owner.

Potential for Conflict of Interest
A conflict of interest that could occur but has not yet materialized.

Pre-Bid Shopping
Negotiations between prime contractors (buyers) and trade contractors (sellers) to obtain lower prices prior to submitting prime contract proposals to owners.

Precedence Diagram
Check: Critical Path Schedule; Critical Path(s); Arrow Diagram
Also known as the activity-on-node method. This method uses a node (geometric shape) to represent activities with connecting lines to show the logic or sequence of activities.

Preconstruction Phase
All required phases prior to the start of construction.

Predesign Phase
Also: Feasibility Phase
The phase prior to the start of design.

Preliminary Design Phase
Applies to engineering projects; the initial design effort following signing of the owner/engineer agreement. It is followed by the Final Design Phase.

Prime Contract
A contract held by an owner.

Prime Contractor
A contractor who has a contract with an owner.

Private Sector
The domain where projects are funded with capital other than from taxes.

Product Data
See: Shop Drawings

Professional CM
See: Agency CM, Pure CM

Professional Liability Insurance
Also: Errors and Omissions Insurance
Insurance provided by design professionals and construction managers that protects the owner against the financial results of negligent acts by the insured.

Professional Services
Services provided by a professional, in the legal sense of the word, or by an individual or firm whose competence can be measured against an established standard of care.

Professionalism
Essentially; considerate, courteous, ethical behavior when dealing or communicating with others on a construction project.

Program Management
A substitute term for construction management, sometimes used when two or more projects are concurrently managed by a CM for the owner, or when the CM only functions as an advisor to the owner.

Program Schedule
Also: Master Schedule
A schedule that spans from the start of design to occupancy; includes the signal activities which control the progress of the project from start to finish.

Program Team
The owner, A/E, and CM represented by their Level 1, 2, and 3 Persons. Used interchangeably with Project Team.

Progress Meeting
Also: Site Coordination Meeting, Coordination Meeting
A meeting dedicated essentially to contractor progress during the construction phase.

Progress Payment
Partial payments on a contractor's contract amount, periodically paid by the owner for work accomplished by the contractor to date.

Project Budget
The target cost of the project established by the owner and agreed to be achievable by the team. The Project Budget usually includes the cost of construction and the CM fee, plus any other line-item costs (land, legal fees, interest, design fees, CM fees, etc.) that the owner wishes to have included in the budget.

Project Costs
Costs expended on a project and which debit the line items that comprise the Project Budget.

Project Manager
See: Control CM, Level 2 Manager, Level 2 Person

Project Manual
Check: CM Project Manual
Written information that augments the drawings. The Project Manual contains the General Conditions, Supplementary and Special Conditions, the Form of Contract, Addenda, Change Orders, Bidding Information and Proposal Forms as appropriate, and the Technical Specifications.

Project Meeting
A meeting dedicated essentially to contractor performance and progress payments, involving supervisors from contractor home offices and the team's Level 2 and 3 Managers.

Project Team
Also: Construction Team; *Check:* Program Team
Consists of the architect/engineer, construction manager, and owner, represented by their Level 1, 2, and 3 Persons, plus the designated leaders of contracted constructors.

Public Sector
The domain where owners fund projects with monies that come in whole or in part from taxes.

Pure CM
See: Agency CM, Professional CM

Quality
The value levels of material and equipment selected by the A/E. Conformance to the technical specifications during construction.

Quality Assurance
The procedure established by the Project Team to inject and extract the level of quality designated by the owner.

Quality Control
That part of the Quality Assurance procedure that determines if specified quality is attained.

Quality Engineering
That part of the Quality Assurance procedure where the required level of quality is accurately inserted into the construction documents by the A/E.

Questionable Practices
Practices, standard or otherwise, that are not totally productive or are unfriendly or unfair to those parties that the practices interface.

RFP (Request For Proposal)
The second request for uniform detailed information from prospective CM practitioners being screened for a project.

Record Drawings
Check: As-Built Drawings
A set of contract document drawings, marked up as construction proceeds, which show the exact location, geometry, and dimensions of all elements of the constructed project as installed.

Reimbursable Expense
Also: Reimbursable
Charges to the owner covering costs for services that could not or intentionally were not quantified at the time the fee arrangement was made.

Samples
See: Shop Drawings

Schedule of Values
The breakdown of a lump sum price into subitems and sub-costs for identifiable construction elements, which can be evaluated by examination for contractor progress payment purposes.

Schematic Design Phase
The initial Design Phase on an architectural project when the A/E delineates the owner's needs in a general way.

Shop Drawings
Also: Product Data; Samples
Detailed information provided by material and equipment suppliers demonstrating that the item provided meets the requirements of the contract documents.

Short Term Contractor's Activity Plan
Also: Short Term CAP; CAP; Short Interval Schedule
A field-based schedule that plans contractor activities on a day-to-day, week-to-week basis from milestone to milestone.

Special Conditions
Also: Supplementary Conditions
Amendments to the General Conditions that change standard requirements to unique requirements, appropriate for a specific project.

Special Consultants
Experts in highly specialized fields not inherent to an owner, A/E, or CM.

Specifications
Detailed statements covering procedures, and quantitative and qualitative information pertaining to material, products, and equipment to be incorporated into a project.

Start-Up
Check: Commissioning; Beneficial Occupancy
The period prior to owner occupancy when mechanical, electrical, and other systems are activated and the owner's operating and maintenance staff are instructed in their use.

Static Decisions
Decisions that are made or can be made under the full influence of the project team's checks and balances.

Static Risks
Risks inherent to the project-delivery process which occur or can occur by accident and have no opportunity for gain in the manner of their disposal.

Study and Report Phase
Check: Feasibility Phase
Principally applicable to engineering projects. Includes the investigation and determination of a situation(s) and the recommendation of design solutions to an owner's needs.

Subcontractor
A contractor who has a contract with a prime contractor.

Substantial Completion
The date on which a contractor reaches a point of completion, when subsequent interfacing contractors can productively begin work or the owner can occupy the project, in whole or in part, without undo interference.

Sub-subcontractor
A contractor who has a contract with a subcontractor.

Superintendent
A job title usually reserved for the administrative level person who supervises the work of an on-site contractor.

Supervision by a Field CM
To coordinate and guide, but not inspect, the performance of construction resources contracted for by the owner.

Supplementary Conditions
Supplements or modifies the standard clauses of the general conditions to accommodate specific project requirements.

Synergism
Actions by two or more persons to achieve an end result that could not be achieved as well by one of the persons.

Team
Check: Project Team; Construction Team

Technical Inspection
Check: Quality Assurance; Quality Control
Matching technical specification criteria with visual or mechanical tests on the project site,

or in a remote location or laboratory, to ascertain conformance.

Technical Review
The critique of design solutions, or criteria used for design solutions, by a party other than the one providing the solutions or criteria, to determine adequacy and suitability of purpose.

Technical Specifications
Written criteria that augment the drawings pertaining to the technical construction of the project that cannot be conveniently included on the plans.

Tenure
The duration, term, or length of time required by agreement or precedent for performance of services.

Testing
Applying standard procedures to determine if prescribed technical criteria have been met in performance.

Timeline
A synonym for scheduling of activities in the context of time.

Timely Performance
Compliance with a time requirement.

Trade Contractor
Also: Performing Contractor
A contractor that specializes in providing/installing specific elements of the overall construction requirements of a project.

Up-Front Services
Free or reduced-rate services provided to prospective clients in the interest of obtaining a contract. Often rationalized as a part of a firm's selling or public relations program.

Value
The intrinsic worth of something determined on an individual basis.

Value Engineer
Also: Certified Value Specialist
A person qualified to perform value engineering services for a client.

Value Engineering
A technical review process; the close matching of engineering design to the value an owner derives from the design.

Value Management
Check: Contractability; Constructability; Designability; Value Engineering
The matching of project decisions and directions with the expressed requirements of the owner, from an owner value derived perspective.

Value Manager
A person qualified to perform value management services for a client.

Warranty
Also: Guarantee
Assurance by a providing party that the work, material, and equipment under warranty will perform as promised or as required by contract.

Warranty Phase
Also: Guarantee Phase; Warranty/Guarantee Phase
See: Occupancy Phase

Work-Scope
See: Bid Division

Work-Scope Description
See: Bid Division Description

XCM
See: Extended Services CM

INDEX